T0093990

Bizarre Medicine

Bizarre Medicine

Unusual Treatments and Practices through the Ages

Ruth Clifford Engs

An Imprint of ABC-CLIO, LLC
Santa Barbara, California • Denver, Colorado

Library of Congress Cataloging-in-Publication Data

Names: Engs, Ruth Clifford, author.
Title: Bizarre medicine : unusual treatments and practices through the ages / Ruth Clifford Engs.
Description: Santa Barbara : ABC-CLIO, [2022] | Includes bibliographical references and index.
Identifiers: LCCN 2021019878 (print) | LCCN 2021019879 (ebook) | ISBN 9781440871245 (hardcover) | ISBN 9781440871252 (ebook)
Subjects: LCSH: Medicine—Miscellanea. | Medicine—Anecdotes.
Classification: LCC R706 .E54 2022 (print) | LCC R706 (ebook) | DDC 610—dc23
LC record available at https://lccn.loc.gov/2021019878
LC ebook record available at https://lccn.loc.gov/2021019879

ISBN: 978-1-4408-7124-5 (print)
978-1-4408-7125-2 (ebook)

26 25 24 23 22 1 2 3 4 5

This book is also available as an eBook.

Greenwood
An Imprint of ABC-CLIO, LLC

ABC-CLIO, LLC
147 Castilian Drive
Santa Barbara, California 93117
www.abc-clio.com

This book is printed on acid-free paper ∞

Manufactured in the United States of America

This book is a reference, not a medical or diagnostic manual. No portion of this text is intended to supplement or substitute medical attention and advice. Readers are advised to consult a physician before making decisions related to their diagnosis or treatment.

This reference work is dedicated to the memory of
the tens of thousands of lives unnecessarily lost
due to the COVID-19 pandemic.
May we become educated and wise enough to
learn from history,
look to and accept science, and
be wary of bizarre medicine.

Contents

List of Entries

Preface

This work is aimed at a variety of students and the general reading public, including those interested in health and the medical sciences, history, sociology, and anthropology, among others. The purpose of this reference work is to briefly acquaint the reader with some information concerning common maladies from over the centuries. However, the focus is on unusual and bizarre treatments to manage or cure maladies from antiquity into the early twenty-first century primarily in the United States and English-speaking Western societies.

Each entry begins with a brief discussion of the symptoms and cause of the ailment. Some historical background and, if known, how the individual or society perceived the condition and the percentage of population affected are described. Following this, the derivation of the malady's name when it was first identified and its history over time are presented. Current treatments are briefly mentioned. Examples of its influence on military campaigns and its representation in the arts are mentioned.

In the second part of each entry, any unusual treatments from antiquity to the present for a particular condition are discussed. The reason for using a bizarre treatment, if known, is mentioned along with possible outcomes. In many cases, remedies were based upon trial and error or upon information discussed in the introduction of various theories. Some individual illnesses have been combined into an overall category. For example, the entry "Sexually Transmitted Infections" discusses several conditions such as syphilis, chlamydia, and human immunodeficiency virus (HIV). These maladies are listed in the body of the reference and point to the entry where they can be found.

SOURCES OF INFORMATION

For in-depth information, the reader will find an extensive bibliography that is from scholarly, governmental, professional, and academic internet sites. Some sources are only available in print copies but can be found in libraries. Even advertisements from the nineteenth into the twenty-first century for bizarre remedies are used as a source; most are now found on the internet. Many primary source items from the past are from archival and public domain sites, such as hathitrust.org, archive.org, mayoclinic.org, and gutenburg.org. Current information from commercial sites, such as WebMD or Healthline, is also utilized. To find many references, type the title into your web browser.

Frequently used sources, for many unusual treatments and many conditions throughout this reference work, include the ancient Egyptian *Ebers* and the *Edwin Smith Surgical Papyri* (c. 1500–1600 BCE), *The Hippocratic Corpus* (c. 5–4 BCE), and Celsus's (c. 25 BCE–c. 50) *Of Medicine*. From the early Middle Ages, *The Seven Books of Paulus Aegineta* (c. seventh century) and from the early modern era *The Works of Thomas Sydenham, M.D.* (1676) and William Buchan's *Domestic Medicine* (1774) discuss many bizarre remedies. In the nineteenth century, Alexander Macaulay's (1831) *A Dictionary of Medicine* and Henry Hartshorne's *Essentials of the Principles and Practice of Medicine* (1867, 1874) list numerous unusual treatments. Useful governmental organizations include the Centers for Disease Control and Prevention, the National Institute of Health, Statistics Canada, Public Health England, and the Australian Institute of Health and Welfare. The World Health Organization and various nongovernmental associations such as the American Heart Association also provide up-to-date information and data. Finally, various online dictionaries including the *Merriam-Webster Dictionary* and the *Online Etymology Dictionary* provide information for the derivation of the name of the disease.

Lastly, a caveat. This book attempts to describe weird, bizarre, and unusual treatments for various illnesses and conditions throughout the ages. This reference work is not to be used for diagnosis or treatment of any condition. Although some remedies are still used as part of folk or alternative systems, many are ineffective, and some are toxic or deadly. Therefore, do not attempt any cure or unusual procedures without first talking to your physician or health-care professional.

Acknowledgments

Reference books pull from many primary and secondary sources. They are also validated by scholars with areas of expertise, experience, and knowledge. Therefore, I want to thank the following individuals with doctoral level degrees from medicine and other disciplines for constructive comments and suggestions. These include Lindsay Franz-Waltsack, Priscilla W. Gabriel, Shirin Hassan Gilbert, Adam T. Spaetti (1977–2020), John Strobel, and Michael Wenzler, as well as Jeff Graf, Indiana University Bloomington, reference librarian, for help with compiling the extensive bibliography.

Various libraries of the Indiana University system have been accommodating in my research efforts for this publication. These include the Wells Library and Lilly library on the Bloomington campus along with the Ruth Lilly Medical Library on the Indianapolis campus. The Interlibrary Loan Department and the Auxiliary Library Facility (ALF) during the 2020 COVID-19 crisis, when library facilities were closed or on limited hours, have also been most supportive.

Various staff from ABC-CLIO have been most helpful. First, I need to express gratitude to my editor Maxine Taylor who was able to help me sort out various aspects of this reference and always had good ideas for improving it. I would also like to thank Art Editor Robin Tutt, Senior Production Editor Nicole Azze, Project Manager Kousalya Krishnamoorthy, Copy Editor Moushumi Dutt, and the Editorial Team at Amnet for their input.

Last, but not least, I am eternally grateful to my husband, Jeffrey Franz, for giving me the space, encouragement, and support to work on this project.

Introduction

Bizarre medicine and unusual practices and treatments have been found throughout history. Sometimes they were carried out by mainstream healers, and at other times they were found in alternative systems, such as folk medicine, or even marketed or administered by well-meaning quacks and outright charlatans. Some nostrums were used as cure-alls, while others were used for specific conditions. In this reference book, current and past common illnesses from antiquity through the present and their unusual treatments are depicted. Treatments are not always useless, but just strange or odd by twenty-first-century standards. A few may even help, but most are ineffective, and some are harmful or deadly.

Some of these remedies have spanned human history, while others are of modern origin. What was considered normal in one era was often considered bizarre in another. For example, inserting crocodile dung into the vagina was an accepted practice to prevent pregnancy in ancient Egypt but was not accepted in later cultures. Some ancient treatments, such as using leeches for bloodletting, have now come back into fashion. Leech therapy was common for centuries to reduce an "imbalance of body fluids" but went out of favor in the late nineteenth century when "humoral theory," discussed later, was discredited. Leech therapy emerged again in the early twenty-first century for use in microsurgery to prevent swelling.

DIFFERENCES BETWEEN FOLK AND CONVENTIONAL TREATMENT

Remedies prescribed by clinicians in previous centuries sometimes became folk medicine or were incorporated into alternative healing systems when they were abandoned by modern medicine based upon new science and technology. These older treatments were often based on "trial and error" and unproven healing theories and philosophies. Conventional treatments by the mid-twentieth century had extensive clinical trials to determine the efficacy (effectiveness) and safety of the treatment. When this was demonstrated, governing bodies such as the FDA (Federal Drug Administration) in the United States permitted their use.

Folk and most alternative remedies, on the other hand, have generally not been tested by the scientific method using clinical trials with experimental and control groups. For the most part, they are based upon personal belief or a philosophical system. In the late twentieth century, for example, some Asian medicine traditions were introduced into Western culture with a different outlook compared to Western medicine but were rarely researched. In these alternative or folk systems, positive results from a remedy tend to be backed up by testimonials. Some patients may not have had the condition or, if they did, their recovery was likely due to the "placebo effect."

PLACEBO EFFECT

In the placebo effect, people experience a positive result from an inert or "look-alike" substance or medication—sometimes called a "sugar pill"—or a sham procedure or treatment. This positive outcome is due to the belief in the effectiveness of the remedy. The placebo effect appears to be a psychological phenomenon that in some cases can heal or reduce symptoms of the illness. For example, some people believe that magnetic or copper bracelets help reduce arthritic pain, but numerous studies have found no scientific validity for this. Their pain relief is likely due to the placebo effect. In addition, many cures in the past were undoubtedly due to this phenomenon.

HUMORAL THEORY OF DISEASE

The major theory of disease and treatment throughout much of Western history was "humoral theory." This theory was proposed by the Greek physician/philosopher Hippocrates (c. 460–370 BCE) who is considered the father of Western medicine. He and his followers wrote numerous manuscripts that were assembled together and referred to as the *Hippocratic Corpus*. (In this current work, the mention of Hippocrates refers to this corpus as only some manuscripts have been attributed directly to him.)

Humoral theory was based upon Greek scholar Aristotle's (384–323 BCE) doctrine of four basic qualities: hot, cold, wet, and dry, which Hippocrates applied to the "four elements, four humors, and four temperaments." The theory proposes that all life contains four basic elements—earth, air, fire, and water—which in humans are related to four basic fluid humors: blood, phlegm, black bile, and yellow bile. Each humor is focused upon an organ: brain, lungs, spleen, and gall bladder in addition to a particular personality type: sanguine, phlegmatic, melancholic, and choleric. If a person became sick, it was believed that an imbalance of the humors existed in the body. Therefore, treatment was aimed at getting rid of the excess, or bad humor, by bloodletting, vomiting, diarrhea, or sweating. In the second century, Greco-Roman physician, Aelius Galenus, also known as Galen of Pergamum's (129–200), expanded these works that became the primary authority for diagnosis and treatment of disease in the West.

However, after the fall of the Western Roman Empire in the late fifth century and during the early Middle Ages—sometimes referred to as the Dark Ages (c. 500–1000)—humoral theory and the healing practices of ancient Greece and Rome largely disappeared but were preserved in the Arab world by scholars such as the Persian Avicenna (980–1037) and the Andalusian Jew Maimonides (1135–1204) who translated Galen's and other works. However, it is not known if these

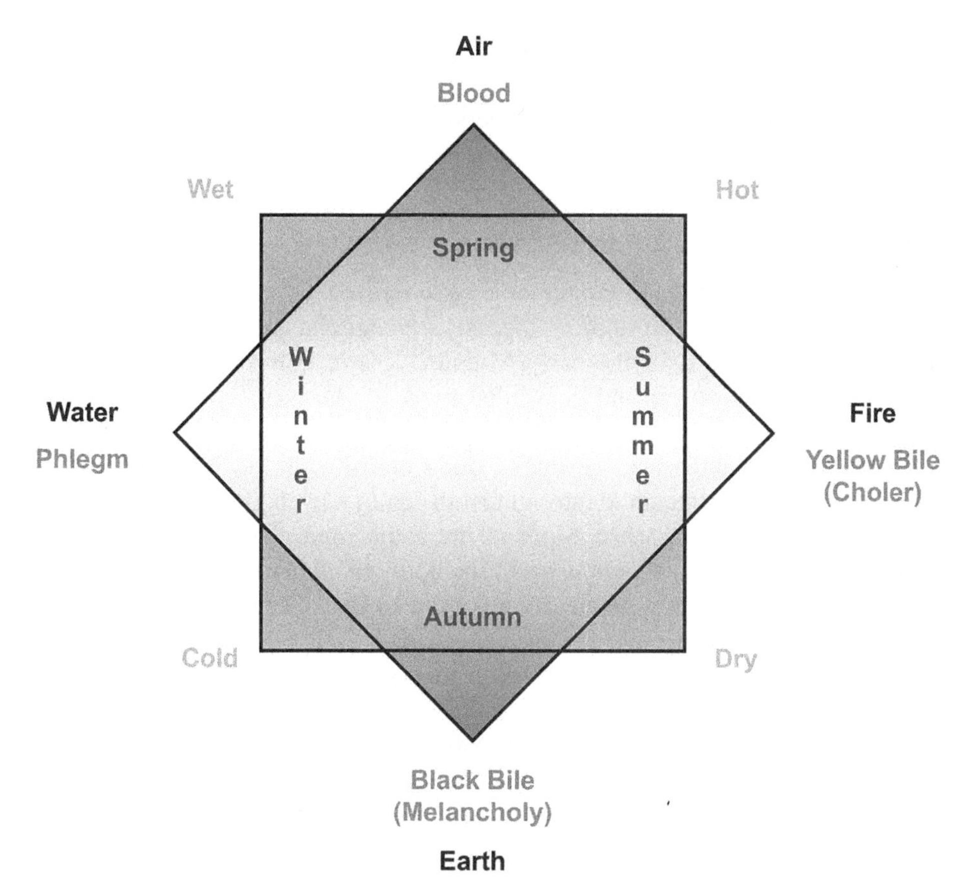

Diagram of the humoral system used from the time of Hippocrates into the nineteenth century upon which treatments of bloodletting, cupping, leeches, and purges and emetics were based. These procedures were used to balance the humors.

written sources reflected common practices, if they were available only to the elite, or if they were accessible to everyone.

In the early eighth century, when the North Africa Islamic Moors conquered Spain they reintroduced humoral theory and Galenic practices. Persian philosopher and physician Avicenna (980–1037) wrote *The Canon of Medicine* based upon these translations. The text was introduced into Europe and was used as the standard medical authority in the Islamic world and in Europe up into the eighteenth century. During this era, the accuracy of the text was not questioned, and clinicians closely followed it. Since Galen considered blood the most dominant humor, bloodletting through venesection, leeching, or wet cupping—placing a hot cup on the skin to cause a vacuum after scratches had been made to draw out the blood—became a major treatment for many diseases.

Humoral theory and treatments to rebalance the humors were the bases of mainstream medical management into the mid to late nineteenth century until French scientist Louis Pasteur (1822–1895) demonstrated that some diseases were caused by microscopic organisms. However, even Pasteur's theory was controversial as it opposed the accepted hypothesis of "spontaneous generation" where living organism emerged from nonliving things such as maggots arising from rotten meat. Acceptance of spontaneous generation disappeared by the twentieth century, but humoral theory remains popular among some alternative medicine proponents today.

OTHER THEORIES OF DISEASE

Other theories of disease and remedies also existed. In ancient Egypt, supranational spirits were thought to bring disease, so magic incantations were generally part of any treatment. In the early Middle Ages (c. 500–1000), the Christian Church had much power and controlled most of Western Europe. The church preached that God could heal all sickness. However, it was also believed that illness and death from disease was due to God's punishment for sinful and immoral behavior. From the fifteenth to the eighteenth century, even astrological influences were thought to cause disease. Signs of the zodiac and planets were believed to have governance over different parts of the body. In addition, it was believed that nostrums such as herbs to be effective needed to be collected when their associated planets were visible.

Another theory of disease, "contagion theory" arose out of the plague years, which began in the mid-fourteenth century. Italian Renaissance physician Girolamo Fracastoro (c. 1478–1553), who is known for naming syphilis, believed that infectious disease or epidemics—often called "fevers"—could be spread by infected people, clothing, and goods. He believed that the spread of these diseases could be slowed down by isolating people. Earlier quarantine laws in the port city of Dubrovnik, Croatia, in 1377 had helped to repress Black Death and restore social order in the community. The laws required people and goods from plague-infested areas to remain outside of the city and be isolated for 40 days on an island before they were allowed into the port. Isolation hospitals were also established to care for the sick.

This contagion theory, however, was not accepted by many clinicians. In opposition to this theory was the "miasmas theory" that proposed that infectious diseases were caused by poisonous vapors or miasmas released by rotting organic material, dirty soil, and stagnant water. Although quarantine had been found to be successful in halting some epidemics, it was not effective for others. However, based upon miasmas theory, public health measures such as cleaning up filthy streets, swamps, and water supplies were successful and began to be practiced by the late nineteenth century. Other healing systems believed that disease resulted when the body's "energies" or "vibrations" were out of balance, so the purpose of a remedy was to bring the energies back into balance again. Based upon these and other theories of illness, bizarre and unusual treatments were often carried out.

SOCIETIES' REJECTION OF SCIENCE

Over the centuries, scientific knowledge and medical advances have often been rejected or even repressed by religious, medical, or political groups or their leaders. This has resulted in the continuation of ineffective bizarre treatments and the slow adoption of methods based upon science. For example, the Christian Church of the Middle Ages generally forbade autopsies and dissection, in particular on men, to determine the cause of death, thus keeping physicians in the dark about many diseases, as the Church felt that a whole body was needed for resurrection. However, as science advanced in the seventeenth and eighteenth centuries, old theories, such as bloodletting for many ailments to balance the humors, were not supported by research and gradually disappeared from medicine. However, science is still sometimes rejected by those in power. In the United States during the worldwide COVID-19 pandemic of 2020–2022, some politicians ignored science. Scientific results warned about the seriousness of the novel SARS-CoV-2 virus and its potential to spawn an international pandemic and advocated standard public health measures to contain the spread of the virus. These included wearing face masks, avoiding crowds, washing hands, and staying several feet away from others. However, many viewed this advice as a political maneuver and ignored the advice of the scientific community. This lack of early public health action led to serious health, social, and economic consequences on a national and international level.

Appendicitis

Appendicitis often begins suddenly with abdominal bloating and mild pain around the navel that moves to the lower right side of the abdomen with localized pain and tenderness. In addition, abdominal rigidity, stiffening of abdominal muscles in response to pressure over the inflamed appendix, fever, inability to pass gas, constipation, nausea, and vomiting sometimes occur, and lab tests frequently find a high white blood count. If the fever becomes more than 101 °F (38.3 °C) along with an increased heart rate, it may be a sign that the appendix has burst, which can lead to a life-threatening peritonitis—infection in the abdomen. In a few people, appendicitis becomes chronic or even goes into remission.

The vermiform (looking like a worm) appendix is a three-and-a-half-inch-long tube of tissue that extends from the cecum—the first part of the large intestine. Appendicitis is caused by calcified stones, foreign objects, or parasites blocking the lumen, the opening into the appendix. The appendix until recently was thought to be a useless organ as most mammals did not have one. However, in 2007 some researchers hypothesized that the appendix might store beneficial microorganisms that are released into the gut when the gastrointestinal tract loses the bacteria that supports digestion and aids the immune system.

The lifetime chance of developing appendicitis ranges from 7 to 14 percent. It is more common in men than women and among those between ages 10 and 30. The incidence has been declining in Western nations but increasing in Asian nations perhaps due to change in diet. In the United States, appendicitis is the most common cause of sudden abdominal pain requiring surgery with roughly 10 cases per 100,000 people per year, and the mortality rate is less than 1 percent.

Appendicitis, as a separate disease, was not recognized until the late nineteenth century. Prior to this time, the malady was termed "iliac passion" (inflammation of the cecum) or "perityphlitis" (inflammation of the connective tissue about the cecum and appendix). The term "appendicitis" was first used in 1886 when American physician Reginald H. Fitz (1843–1913) reported, based upon his research, that all inflammation of the right iliac fossa was from the vermiform appendage on the cecum. The word "appendicitis" is from the Latin *appendix*, meaning "something attached," and Greek *-itis*, meaning "inflammation."

Pain in the abdomen has been described since antiquity, but the cause was generally not known. An Egyptian mummy of the Byzantine era, for example, exhibited adhesions in the right lower quadrant of the abdomen, suggestive of appendicitis that had healed. The Greco-Roman physician Galen (129–c. 210 CE) only dissected monkeys and did not find an appendix as they do not have one.

Since physicians throughout the Middle Ages relied upon Galen and other ancient healers as the authorities, and not on their own experiments or observations, most ignored the organ; it was also illegal to dissect humans. But physicians did describe symptoms that today would likely be classified as appendicitis. Scottish physician William Buchan (1729–1805), for example, described iliac passion that had the classic symptoms of appendicitis. He believed the condition was caused by wet feet; carrying heavy weight; and eating unfermented bread made of peas, beans, rye, and "other windy ingredients," along with unripened fruit.

It took centuries for appendicitis to be identified for what it was. In 1839, two physicians of Guy's Hospital, London, England, clearly described the symptoms of appendicitis and claimed that the appendix was the cause of most inflammatory conditions of the lower abdomen. However, this was rejected by more notable physicians of the day, and it was not until Fitz's paper that this hypothesis was finally accepted.

Until the advent of anesthesia and antiseptics, surgery was the last treatment for most bowel issues as the death rate was high. However, in the late 1800s through the early twenty-first century, surgery became the primary treatment for all appendicitis cases even though a mildly inflamed appendix sometimes gets better on its own. On the other hand, some physicians began to prescribe antibiotics when they were developed in the mid-twentieth century for uncomplicated cases, which has proved to be rather successful. But for cases where it appears that the appendix might burst, appendectomies are immediately performed. They are done either as open surgery with one abdominal incision or by laparoscopic surgery through a few small abdominal incisions. In addition, by the late twentieth century, many tests had been developed to diagnose the seriousness of an appendicitis attack, including ultrasound, computerized tomography (CT) scan, or magnetic resonance imaging (MRI).

Appendicitis may have influenced the course of history over the centuries. In the early twentieth century, for example, one of the best-known Western artists and leading American impressionist, Frederic Remington (1861–1909), died at the peak of his career after surgery for a ruptured appendix, which may have meant a loss of further works. The famous Hungarian American escape artist and magician Harry Houdini, born Erik Weisz (1874–1926), died from a ruptured appendix when a Canadian university student punched him in the abdomen when he was not prepared for it. Some of the secrets of his illusions are still unknown.

UNUSUAL TREATMENTS

Many physicians considered abdominal pain to be caused by obstruction, so they often recommended cures to remove the offending object. Ancient Greek physician Hippocrates (c. 460–370 BCE), for example, gave an air enema to distend and dislodge the item in the intestine. In the sixteenth century, some physicians insisted that patients were cured "of the most deplorable iliac passions" by drinking three pounds of quicksilver (toxic liquid mercury) in hot water; this was still used by some into the early nineteenth century and likely led to death. Physicians in the seventeenth century suggested that the patient swallow lead musket balls for iliac passion to drive out the obstruction. The patient had to walk around to help the balls fall through the intestines, which could have led to intestinal perforations and killed the patient. Bleeding from a vein at the elbow and ankle was practiced throughout the mid-nineteenth century along with purgatives and enemas to balance the humors. The patient was also forced to sit in a bath of hot oil in which various drugs were dissolved. These were not helpful, and many patients did not survive these remedies.

Renowned English physician Thomas Sydenham (1624–1689) recommended that tobacco smoke be blown through a large bladder into the rectum. He would also order the patient—in order to free the stomach and bowels from "sharp humors"—to ingest wormwood (used as a flavoring in alcohol beverages) and lemon juice and at the same time to keep a live puppy on the belly in order to transfer living energy to the patient. This was not effective although harmless. On the other hand, Dutch physician Herman Boerhaave (1668–1738), besides bloodletting, laxatives, enemas, and purgatives, recommended that young living animals be split open and kept on the abdomen. These were not effective, and laxatives may have precipitated a ruptured appendix.

Buchan in the eighteenth century ordered purgatives and oily enemas and had the patient drink several times a day a small amount of Barbados tar (from Barbados) mixed with "an equal quantity of strong rum." He also recommended bathing the feet and legs in warm water and applying warm cloths to the stomach and bowels. These were not effective, but the rum may have helped quell the pain.

In the nineteenth century, American physician Henry Hartshorne (1823–1897) in his medical text recommended, for inflammation of the cecum, bleeding from a vein and leeching over the painful area. He then ordered a poultice of flaxseed or cornmeal to be placed over the abdomen with oiled silk to retain moisture. He also endorsed tobacco and ice-cold enemas. After these procedures, he had the patient lie still in bed and gave opium to relieve severe pain. In addition, he suggested that an electric current should be run through the abdomen to restore peristalsis. Opium may have quelled pain, but the other treatments were ineffective but harmless.

Indian American physician Dinshah P. Ghadiali (1873–1966) developed the "Spectro-Chrome Therapy" (method of healing using different color light waves), which was a popular alternative treatment in the mid-twentieth century. For appendicitis, he advised green, blue, and indigo light. He also required his patients to be on a liquid diet of water for several days until the pain subsided. Cures were likely due to the placebo effect or they were mild cases that went into remission.

Arthritis, Rheumatism, and Gout

Arthritis, sometimes called rheumatism, causes painful or swollen joints, leading to a lack of mobility, and is a chronic condition. It includes several types with the most common being osteoarthritis (OA), rheumatoid arthritis (RA), psoriatic arthritis (PsA), and gout. Osteoarthritis is a degenerative joint disease in which the cartilage that cushions the bones wears down. This causes pain, stiffness, and loss of mobility. It is the most common form of arthritis and afflicts millions of people worldwide. It is more prevalent among older women with northern European ancestry and is associated with diabetes, obesity, joint injury, and repeated stress. However, by age 60 most people have some degenerative joint changes. The first use of the term "osteoarthritis" in English was in 1878. The term is derived from the Greek words *osteo*, meaning "bone"; *arthro*, meaning "joint"; and *-itis*, meaning "inflammation." Non-English-speaking nations use the term OA ("osteoarthrosis") as they contend it is noninflammatory.

In 2012, this ancient condition was even found in the joints of a 150-million-year-old dinosaur. Likewise, Neanderthal remains from over 35,000 years ago show arthritic changes. For centuries, arthritis has been recorded in written material. The Egyptian *Ebers* and *Edwin Smith papyri* (c. 1550 BCE) were the first to clinically document arthritis and gout as separate ailments. Greek physician Hippocrates (460–370 BCE) also described the maladies. In the eighteenth century, Scottish physician William Buchan (1729–1805) described OA as "chronic rheumatism" and noted that it was incurable and found primarily among the elderly.

On the other hand, rheumatoid and PsA tend to be found among individuals during their midlife. Both are autoimmune conditions and likely have genetic and environmental components. The body's own immune system attacks joints and various organs causing inflammation and pain. Over time, constant inflammation can break down the bone and cartilage and cause permanent damage. In RA, joints of fingers and hands often become deformed. Women are up to three times more likely to develop RA than men, and it is found in about 1 percent of the population globally. With PsA, red scaly rashes may appear on the scalp, back, near elbows, knees, and other joints. The prevalence of PsA is under 1 percent, and it is twice as common among females. Without treatment, patients can become crippled throughout life due to these maladies, which can affect their work, social, and personal life.

The first-known use of the term "rheumatism" in English was in 1670 via Latin, from the Greek *rhein* "to flow," as the disease was thought to be caused by the internal flow of watery humors. Buchan in the eighteenth century believed that acute rheumatism was caused by wet clothes, damp beds, lying on the ground at night, sudden change in weather, and going from a hot to a cold environment. English rheumatologist Alfred Baring Garrod (1819–1907) coined the term "rheumatoid arthritis" in 1859. The term "psoriatic arthritis" was first used in English in the late seventeenth century from the Greek *psōra*, meaning "itch."

Gout is an extremely painful inflammatory arthritis that tends to arise in the large joint of the big toe but can be found in other joints too. It occurs when a high level of uric acid in the blood forms crystals—tophi—in the joint, which causes excruciating pain and swelling. A gout attack generally strikes in the middle of the night with symptoms worsening over the next few hours. The pain diminishes in a week or so. Gout tends to be more common in men than women. Since antiquity it has been called the "disease of kings" and the affluent as it has been associated with excessive alcohol, rich food, and heavy meat consumption. These high purine substances oxidize into uric acid. Overweight, high blood pressure, and a sedentary lifestyle in recent years have also been found to be associated with the affliction. Gout is rarer in Asia and cultures whose diet is based on rice and vegetables. However, the condition has been increasing worldwide as more people are adopting a Western lifestyle.

In antiquity, the malady was called "podagra." The term "gout" was first used in English by a Dominican monk in the thirteenth century. It is derived from the Middle English *goute*, from the Latin *gutta*, meaning "drop," as it was believed that diseased particles dropped from the blood into the joint. French physician Guillaume de Baillou (1538–1616) was one of the first to recognize that rheumatism was an affliction of joints of the entire body while gout was a disease of one joint.

Today, many treatments for arthritic conditions are available. They include aspirin, over-the-counter nonsteroidal anti-inflammatory drugs (NSAIDs), steroids such as prednisone, and artificial joint replacements for severely damaged joints. By the early twenty-first century, pharmacologists had developed antirheumatic drugs including biologics, and stem cell replacement therapy for cartilage is on the horizon. Physicians have also recommended exercise; weight loss; and avoidance of rich foods, red meat, wine, and beers for the management of arthritic afflictions.

Gout and arthritic conditions have shaped the course of history. For example, King Charles I of Spain (1500–1558) developed gouty arthritis—perhaps RA and gout. The pain in his inflamed joints prevented him from leading his army into battle and became so severe that he abdicated the throne in 1556. British statesman William Pitt (1708–1778) was absent from the parliament due to a gout attack when it taxed the colonists for imported tea to pay for British troops that had been in North America during the French and Indian or Seven Years' War (1754–1763). This precipitated the Boston Tea Party in 1773, the American Revolution, and the formation of the United States of America.

UNUSUAL TREATMENTS

Many remedies for joint pains have been around for centuries, including the use of herbal and other concoctions. The earliest-known reference for gout treatment is from the *Papyrus Ebers*. Healers used sandalwood incense and had patients ingest colchicine (derived from the autumn crocus plant) that is still used today. It acts as an anti-inflammatory medication but can be toxic. For stiff joints, healers also mixed the fat of serpent, mouse, and cat and placed this on the joint. In the seventeenth century, a folk treatment for gout and arthritis was the "spirit of skull," "Goddard's drops," or "king's drops." These drops were the distillate of a mixture of dried snakes, pieces of skull from a recently hanged man, and ammonia boiled in a glass container. These were not effective against gout other than reducing the pain due to placebo effect. English physician Thomas Sydenham (1624–1689), however, during the same era treated rheumatism and gout by having his patients abstain from wine and meat and drink a concoction of sassafras, china-root herb, and shavings from a hartshorn (male deer) and anise seed boiled in fountain water. Abstinence from wine and red meat may have helped. Physicians treated OA and RA in late nineteenth century with cod liver oil, potassium iodide, arsenic, and turpentine. The cod liver oil may have helped reduce the pain due to the anti-inflammatory effect of omega-3 fatty acids, but arsenic and turpentine were toxic.

Advertisement in the nineteenth century for patent medicine to cure gout and arthritis, which likely contained alcohol and opium. (Wellcome Collection. Attribution 4.0 International (CC BY 4.0))

Apitherapy, or bee sting therapy, has been practiced since antiquity to help relieve arthritic pain. By 2010, bee venom was found to increase levels of anti-inflammatory hormones. Now called bee venom therapy (BVT), this alternative therapy can lead to a decrease in swelling, pain, and stiffness in some people with RA. However, patients have died from allergic reactions to the venom, making the treatment controversial. In the 1920s through the late 1900s, physicians gave gold injections for RA, which may have relieved some pain due to its anti-inflammatory properties. However, it also had many side effects and is rarely used today.

Electric treatments have been used for centuries. For example, Hippocrates and other ancient Greeks used the electric "torpedo" fish and electric eels to treat gout. The patient stood on a live fish in a moist place until his whole foot and leg up to the knee was numb, which took away pain. In the sixth century, Alexander of Tralles (c. 525–605) recommended cooking a live torpedo fish in oil with a narcissus plant. The resulting oily mixture was applied to the joint several times a day to prevent or treat pain and swelling. Oily concoctions from electric fish were rubbed on painful joints into the mid-nineteenth century but were likely ineffective. By the late nineteenth century, once electric batteries had been invented, electric baths were advertised to help rheumatism, gout, and sciatica. By the late twentieth century, small, portable transcutaneous electrical nerve stimulation (TENS) units that delivered mild battery-operated electric current were used as a counterirritant to pain and were often helpful.

Bleeding, leeching, and creating blisters were used as a treatment for joint pain in ancient Greece to draw out bad humors. These procedures were still used into the mid-nineteenth century. Frequent bleeding was done on younger individuals, but English physician Sydenham, in the eighteenth century, warned that if a patient had rheumatism for years, bleeding should only be done every few weeks. For joint pain, physicians throughout the early nineteenth century also applied blistering agents to the affected part. In addition, they recommended dry cupping (heated cups placed on the skin, which created a vacuum) on the back or pouring hot water over the sore joint or muscle. For gout, some physicians had the patient soak his feet in water inhabited by leeches. These treatments may have relieved pain due to placebo effect.

Topical poultices, plasters, and liniments have been popular treatments for joint pain throughout history. In ancient Egypt, cumin powder mixed with wheat flour and water was applied to aching joints, and in the eighteenth century, Buchan used a plaster of burgundy pitch (resin from Norway spruce) placed on soft leather and laid on the joint for pain relief, which may have offered temporary pain relief. Sydenham alternated a poultice of white bread, milk, and saffron with bloodletting on the side of the pain. American physician Henry Hartshorne (1823–1897) recommended oil of turpentine, oil of sassafras, ammonia, laudanum, and chloroform liniments be rubbed on the painful joint. Some of these substances may have temporarily killed the pain. Immersing hands in warm wax has been and is still used to soothe away arthritis aches and pains.

An oily liniment used by physicians from the Middle Ages until the late eighteenth century was "man's grease." This oil was the boiled-down fat cut from an executed criminal and rubbed on inflamed joints. It was a popular treatment in England but was ineffective other than for the placebo effect. In Chinese medicine, the oil from the fat Chinese water snake, which contains a high level of omega-3 fatty acids, was used to treat arthritis pain. It is an effective anti-inflammatory medicine and likely helped relieve some pain. However, unscrupulous patent medicine salesmen manufactured fake snake oil out of mineral and other oils, which was ineffective. In the early twentieth century, a radioactive radium cream liniment was advertised to drive out pain from aching joints. However, since radium was extremely scarce and expensive, only a few people died from using this deadly ointment. (Radium-branded products in the United States were outlawed in the 1938 Federal Food, Drug, and Cosmetic Act.)

The placebo effect likely reduced pain in some remedies. Wearing a copper bracelet has been a popular folk cure since antiquity and is still advertised along with magnets and magnetic bracelets. However, research has shown that neither copper bracelets nor magnetic wrist straps have any effect on arthritis. By the early twenty-first century, foot detox pads were advertised as a treatment for arthritic and other contentions to draw out "toxins." However, this treatment has no scientific evidence. The brown color on the pads after use is a result of chemical reaction with moisture.

Light and color therapy was attempted to relieve gout and joint pain in the late nineteenth century. In the 1870s, a blue light fad emerged, wherein light baths through blue glass were advertised as curing gout; it was ineffective. However,

American physician John Harvey Kellogg (1852–1943) developed a light bath box with an ultraviolet (UV) "arc-light" as a cure for joint ailments and gout. Since vitamin D deficiency has been linked to autoimmune diseases such as RA, this treatment may have been effective in some cases. The Spectro-Chrome Light Therapy alternative system of the mid-twentieth into the twenty-first century recommended lemon and magenta light for gout. For osteo- and rheumatoid arthritis, a green and magenta light on all body areas, and for acute conditions, blue and indigo light was used on the joints. Other than the placebo effect, this was not effective.

B

Back Pain and Spine Issues

Back problems include pain and stiffness in the lower back, mid-back, or neck. If pain, tingling, or numbness runs from the lower back to the buttock or down to the toes it is often called "sciatica." Pain and numbness can also radiate from the neck to the fingers. Misalignment or curvature of the spine also causes back pain and other issues.

Up to 80 percent of adults experience some type of back pain during their lives that can be the result of muscle strain, including lumbago in the lumbar area. This pain can affect personal relationships and job performance and are a major problem for many people. A major low back concern is sciatica. Sciatica results from irritation of the sciatic nerve roots in the lumbar and lumbosacral spine area and is a symptom of nerve root (radicular) pain and not a disease in itself. It is generally caused by a herniated or "slipped" disk, degenerative disk disease, a bone spur, narrowing of the spine (spinal stenosis), fractures, or other issues, such as pregnancy, which compresses part of the nerve and causes pain along the sciatic nerve pathway. Sciatica is a relatively common condition, and at least 40 percent of the global population suffers at some point from the malady. It causes work, social, and quality of life issues but is rarely fatal. Aging, obesity, sedentary lifestyle, chronic back pain, and smoking are also associated with sciatica. Acute back pain and sciatica generally last from a few weeks to six months regardless of treatment.

The first-known use of the word "sciatica" in English dates to the mid-fifteenth century from the Late Latin *sciaticus* and from the Greek *ischias* or pain arising from or around the hip and thigh. "Lumbago" is from the late seventeenth century, from the Latin *lumbus*, meaning "lower back or loin." Back aches and sciatica likely arose when humans became bipedal, and they have been recorded since antiquity. Greek physician Hippocrates (460–370 BCE), for example, noted that "ischiatic" pain in the hip and into the foot mainly affected men aged 40–60 years and that in younger men it usually lasted 40 days and usually resolved on its own. Greco-Roman physician Galen (129–210), based upon Hippocrates's humor theory of disease, declared that toxic humors caused back pain. His theories dominated western European medicine into the eighteenth century although others in this era thought sciatica and backache were caused by exposure to cold damp surroundings. In addition, the Italian anatomist Domenico Cotugno (1736–1822) wrote the first book on sciatica in 1764 and thought it was due to excess fluid surrounding the nerve. By the early nineteenth century, sciatica was thought to be an inflammation of the sciatic nerve due to a rheumatic condition, but by the late nineteenth century, low back pain became associated with injury. Neck pain and numbness down the arm were also thought to be due to similar causes.

In addition to lower back and neck issues, several spinal deformities can lead to back pain, breathing, and stomach problems. In kyphosis, the upper spine is curve outward. Scheuermann's kyphosis ("round back") appears most often among teenage boys caused from misshapen vertebra. If it is severe, the condition is known as "hunchback" or "humpback." Some infants are born with congenital kyphosis, and even with treatment, these curvatures sometimes can become permanent. About 1/10,000 newborns are born with the affliction. In the elderly, especially those with osteoporosis (thinning bones), small fractures may gradually lead to a "dowager's hump," which may also cause a chronic dull pain in the lower back. Occupations where bending is common are a risk factor over time for kyphosis. The word "kyphosis" is from the Greek *kyphos*, meaning a "hump," and was first used in English in the mid-nineteenth century. Another type of kyphosis, Pott's disease (coined in the early nineteenth century), is a tuberculous arthritis between the vertebra and named after Percival Pott (1714–1788), a British surgeon. About 1 percent of people with TB develop the affliction, and it is now more common in Asia and sub-Saharan Africa.

In opposition to kyphosis is lordosis ("swayback"). The lumbar region is bent inward and is often found in late pregnancy. A common back issue among children is idiopathic (unknown cause) scoliosis—a sideways curvature of the spine. It is present in around 3 percent of children aged 10–16 years and is more common in girls and in women of all ages. It may also be caused by neuromuscular illness and can be congenital. During the Victorian ages, it was thought to result from poor posture, bad fashion choices, and female occupations. However, more than 60 percent of people over the age of 60 have some degree of degenerative scoliosis. Severe curvatures can lead to differences in leg lengths, back and hip pain, breathing difficulties, balance, and other issues. The term "scoliosis" is from the Greek *skolios*, meaning "bent," and "lordosis" from the Greek *lordos*, meaning "curving forward." Both terms were first used in English in the early eighteenth century. These back misalignments and degenerative changes have been found in the spines from antiquity. Hippocrates, for example, introduced the terms "kyphosis" and "scoliosis," and Galen documented "lordosis."

Management of these back conditions has been a concern since antiquity. Today various treatments are available but are not always successful. They include physical therapy, exercise, massage, chiropractic care, vitamin D and calcium supplements, moist heat, spas, bracing for spinal misalignment, and various over-the-counter and prescription medications. As a last resort, spinal surgery that sometimes fuses the spine, adds rods, removes herniated disks, or adds stem cells may, or may not, be successful in removing chronic pain or spinal misalignments.

These maladies have been so common that they have been featured in literature. For example, in William Shakespeare's (1564–1616) *Timon of Athens* (c. 1605), the protagonist, Timon, places a "sciatica curse" on his enemies. In Shakespeare's historical play, *Richard the Third* (1593), he describes Richard as having a deformed back. The remains of the actual English King Richard III (1452–1485) with severe scoliosis were disinterred in Leicester, England in 2012. Both scoliosis and kyphosis are depicted in *The Hunchback of Notre-Dame* (1831), a French novel by Victor Hugo (1802–1885). Norwegian dramatist Henrik Ibsen (1828–1906) in *A Doll's House* notes that a character has "consumption of the spine" or Pott's disease.

UNUSUAL TREATMENTS

Over the centuries, various folk and other remedies have emerged based upon trial and error and the theory of the day. Herbal substances along with other measures have been popular. The Egyptian *Papyrus Ebers* (c. 1500 BCE), for example, recommend rubbing saffron on the lower back or neck to relieve pain. It also suggests rubbing myrrh to relieve backaches and onions to relieve sciatica. For sciatica, Greek physician Dioscorides (40–90) suggests goat dung be placed over the painful area. These may have helped due to placebo effect and were largely harmless to the patient.

Throughout the mid-nineteenth century, physicians used cupping (heated cups placed over the skin, which acts as a vacuum) on the spine, caustic cantharis beetle on the buttock to cause blisters, bloodletting on the foot of the painful side, and leeching over the painful area, which only caused more pain to the patient. In the eighteenth century, Cotugno promoted setons (a thread placed under the skin to cause festering and keep the wound open) along with enemas of opium dissolved in milk and water and emetic wine. He also suggested rubbing the painful area with viper (snake) oil and man's grease (human fat), singing, riding in a carriage, and galvanism (electric current). These likely had little effect other than temporary pain relief with opium and counterirritation, and some could have caused infection.

In the nineteenth century, folk practitioners placed mustard plasters and chloroform-soaked cloths over the sciatic nerve pathway or painful backs, which may have helped with pain, but chloroform could cause liver damage. In rural England, the legs of people with sciatica were smoked in a fern fire, which was likely ineffective and could have caused burns. In the 1880s, an "Electropathic Corset for the Ladies" was advertised for sciatica and lumbago. Most of these methods were abandoned by the twentieth century. However, continuous electric current, for example, was reintroduced in the late twentieth century as a transcutaneous electrical nerve stimulation (TENS) portable unit, which is sometimes successful in reducing pain.

By 1900, since "slipped disks," sciatica, and back pain were now thought to be related to trauma, clinicians recommended strict bed rest for two to six weeks, traction, or even a plaster body cast to heal the injury. This was contrary to earlier treatment and caused additional health issues such as blood clots and muscle weakness, and by the early twenty-first century, it was rarely recommended.

Spinal manipulation and traction have been popular over the centuries for back issues. In antiquity, for example, Hippocrates invented three methods to straighten spines—the Hippocratic ladder, board, and bench. For the ladder treatment, the healer would tie the patient to the ladder either right side up, or with head down, depending upon where the hump was located and shake the ladder back and forth.

The weight of the trunk and the limbs supposedly acted as a pulling force, which straightened the spine. In the early twenty-first century, upside down traction as the "inversion table" to relieve back pain—but did not shake—was advertised. These treatments may have been temporarily effective in some cases to relieve pain. The Hippocratic board had simultaneous traction of the spine along with manual pressure over the kyphotic area. In the Hippocratic bench, a piece of board would act as a lever that pushed upon the spinal deformity and caused pain to the patient and was ineffective.

From the 18th through the mid-twentieth century, to straighten spines, various apparatuses were tried such as back boards, screw chairs, steel bodices, plaster-of-Paris corsets, and heavy metal braces, which were uncomfortable to the patient and largely ineffective.

From the early 1960s through the late 1990s, scoliosis was treated with straight rods surgically implanted along the spinal column. This treatment was replaced by various treatments including fusing the spinal column together, motorized implants, and stem cell injections, which were only marginally successful and were often painful to the patient.

Birth Defects and Congenital Anomalies

Birth defects—also called congenital deformities, disorders, or anomalies—are physical and mental abnormalities that arise in utero. Defects include structural malformations, neuro tube defects, genetic conditions, and heart defects that arise from environmental and genetic factors. Major birth defects are common in about 3–4 percent of newborns. These abnormalities often cause parents grief, anguish, and stress in addition to marital and financial difficulties. However, some of these conditions can be prevented or identified with genetic counseling, DNA analysis, in utero testing, or ultrasound scans and sometimes can be corrected by surgery.

The term for an infant born with any physical or mental anomaly has changed over time. In the sixteenth century, "monstrosity," based upon Latin meaning "abnormality of growth," was common. "Congenital" was first used in the late eighteenth century from the Latin *congenitus*, meaning "to bring forth." "Birth defect" became common by the late 19th, and in the late twentieth century, "congenital anomaly" began to be used. Birth defects were noted in antiquity. For example, Mesopotamia tablets (c. 1700 BCE) declared that newborn malformations were omens of future events, and these infants were often seen as evil. Ancient Hebrews/Christian scriptures believed that birth defects were the result of God's punishment for parental sin. Along with the work of the devil, frights, and witches' curses, these folk beliefs were still prevalent into the eighteenth century.

Structural Birth Defects

These defects are some of the most prevalent congenital malformations. The most common is a cleft lip caused by the lack of tissue fusion during fetal development. In some cases, it is associated with a cleft palate. In a cleft lip, the upper lip is separated in one or two places, and in a cleft palate, a split is found in the roof or the mouth that opens into the nasal cavity. A cleft lip or palate can lead to feeding, speech, dental, and hearing problems in addition to ear infections and psychosocial issues. A cleft lip is about twice as common in males as females, while cleft palate—without cleft lip—is more common in females. Cleft lips occur in about 1 in every 1,000 births while cleft palate occurs in about 1 in every 2,500 births. The term "harelip" for the malformation was used from the mid-sixteenth century as the cleft resembled the mouth of a hare. In the past, it was thought to be caused by the pregnant woman eating the head of a hare, seeing a hare that had been bewitched, or even cavorting with the devil. The first-known use of "cleft lip" was in the late nineteenth century. "Cleft" is derived from the Middle English *clift*, meaning "split," and "lip" from the Old English *lippa*, of Germanic origin meaning "edge of mouth or vessel." Cleft palates have been observed in Egyptian mummies, and the Greco-Roman physician Galen (131–206) was the first to mention clefts in European history. Physicians in the sixteenth century believed cleft palates to be caused by syphilis. Risk factors today for these defects include genetics, the use of tobacco, heavy alcohol or other drug use, and obesity during pregnancy. Plastic surgery today can usually repair these problems.

Another major structural defect is clubfoot (talipes equinovarus), where one or both feet are rotated internally and the foot points down and inward at the ankle. There is an imbalance in muscle forces in the lower leg that results in the foot deformity. Without treatment, individuals with the condition walk on the outside of the foot or their toes. The disorder is found in about 1.5 per 1,000 live births. It occurs twice as often in males than in females and is more common among Pacific Islanders. The common term "clubfoot" and the medical term "talipes equinovarus" were first used around the sixteenth century. "Talipes" is from Latin, from the Greek *astragalos*, meaning "ankle bone," plus *pes*, meaning "foot." *Equinovarus* means the "inability to bend ankle toward the leg." The condition has been known since antiquity. Hippocrates gave the first written description of clubfoot and believed the causative factor was mechanical pressure inside the uterus on the feet. The malady can now be detected in utero through ultrasonography. Treatment with both manipulation and surgery today can often correct the condition so the individual can walk in a normal fashion.

Neural Tube Defects

The neural tube, found in the embryo, is the precursor to the brain and spinal cord. Environmental and genetic factors may lead to malformations in the tube causing severe disability. In hydrocephalus—from Latin meaning "water on the brain"—excess cerebral spinal fluid (CSF) causes pressure on the brain that can result in a large head, brain damage, cognitive impairment, or death without treatment. Hydrocephalus occurs in 1–2 per 1,000 births per year. In 1952, a shunt to drain excess CSF into another part of the body to relieve pressure was developed, but even with treatment, mental and physical impairment can occur. Risk factors during early pregnancy include lack of folic acid, diabetes, infection such as rubella, soaking in hot water bath, fever, and spina bifida. Spina bifida occurs when the meninges that enclose the spinal cord and/or the vertebrae fail to close over the cord. Several levels of severity exist. The most severe is spina bifida cystica where the meninges and/or the cord protrude through the opening in the vertebrae in the lower back. It can impair bowel, bladder, and leg functioning and sometimes cognitive impairment. Until the mid-twentieth century, it was a fatal disease with most children dying within three years. However, both pre- and postnatal surgery can now help lessen the disability and enable the individual live into adulthood.

Genetic Conditions

Many congenital genetic conditions are not apparent at birth and may take weeks or months to develop. A major congenital disability usually seen at birth is Down or Trisomy 21 syndrome. The newborn often exhibits a flat broad face, slanting eyes, a short neck, extra skin at the back of the neck, a large tongue, low small ears, and a single deep crease across the palm. Risk factors include older mothers and genetics. Individuals with Down syndrome can be mildly or profoundly mentally disabled, and the condition occurs in roughly 1 per 1,000 births in all ethnic and socioeconomic groups. In the past, many have died in a few days due to various heart and other organ defects. Some survived to be 6 or 7 years of age. However, because of improved health care, life expectancy rose from 9 years in 1900 to more than 50 years in the early twenty-first century. The syndrome was first classified as a separate condition in 1866 by British physician John Langdon Down (1828–1896). Since these newborns had features similar to those of the "Mongolian race," he called it "Mongolism." The cause was not found until 1959, when French physician Jérôme Lejeune (1926–1994) identified an extra copy of chromosome 21 in the cells of people with the disorder, which gave it the name of Trisomy 21 (three 21 chromosomes). The term "Down syndrome" was not used until the early 1970s and named after Down when "Mongoloid" was deemed offensive to people of Asian descent.

Attitudes toward Down syndrome changed over the centuries. Archaeological and cultural evidence from the past suggests that children with the syndrome in many cases were accepted as part of the family and community. However, from the early nineteenth into the mid-twentieth century, parents were sometimes ashamed of "defective" newborns and would keep them hidden in the household or have them institutionalized. They were often considered a sad burden on the family, and there was little support for families or education for these children. By the late twentieth century, special education, sheltered workshops, and group homes evolved to help individuals with Down syndrome to be productive members of society. However, almost all need social support for life.

Congenital Heart Defects

A newborn with a serious heart defect often has blue skin (cyanosis), problem in breathing, pounding heart and weak pulse, clubbed fingers, sleepiness, problem in feeding, and failure to thrive. Around 1 percent of newborns will have a heart defect, and about 25 percent of these will have a critical congenital heart defect

(CHD) for which early surgery is needed for survival. A common defect found in about 20 percent of all heart abnormalities is a hole in the wall of the ventricles (two lower chambers) called a *ventricular septal defect* that is generally non-life-threatening. Another common anomaly is a narrowing of the pulmonary valve and the main pulmonary artery (*pulmonary stenosis*). Tetralogy of Fallot is a critical CHD that involves four abnormalities occurring together: pulmonary stenosis, ventricular septal defect, ventricular hypertrophy (thickening of right ventricle), and an enlarged aortic valve, which seems to open from both ventricles. In the 1920s, the malady was named after French physician Etienne L. A. Fallot (1850–1911) who discovered it. The cause of these defects is largely unknown but is likely influenced by genetic and environmental factors. Some heart defects can be determined in utero or by oxygen levels at birth. Surgery is generally needed for critical CHD, but some congenital heart defects may not show up for months or years.

Congenital anomalies have been illustrated in the arts, literature, and drama. For example, a clubfoot was illustrated in ancient Egyptian paintings. Also, the Greek god, Hephaestus a blacksmith, is shown in artwork with a clubfoot. By the sixteenth century, society became more sympathetic toward disabilities that were reflected in the arts. For example, Dutch painter Hieronymus Bosch (c. 1450–1516) illustrates a person with a clubfoot in *Procession of Cripples*. Genetic anomalies have also been featured in the arts. In the sixteenth century, two children with Down syndrome are shown in the *Adoration of the Christ Child* (c. 1515) by an unknown artist. Even heart defects have been illustrated in paintings. Italian baroque artist Giovanni Lanfranc's (1582–1647) painting, *St. Luke Healing the Dropsical Child* (1620), shows a young child with edema (swelling) and cyanosis who likely had a heart condition. English author Mary Webb's (1881–1927) *Precious Bane* (1924) was turned into a BBC series in 1989. It featured a female protagonist who describes her life blighted by having a "harelip." Finally, in the early 2010s, the popular BBC series *Call the Midwife* in one episode featured a newborn with a serious heart defect who dies.

UNUSUAL TREATMENTS

Until the twentieth century, many infants with birth defects died soon after birth or at an early age and treatment was not attempted. Cures in antiquity consisted of sacrificing a deformed infant to the gods or leaving the infant to die outside in the elements to protect the community as it was believed these babies harbored evil spirits. In Sparta, for example, the decision to keep a child was made by the state, and many deformed or weak infants were abandoned on a hillside. In Greece, the philosopher Aristotle (384–322 BCE) advocated exposure to the elements in the case of a congenital deformity, and in Rome, it was the obligation of the father or head of household (*pater familis*) to expose a child who was visibly deformed. However, the upper classes often kept their malformed and mentally disabled infants who in the Middle Ages sometimes became the family and community "fool" or "simpleton" for entertainment. With the Christianization of Rome and Europe, neonaticide was forbidden. However, killing unwanted or deformed newborns continued into the nineteenth century in some Western nations, and in some Asian nations, some female newborns were continued to be killed into the twenty-first century as male babies were desired.

Structural Deformities

Infants with a cleft lip or palate often died as it was difficult for them to nurse. By the tenth century, however, successful surgery to repair the disfigurement was first recorded. In Saxon Britain around 950, the healer cut the edges of the cleft and stitched them together with silk. Then the healer smeared a mixture of pounded mastic gum, egg whites, and vermilion (it has antiseptic properties) on the repair. Around the same time, an Arab healer inserted a garlic clove, which also has antiseptic properties, into an incision he made in the divided lip for a day. It was then removed, and the two parts were held together with a bandage moistened with butter. Other Arab surgeons cauterized the fissure, removed the scabs, and sutured the raw edges together. By the early seventeenth century, surgery was performed on older children and adults. Skin was cut from the two sides of the gap, and several gold pins were pushed from one side to the other. The pins were kept together by thread in a figure 8 pattern for several days. These early surgeries were often successful as long as they did not become infected.

Of all birth defects, clubfoot has had somewhat effective treatment since antiquity. Hippocrates considered clubfoot curable for most cases. He described methods for repeated gentle manipulations of the foot with his hands followed by applying strong bandages to maintain the correction. He also explained that treatment should begin soon after birth before the abnormality of the bones became fixed. After he had made the correction, the child wore special shoes to maintain the correct position and prevent the deformity from recurring. However, if the treatment did not work, Hippocrates often amputated the foot.

Hippocrates's techniques were lost during the "Dark Ages," and by the sixteenth through the early eighteenth century, barber surgeons and bonesetters treated clubfoot and other deformities. One clinician first bathed the foot in hot baths and applied hot poultices to "soften the parts." Then with a great amount of force, he pushed the foot into its proper shape and applied fabric soaked in a mixture of egg white and flour to stiffen the bandage. A wooden sole was fastened to the foot of the bandage along with malleable wooden splints. Another treatment included months and even years of warm baths, massage, manipulation, stretching, and the use of splints. These methods sometimes helped remedy the defect. Other physicians tried mechanical devices for maintaining the correction that looked like medieval torture instruments and were not effective. In the nineteenth century, with the advent of aseptic methods and anesthesia, physicians tried various surgeries to correct the defect such as cutting a tendon, which were also met with mixed successes.

Neural Tube Disorders

Few treatments were attempted for these disorders. Scottish physician William Buchan (1729–1805), for hydrocephalus, remarked that no medicine had been found to "carry off dropsy of the brain." But since it was regarded as an imbalance of body fluids, to rebalance the system he administered purges with rhubarb or jalap along with calomel (contains toxic mercury) and placed blistering plasters generally made with cantharis (Spanish fly) to the neck or back part of the head to draw out the fluid. He also recommended diuretics to promote secretions in hopes of eliminating the swelling. These were not effective as swelling was in the brain only and likely hastened death of the child. For spina bifida, British physician Alexander Macaulay (1783–1868) recommended putting permanent pressure on the bulge. He also evacuated the spinal fluid from the bulge with small punctures and then induced inflammation "to obliterate the cavity," which likely led to early death. Some clinicians attempted to open or cut off the bulge, which usually led to immediate death as they had limited knowledge about the condition.

Genetic Conditions

Most genetic congenital anomalies are not observable at birth. However, it is usually obvious in Down syndrome (DS). In the Christian Middle Ages through the early nineteenth century, although some DS newborns were likely killed at birth and recorded as "stillborn," many were raised by their families. In contrast, during the late nineteenth century, educators advocated strict control and confinement of persons with DS and other mental disabilities to "protect society," as they were concerned they would produce more defective children. In addition, clinicians encouraged parents to institutionalize these children away from their families and communities into the mid-twentieth century, believing it was better for the child.

Congenital Heart Defects

Most infants with serious congenital heart defects likely died within a few weeks or months. Treatment by the nineteenth century included warmly wrapping the infant to guard against bronchitis that could lead to death. In an attack of dyspnea (difficult breathing), some physicians practiced bloodletting and administered a saline cathartic to balance the humors. These treatments did not correct the underlying issues. But clinicians did warn against the use of digitalis that produced stronger heart contraction, as it could kill the infant.

Bites: Mammals and Reptiles

Vertebral animals including mammals and reptiles can bite humans and cause serious injury or death. People for centuries have been terrorized by the bites of mammals such as dogs, wolves and bats, and reptiles such as snakes. Spiders have also been feared and are discussed in the "Bites and Stings: Insects and Arachnids" entry.

One of the most dreaded bites throughout history has been that of a rabid or "mad dog." If the bite, or even a scratch, is contaminated with saliva that contains the rabies virus, the affliction can be transferred to another animal. The disease, also called hydrophobia, is caused by a *lyssavirus* (RABV) that infects the central nervous system and causes lethal encephalitis (brain inflammation). Symptoms of rabies include fatigue, confusion, numbness around the bite, hallucinations, fear of water, foaming at the mouth, aggressiveness, paralysis, coma, and death. It is usually fatal once symptoms develop. However, Louis Pasteur (1822–1895), a French biologist, in 1885 developed a vaccine against the disease.

Each year, about 55,000 people in the United States have a potential rabies exposure. Before 1960, bites from rabid dogs caused most human cases, but as mass pet vaccinations and dog muzzle laws were enacted in developed nations, dog rabies declined and bats, foxes, and other wild mammals became responsible for most cases. Human deaths also declined due to availability of postexposure inoculations. In the United States, bats are now responsible for roughly 7 in 10 rabies deaths. Globally, rabid dogs cause about 98 percent of the 59,000 deaths from rabies each year. About two-thirds of these are in Asia, and most victims are children.

The first use of the term "rabies" was around 1598. It is derived from the Latin *rabere*, meaning "to rage" or "to rave." "Hydrophobia" was first used in the mid-sixteenth century from Late Latin, from the Greek *hudrophobia*, meaning "fear of water." Rabies has been described since antiquity. Greek physician, Hippocrates (c. 460–370 BCE), for example, refers to signs of rabies. Explanations for this disease were proposed by some. For example, Roman philosopher Pliny the Elder (c. 24–79) thought dogs become rabid by tasting the menstrual blood of a woman, and they were unusually susceptible to rabies when Sirius the Dog Star was seen in the summer. In the Middle Ages, tales of rabid beasts attacking humans emerged. In 1271, for instance, rabid wolves reportedly invaded a village and attacked livestock and people with several deaths occurring among those bitten. By the sixteenth century, the disease became associated with bats. The current management for exposure to a rabid animal is postexposure prophylaxis (PEP). This includes the administration of rabies immunoglobulin and a series of immunization injections. Throughout history, rabies has been featured in the arts. Roman poet, Grattius Faliscus (63 BCE–14 CE), in his "On Hunting" poem describes rabies. An outdoor scene with a mad dog biting a man was found in an Arabic translation of the *Materia Medica* by Dioscorides (c. 40–90). In more recent times, the children's novel *Old Yeller* (1956) involves a frontier dog that becomes infected with the scourge from a wild rabid wolf.

In addition to animal bites, snakes (also called vipers or serpents) have also been feared for centuries due to deadly bites. Globally about 3,000 snake species are found. Of the several hundred venomous species, most are not deadly, but their bite can cause redness, swelling, severe pain at the site of the bite, and nausea and vomiting. However, the bite of some venomous snakes, such as the black mamba in sub-Saharan Africa, can lead to death within half an hour. In India and South Asia, the Indian Cobra's bite delivers enough neurotoxins to kill about a half of its victims. Due to a lack of antivenom, around 20,000 people die each year due to venomous snake bites in South Asia. In developed nations, most snake bites are managed with antivenom.

In North America, five different poisonous snakes are found. People in all age groups are bitten by snakes; however, hikers, gardeners, hunters, and those in outdoor occupations are most at risk. The venomous copperhead bites the most people, but it is rarely fatal. Most fatal bites are from the eastern and the western diamondback rattlesnakes. In the United States, it is estimated that 7,000–8,000 people per year receive venomous snake bites, and about 5 of those die.

The first use of the term "snake" in English was before the twelfth century and is from the Proto-Germanic *snako*. "Serpent" is from the Latin *serpens*, meaning "to creep." In Modern English, "snake" is gradually replacing "serpent" in popular use. Snakes are an ancient symbol and are found in mythology, legends, and religions of many cultures and have been featured in the arts. They often represent the dual expression of good and evil. Snakes were symbol of healing in ancient Greco-Roman medicine, and even today, the ancient Rod of Asclepius—a staff with a snake curled around it—is used as a symbol of medicine in many nations. Snakes have been featured in English literature. For instance, playwright William Shakespeare's (1564–1616) *Anthony and Cleopatra* (1607) describes Cleopatra committing suicide with an asp (Egyptian Cobra), which is likely a myth. American writer Mark Twain's (1910–1835) *The Adventures of Huckleberry Finn* describes the runaway slave, Jim, being bitten by a rattlesnake and drinking whiskey to eliminate the snake's poison. British novelist J.K. Rowling's (1965–) Harry Potter books (1997–2007) portray the evil Voldemort's giant python-viper hybrid killer and the house of Slytherin with a snake as its crest.

Besides snakes, other reptiles can be dangerous. In North America, some venomous reptiles include the Gila monster and the Mexican beaded lizard. These desert-dwelling reptile bites are generally not fatal but can cause pain, swelling, severe bleeding, or allergic reactions. However, bites from crocodiles or alligators found in tropical areas are usually fatal. If the victim survives, there is a high risk of infection. The first use of the word "reptile" was in the fourteenth century from the Middle English *reptil*, from Late Latin. Reptiles have been associated since antiquity with religion and popular culture genres. For example, the crocodile-headed god of ancient Egypt represented power, protection, and fertility, while children's fiction such as *Peter Pan* (1904) and *Solomon Crocodile* (2011) featured both a bad and a good reptile. An alligator is even a mascot for an American athletic team, Florida Gators, representing the University of Florida. Although humans are often terrified by snakes and other reptiles, today modern medical treatment helps most people recover from the bites of small reptiles.

UNUSUAL TREATMENTS

Illnesses and death were associated with bites of animals and reptiles since antiquity. In terms of rabid dog bites, numerous unsuccessful cures have been attempted over the millennia. Roman physician Celsus (c. 25–c. 50) treated the animal bite as soon as possible by cupping (warm glasses to draw out the canine poison) after the bite had been cut, or he cauterized the bite with a hot poker, which may have sometimes been effective in destroying the virus, if the bite was superficial. He also cut open a live chicken and applied it to the wound to draw the poison into the chicken, which was not effective. Once victims exhibited the disease, he recommended holding them under water to force them to drink it, which resulted either in drowning or inevitably the death from the malady. This unsuccessful practice survived into the nineteenth century, when victims were tied in a sack and lowered into a well until they had drunk enough water to overcome the hydrophobia; they died anyway. Early healers also suggested that the hair of a rabid dog should be laid on the wound or ingested by the patient—"eating the hair of the dog that bit you."

To counteract the rabies poison, Greek physician Dioscorides in the first century recommended drinking a mixture of the ash of burnt river-crabs, gentian root, and wine for three days. Some healers suggested that rabies victims should consume the liver or head—either raw or cooked—of the dog that bit them, poultry dung mixed with vinegar, the ash of a mouse's tail, or pulverized beavers' testicles called castoreum. Others recommended that a poultice should be placed on the wound made of burnt pig dung dissolved in olive oil; none were successful and the patient usually died.

Magic amulets were also used to unsuccessfully treat rabies. For example, Belgium's St. Hubert's (c. 656–727) magical golden key was recast into an iron key that was heated and pressed on the bite as a cure. By the eighteenth century, Scottish physician Alexander Macaulay (1783–1868) noted that popular remedies including large amounts of bloodletting, cold and warm bathing, wine, opium, turpentine, vinegar, and potash all used to treat rabies were useless. He recommended, instead, that the bite wound should be cut out as soon as possible to prevent the poison from entering the body, which in some cases may have helped. Another Scottish physician, William Buchan (1729–1805), admonished physicians for carrying out the common practice of bleeding rabies patients to death or suffocating them "between mattresses or feather beds" when they became violent.

In North America by the nineteenth century, physicians prescribed "madstones." These calcified hairballs found in the stomachs of ruminants were thought to have curative powers by drawing the madness out of the bite wound. The cure-all "man's grease" (fat boiled down from an executed prisoner) was rubbed on the bite. American physician Henry Hartshorne (1823–1897) even suggested euthanasia with chloroform to reduce the suffering of painful spasms in the dying rabies patient.

For snake and other reptile bites, if ancient Egyptian healers thought the patient would survive the venom, they used a combination of magical incantations and other remedies. If they did not think the patient would live, they did not treat them. Healers cut into the wound many times, applied red natron salt used in mummification, and bandaged it to reduce the swelling and minimize the absorption of the venom. For treating crocodile bites, or other large bites with no venom, the Egyptian *Papyrus Ebers* (c. 1500 BCE) recommended bandaging the wound with fresh meat the first day. Following this, healers washed the lesion with natron and placed honey and goose fat in the injury. The meat may have helped coagulate the blood and honey has some antibacterial properties, which may have slightly reduced the chance of infection.

For snake bites, Dioscorides in the first century applied goat's dung mixed with wine to the bite, which likely introduced bacteria. He also recommended numerous substances to be consumed with wine including garlic, dried stag testicle, the

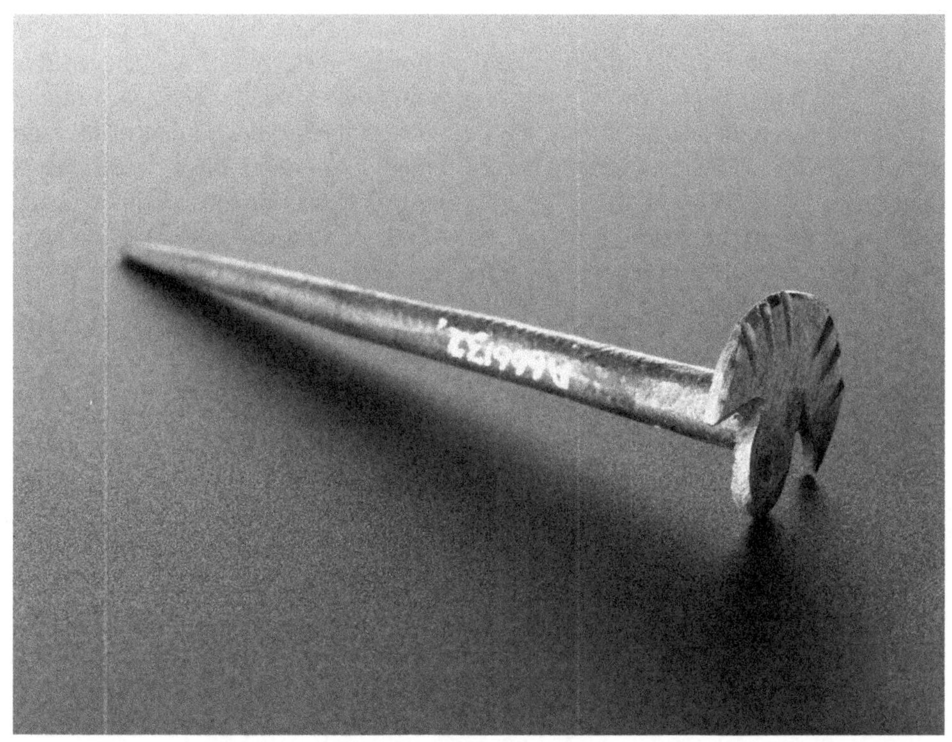

St. Hubert's Key was a treatment for rabies from the eighth to the eighteenth century. The heated key would be placed where the bite occurred. Sometimes it would cauterize the wound and kill the rabies virus, thus preventing the infection. (Science Museum, London. Attribution 4.0 International (CC BY 4.0))

brains of domestic fowls, or curdled milk from the stomach of an un-weaned rabbit or deer. Paulus Aegineta (c. 625–c. 690) in the seventh century first placed a tourniquet above the bite and then bled the patient by cupping.

In the eighteenth century, Buchan rubbed grease from a killed snake into the bite wound but thought the best remedy was sucking the bite to remove the poison. The ancient practice of cutting into the bite and sucking out the poison with the mouth was considered safe unless the person had a sore in the mouth. But this method could introduce bacteria and cause a wound infection. This was the primary treatment suggested by health personnel into the 1950s until antivenom was in general use. First aid kits for snake bites are still sold today and include a tourniquet, a blade for cutting the wound, and a suction device to suck out the venom but are generally not effective.

Another panacea for reptile bites to counteract the poison was ammonia, which was first introduced in France in 1747. Other physicians tried ammonia mixed with other items. American Edward John Waring (1819–1891), for example, in the late nineteenth century applied a liniment composed of equal parts of liquid ammonia, olive oil, and tincture of opium, which may have reduced the pain of a snake bite. He also recommended the patient drink a few drops of liquid ammonia in water. Some Australian physicians injected liquid ammonia into a vein, which could have killed the patient. Physicians also placed caustic material such as nitric acid or inserted a silver nitrate stick into the wound. They also kept the patients up walking and did not allow them to lie down or go to sleep as they believed they would die if they slept. However, walking would have spread the venom faster. A treatment recommended by many physicians was having the patient drink large amounts of whiskey. In addition to this remedy, American physician Elijah B. Hammack (1826–1888), also had patients chew and swallow tobacco juice, which had no effect on the wound. Over the centuries, some treatments were the same for both animal and reptile bites including cauterization, bloodletting, and plying the patient with large amounts of alcohol. However, these had little effect.

Bites and Stings: Insects and Arachnids

A variety of arthropods (a class of joint-legged invertebrate animals) can bite or sting humans and other animals. Arthropods include insects such as mosquitoes, bees, hornets, wasps, and ants along with arachnids such as spider and scorpions. Other arachnids include mites, lice, and ticks. Their bites cause local skin reactions such as redness, itching, stinging, swelling, and occasionally serious allergic reactions. In addition, they are also major vectors of zoonotic and other diseases. Ticks, for instance, carry numerous illnesses, including Lyme disease and Rocky Mountain spotted fever, and mites carry scrub typhus. Fleas and lice are vectors for bubonic plague and typhus, respectively, which are discussed in detail in separate entries. However, it has only been since the late nineteenth century that it has been known that many insects and arachnoids carry diseases.

Of all insects, mosquitoes cause the most bites and are the major disease vectors worldwide. Mosquitoes spread malaria, yellow, and dengue fever, which together are responsible for several million deaths and hundreds of millions of cases every year. (Malaria and yellow fever are discussed in detail in separate entries.) Dengue results in 20 million cases a year in more than 100 countries and causes a high fever, headache, vomiting, and muscle and joint pains. Those at highest risk are the young and the old.

Mosquitoes transmit several deadly encephalitis viruses that can cause brain damage or death. These include Japanese encephalitis found in Asia, and West Nile virus—which originated in Africa—now found around the world. In North America, some mosquitoes transmit Eastern Equine, Western Equine, and St. Louis encephalitis, among others, with 1 per 200,000 per year being infected in the United States. Murray Valley encephalitis is found in northern Australia. Most encephalitis viruses are cycled between birds and mosquitoes and transmitted to mammals. Mosquitoes are also vectors of emerging diseases including chikungunya, a viral infection that causes fever and severe joint pain, and Zika, which can spread from mother to developing fetus and cause serious birth defects. A Zika epidemic occurred in the Americas during 2015–2016, which affected some infants, particularly in Brazil.

Worldwide numerous venomous spiders exist that can cause pain and death. For example, the black widow, found in North America—other than Alaska and northern Canada—causes pain and swelling that can spread to other parts of the body, stomach cramps, and sweating. The bite of the brown recluse spider—found in the Midwest and southern areas of the United States, Mexico, and Central and South America—results in pain, a blue or purple center in the lesion that can form a large ulcer, and destroy flesh that may take several months to heal. No venomous spiders exist in the United Kingdom, and few are found in Europe. However, in Australia, several highly poisonous spiders are found, including the Sydney Funnel-web, and the Red-back Spider, which have caused deaths. Though most people are afraid of spiders, on the whole, they are not a large threat to humans.

Scorpions, which are also greatly feared, cause a stinging or burning sensation at the injection site. The venom can cause slurred speech, drooling, muscle twitches, abdominal pain, and cramps, which generally dissipate in a few days and rarely results in death. Two deadly scorpions are the Brazilian yellow scorpion and the Palestinian death stalker. Antivenin exists for both species' stings. However, bees, wasps, and hornets are deadlier and in the United States kill an average of 62 people per year from allergic reactions. Most of these stings can be managed today by various antihistamines.

Words in English for common arthropods have emerged since the Middle Ages. The first use of the term "arthropod" was during the mid-nineteenth century. It is derived from the New Latin *Arthropoda*, from the Greek *arthr*, meaning "joints," plus *pous*, meaning "foot." "Insect" is used from the early seventeenth century from the Latin *insectum*, meaning "animal with a notched or divided body." Some insects' names are from Old English. These includes "bee" from *beo*, which was also used as a metaphor for a "busy worker" since the 1530s; "wasp" or *wæps*, from the Latin *vespa*; louse from *lūs*; and flea from *flēa* in the twelfth century. Hornet is likely from the Old Saxon *hornut*. "Spider" was first used in in the late fourteenth century from the Germanic *spinne*. "Tick" is from the Middle English *tyke* from this same century.

Although humans have been bitten by insects since prehistory, most societies have considered these bites and stings to be a minor nuisance other than those of spiders and scorpions. In sixteenth- and seventeenth-century southern Italy, the bite of a "tarantula" (wolf spider) or even a scorpion was popularly believed to be highly venomous and to cause a dancing mania. The person, generally women from the lower classes, would fall into a stupor. When they heard their favorite tune, they jumped up, often ripped off some of their clothing, and danced until they fell to the ground exhausted and "cured" of their condition. This generally happened during the summer or harvest and led to bands of roving musicians traveling from village to village to play lively music, which became known as the tarantella. The dancing allowed women to act out in a repressive society.

Popular culture over the centuries has fostered the fear of insects and arachnoids. Today medical management of stings and bites includes effective antivenin and antihistamines to prevent serious toxic and allergic reactions, and death is now rare. For disease-carrying arthropods, immunization and prevention by use of insect repellant and other means is important. However, there is no cure for many mosquito-borne viral afflictions.

Spiders and biting insects have been featured in the arts. Spiders, in particular, have been highlighted in music, poetry, and fiction for centuries. A spider is a central character in the children's poem "Little Miss Muffat," the song "The Itsy-Bitsy

Spider," and the book *Charlotte's Web* (1952) by E.B. White (1899–1985). Spiders also influenced the comic book and films about the superhero *Spiderman* in the late twentieth and early twenty-first centuries. Other insects in literature include the Brothers Grimm's "The Louse and the Flea" and "The Queen Bee" in their collection of fairy tales first published in 1812. A poem "The Flea" by John Donne (1572–1631) describes how the blood of the poet was mixed with the woman he lusted for after both were bitten by the same flea. Scottish poet Robert Burns (1759–1796) wrote "To a Louse, on seeing one on a Lady's Bonnet at Church." Some modern popular culture sci-fi/horror films have depicted evil arthropods such as *The Black Scorpion* (1957) and *The Swarm* (1978). Mutant insects have also been featured in more recent films.

UNUSUAL TREATMENTS

Over the centuries, folk and home remedies have evolved for treating insect bites, and some are still used today. For example, rural individuals sometimes apply mud or urine on a sting or bite to subdue the itching. By the early twentieth century, some people painted "chigger" (mite) bites with nail polish. Today, people smear peppermint toothpaste and place wet aspirin tablets or meat tenderizer on mosquito, bee, and other insect bites, which people claim helps soothe the itching. Many remedies, both folk and professional, were based upon trial and error and may have calmed some itching and irritation.

In antiquity, the ancient Egyptian *Surgical Papyri* prescribed a raw meat poultice for bites, followed by oil and honey. Greek physician Paulus Aegineta (c. 625–c. 690), for bee and wasp stings, rubbed the swelling with clay, cow's dung, or fig juice and applied a paste made of barley flour mixed with vinegar. The cow's dung could have introduced infections but may have reduced the itching. For these stings, Persian physician Rhazes (865–925) suggested a mixture of camphor (think chest vapor rubs) and vinegar sometimes mixed with opium, which likely quelled itching and pain. To prevent stings from bees and wasps, he recommended a mixture of mallows (hibiscus type plant) and oil be smeared on the skin. Home remedies today include the crushed leaves of yellow yarrow, mud, vanilla extract, and some hand lotions rubbed on the skin as insect repellents, which are sometimes effective.

In the eighteenth century, Scottish physician William Buchan (1729–1805) applied honey or crushed parsley on the skin to reduce inflammation from numerous bees, wasp, or hornet stings. He also smeared a mixture of vinegar and warm salad oil on the bites. When the stings were so numerous as to endanger the victim's life, he bled the patient, which likely weakened the patient from loss of blood. He also ordered "cooling medicines" to drink such as cream of tartar, lots of whey, buttermilk, warm water, or thin water gruel, which were harmless but had no effect. Some mid-nineteenth century physicians, who thought bites were acrid or acid, applied alkalis such as ammonia, potassium, salt, and water to the lesions, while others applied alcohol from distilled wine. These may have soothed the itching. American physician Elijah B. Hammack (1826–1888) advocated the folk remedy of using a chewed tobacco plug as a poultice or smearing tobacco juice on the sting, which may have temporarily soothed the itching.

More terrifying than bees and wasps were spider and scorpion bites. Some remedies were utilized for both throughout the ages. In ancient Egypt, for example, if healers thought the patient would survive the venom of these arachnoids, they used a combination of magical incantations and surgical remedies. They would cut the wound many times, apply red natron salt used in mummification, and bandage it to reduce the swelling and minimize the absorption of the venom. Greek physician Dioscorides (c. 40–90) claimed that honey drunk with asses' milk helped prevent the bites of snakes, venomous spiders, and scorpions.

In terms of spider bites, Paulus in the seventh century bathed patients in hot water, which temporarily relieved the pain of the bites. He also applied ashes of figs mixed with salts soaked in wine. In Persia, people bitten by spiders were drenched in milk and then put into a suspended box and violently rotated until they vomited. Since many survived anyway from these bites, these were not effective cures.

In southern Italy during the sixteenth and seventeenth centuries, treatment for the so-called bite of the tarantula spider was music. People would surround the bitten victim while musicians played mandolins, guitars, and tambourines in search of the correct rhythm. Each beat would have a different effect and cause the victim to exhibit different movements. Once the "correct rhythm was found," the person would be cured. Some patients found that rocking in a swing or cradle or having someone administer severe blows to their feet would cure the condition. In the mid-nineteenth century, a few physicians in North America favored bloodletting to treat tarantula and other spider bites along with drinking large quantities of alcohol to counteract the poison. They also placed a tobacco poultice on the bite to draw out the poison. These were not likely effective other than because of the placebo effect.

Some specific treatments were used for scorpion bites. For example, Dioscorides in the first century cut open a house mouse and placed the cut part on the bite to draw out the poison, which could have introduced infection. Some healers poured fig juice into the wound and pounded the scorpion that had stung the person and applied it to the bite. Following this, sulfur, mixed with rosin or turpentine, was placed on the bite, which may have been soothing. Greco/Roman physician Galen (129–c. 210) gave patients an equal amount of opium and henbane seeds (a poisonous plant) mixed with honey to drink. Opium and henbane would have relieved pain but could have caused coma and death in large amounts.

Paulus in the seventh century noted that earlier physicians directed people to immediately place a tourniquet around the limb and apply a dead scorpion to the bite in addition to drinking much wine. The wine may have taken away pain, but the other ingredients were not effective. English physician John Arderne (1307–1392) in the fourteenth century made cuts around the scorpion sting followed by smearing blood from the wound over the bite. In the nineteenth century, British surgeon John Waring (1819–1891), from his experiences in India, placed nitric acid into scorpion wounds to eliminate the poison. None of these remedies were likely effective, and nitric acid could have caused disfigurement.

For bites from insects, such as mosquitoes, Waring applied lemon juice to the lesion. He also rubbed a liniment composed of liquid ammonia, olive oil, and "T. Opii" (opium mixed with alcohol and camphor), which would have stopped the itching. He claimed that lemon juice also took away the itching. For lice, some applied kerosene to the body or head to kill lice, which was effective. Others applied tobacco and various lotions and fumigated the patient with a mercury vapor bath, wherein the fumes of mercury were volatized by heat and mixed with steam. Besides killing the lice, this treatment would have poisoned the patient. Some of these remedies may have helped quell the itchiness and burning of bites and stings, but they would not have cured the diseases that the insects and arthropods carried.

Bowel and Rectal Issues

The most common bowel and rectal issues include constipation and irritable bowel syndrome, hemorrhoids, and colorectal cancer. Diarrhea is also common but is discussed in the "Gastroenteritis, Dysentery, and Diarrhea" entry.

Constipation and Irritable Bowel Syndrome

Most people at some point will suffer acute constipation, in which fecal material is hard and lumpy and difficult or painful to pass. This occurs when there is not enough water and lack of fiber in the colon. Chronic constipation is defined as fewer than three bowel movements a week, which persists for several weeks or longer. Long-lasting constipation may have complications including hemorrhoids, anal fissures, rectal prolapse, and fecal impaction. Chronic constipation is considered a functional disorder, with no evidence of abnormality. About 16 percent of adults and roughly a third of those 60 and older have constipation symptoms. It is a major bothersome issue for people.

Another functional disorder is irritable bowel syndrome (IBS). It is a group of symptoms, and chronic constipation is a major one. IBS was defined in the 1940s, but some physicians by the early twenty-first century argued that chronic constipation and irritable bowel syndrome were different facets of the same functional disorder. People with IBS have chronic constipation, often alternating with diarrhea and abnormal pain. Unlike those with chronic constipation, they tend to be younger. About 10–15 percent of the population worldwide suffers from IBS. Both chronic constipation and IBS are associated with being female, lack of exercise, low fiber diet, taking certain medications including opiates, and some sedatives and antidepressants. The impact of these conditions can range from a mild inconvenience to debilitation, which can influence many aspects of a person's life.

The term "constipation" was first recorded around the fourteenth century and was derived from the Late Latin *constipatio*, meaning "packed together," or a bowel condition where evacuations are difficult. However, throughout human history, healers believed that bowel irregularity was dangerous to health, in particular, constipation, as they feared fecal decay led to physical decay. Around 1500 BCE, the Cyril P. Bryan (1884–1940) translation of the Egyptian *Papyrus Ebers* suggested that all disease is the result of poisoning of the body by material released from decomposing waste in the intestine. In ancient Greece, Hippocrates (c. 460–370 BCE) in *Of Epidemics* discusses constipation and bowel problems, and he believed diseases were caused by imbalance in body humors as did many Western physicians throughout the 1600s.

Other theories were advanced over the centuries concerning the cause of constipation and its consequences. In the West beginning in the seventeenth century, a popular belief was that constipation was caused by moral failing due to gluttony or slothfulness (laziness). In the eighteenth century, German physician Johann Kampf (d. 1753)—like ancient Egyptian belief—proposed that all diseases stemmed from impacted or hard stool. In the late nineteenth century, several researchers theorized that poisonous "ptomaines" arose from constipated feces and produced diseases. Therefore, the faster the stool could be removed, the less likely the person would become sick.

Also, in the late nineteenth century, one physician proposed a theory of "autointoxication" or self-poisoning from one's own retained wastes and bacteria. This concept was the rage into the 1930s, and daily evacuation of the bowels was considered important to prevent disease. These theories were abandoned as they were not supported by research. By the twenty-first century, several physicians hypothesized that a "gut-brain axis microbiome" or a bidirectional interaction between the central nervous system, the enteric nervous system, and the gastrointestinal tract with its numerous bacteria was the cause for constipation and IBS and other health issues. This theory is still under examination. Today to prevent and treat constipation and IBS, a high fiber diet, exercise, and stool softeners are often recommended. Some prescription medications also became available by the early twenty-first century.

Hemorrhoids

Hemorrhoids—also called piles—can cause rectal bleeding, itching, pain, and sometimes fecal leakage. Hemorrhoids are normal structures and are present in all individuals, but the term "hemorrhoids" is now generally used for disease symptoms. In this meaning, they are varicose veins in the anus called external, and in the lower rectum called internal, hemorrhoids. Although their exact cause is unknown, some physicians believe they are associated with chronic diarrhea, pregnancy, obesity, constipation, sitting for long periods of time, anal intercourse, heavy lifting, and genetic influences. About 75 percent of adults occasionally sufferer from the condition, and it is most common in both men and women between 45 and 65 years of age. The major problem of piles is that it impacts the quality of life. However, most cases are self-limited.

The first use of the word "hemorrhoid" in English occurred in the late fourteenth century and is derived from the Old French *emorroides*, from the Latin *hæmorrhoida*, meaning "bleeding veins." The first-known record of this condition is from the Egyptian *Papyrus Ebers*. The Greek physicians Hippocrates and Galen (129–c. 210 CE), who worked in Rome, also wrote on the subject. Scottish physician William Buchan (1729–1805) in the eighteenth century remarked that this ancient condition was caused by too much sweet wine, highly seasoned food, overweight, inactive life, constipation, horseback riding in "thin breeches," and anger, grief, and "other violent passions." In addition, during the nineteenth century, physicians believed masturbation, anal intercourse, frequent train rides, wearing too tight clothes, and spanking could cause them.

Today the treatment of piles consists of over-the-counter ointments, creams, and suppositories that contain local anesthetic for pain, corticosteroid for itching, and a vasoconstrictor to reduce swelling. Surgeons also perform various procedures such as tying off the hemorrhoid with a rubber band, infrared coagulation, and other procedures.

Colorectal Cancer

Although most people in the early stages of colorectal cancer do not have symptoms, in the later stages, bleeding, chronic diarrhea or constipation, change in size or shape of stool, lower abdominal pain, bloating, and weight loss without dieting can occur. Colorectal cancer usually starts as a polyp in the lining of the colon and grows slowly. Around 5 percent of people will develop colorectal cancer during their lifetime, and it is the third most common type of cancer in the world. It is more prevalent in developed countries, where more than 65 percent of cases are found. About 71 percent of these cancers emerge in the colon and about 29 percent in the rectum. The cause of colorectal cancer is unknown although it is associated with older age, genetic factors, poor diet, heavy alcohol consumption, obesity, smoking, and lack of physical activity. Moreover, it is an ancient disease and not just associated with modern life. Rectal cancer has been found in archaeological remains of an Egyptian mummy from around 305–300 BCE.

Hippocrates in his writings first used the terms "carcinos" and "carcinoma" to describe non-ulcer-forming and ulcer-forming tumors. They are from the Greek term "crab," as he observed many tumors had projections resembling crab legs. He believed the disease was caused by an imbalance in the humors, in particular "black bile," which remained a popular theory into the nineteenth century. The Roman physician, Celsus (28–50 BCE), later translated the Greek term into *cancer*, the Latin word for "crab." In earlier times, it was less common because people died at a younger age from other conditions. Throughout history, cancer has been known as an incurable disease.

In the eighteenth century, Buchan believed bowel cancer was caused by celibacy and the religious life. By the nineteenth century, physicians thought that people inherited the potential for cancer, and both the public and physicians viewed it as a death sentence. Cancer patients were often stigmatized, and the disease was considered a taboo topic of discussion. In Western cultures into the 1960s, many physicians did not disclose the diagnosis to their patients. This still occurs in some Asian nations.

Colon cancer was difficult to diagnose as it was hidden until the development of the sigmoidoscope in the 1890s, which allowed for the examination of the descending colon. By the late twentieth century, screening was accomplished with the flexible colonoscope that examined the whole colon. Due to more societal openness about the malady, screening, and early detection, the survival rate for colorectal cancer has increased. Today it is treated by many therapies including surgery, chemotherapy, targeted radiation therapy, among others.

Bowel issues may have affected the course of history; for example, French military leader Napoleon Bonaparte (1769–1821) had many illnesses that may have influenced his decision-making. For instance, before a battle he usually inspected the battlefield. However, he may have had a painful attack of hemorrhoids on the morning before the Battle of Waterloo (1815) and did not inspect it, which resulted in his final defeat. In another example, German Chancellor Adolf Hitler's (1889–1945) physician recommended Hitler take a strychnine compound for constipation and gas and to increase energy. This concoction may have caused his mental aberrations and diabolical behavior.

UNUSUAL TREATMENTS

Various treatments including folk and traditional medicines and those offered by physicians have been found since recorded history for bowel and rectal issues. Some of the earliest are recorded in ancient Egyptian papyri.

Constipation and Irritable Bowel Syndrome

The *Papyrus Ebers* gives numerous recipes for purgatives "to drive out the excrement in the belly." The concoctions were drunk and eaten or made into pills or suppositories. Castor oil from beans of the *Ricinus communis* plant was a common ingredient. However, these beans are highly toxic unless cooked. One remedy called for the patient to chew some beans and swallow them down with beer—the Papyrus does not mention if they were raw or cooked. Wine was mixed with grapevine ashes, manure, and hellebore (a poisonous plant), and goose fat, figs, dates, cumin, and wormwood were combined with other ingredients, which did work to relieve constipation but in some cases may have killed the patient.

Clysters or enemas were another treatment for constipation used since antiquity. The procedure was carried out with metal, ceramic, or ivory syringes until the mid-nineteenth century when rubber bags with hoses and bulbs took their place. Enemas were popular in Western medicine until the late twentieth century when their use declined other than cleaning the bowel before some medical procedures. However, enemas are still popular in alternative and folk systems as "colonic cleansing" but are not effective in permanently curing constipation, IBS, or other conditions.

The purgative, or laxative, calomel (contains toxic mercury) was used from the sixteenth to the early twentieth century for constipation and most other ills that could poison the patient. In the nineteenth century, it was fashionable to drink mineral waters offered at spas in Europe and North America as a cure. The golden age for laxatives was during the 1920s and 1930s. Hundreds of patent medicines were advertised, some of which contained toxic substances such as strychnine. This and other toxic purgatives were banned in most Western cultures by the late twentieth century.

Over the years, various devices were developed to treat constipation. In the seventeenth century, some physicians used the *spatula mundani*, a spoonlike device that extracted "hard excrements."

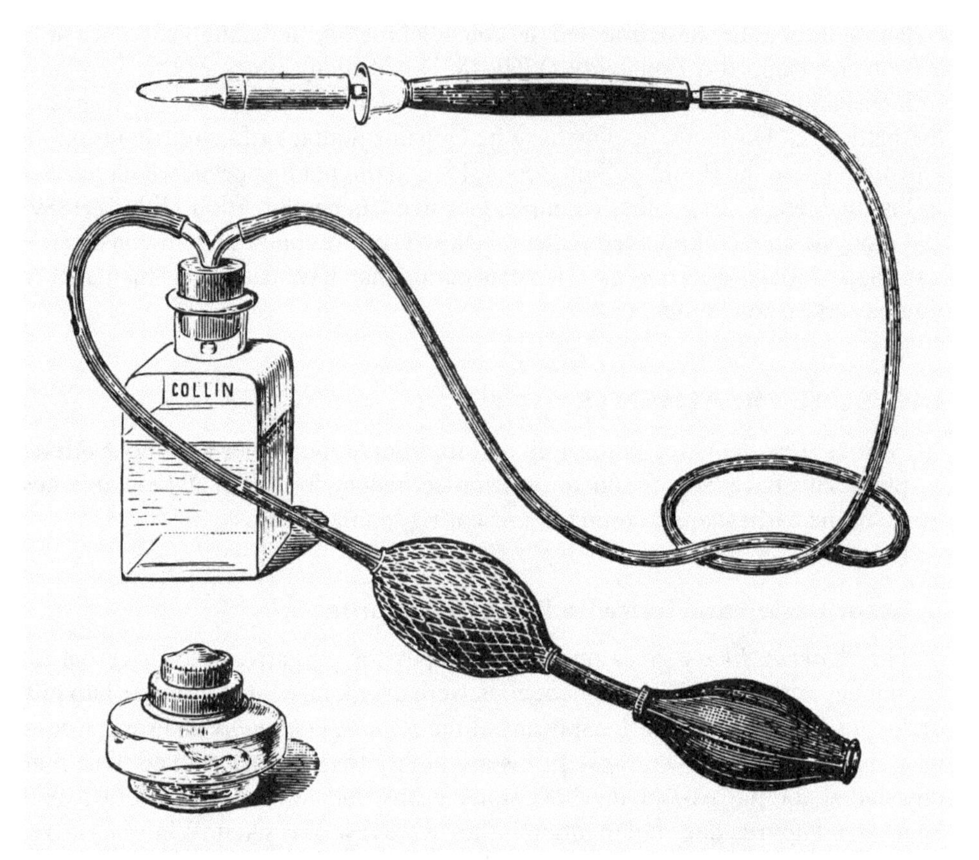

Illustration of cautery tools from the nineteenth century to burn off hemorrhoids and skin conditions and stop bleeding. (© Patrick Guenette/Dreamstime.com)

In the late nineteenth through mid-twentieth century, electrical stimulators and hairbrushes, shocking devices, and rectal dilators—that came in increasingly larger sizes—were marketed. Most were prohibited by the 1980s in Western nations as false advertising and were worthless.

However, one successful treatment for a range of chronic bowel conditions has been fecal microbiota transplantation. A fourth-century Chinese physician Ge Hong records that he administered fecal material from a healthy person as "yellow soup" to patients. Fecal transplantation in Western medicine was not fully recognized as useful until the early twenty-first century when it was found to successfully treat several bowel conditions.

Hemorrhoids

Remedies for hemorrhoids have existed for centuries. The *Papyrus Ebers* suggests mixing onion meal with the tail of a mouse and placing it on the anus. Another suppository was made of cumin mixed with lard to stop itching, which may have worked as a counterirritant. Another treatment was a poultice made of a goose egg and the guts of a goose. In 460 BC, Hippocrates discussed a treatment similar to a modern rubber band ligation in which a cord is tied around the protruding pile until it later drops off.

Leeches have also been used to treat hemorrhoids for centuries, which may have been somewhat effective in reducing the swelling. To prevent a leech from crawling up the anus, a thread was attached. Cautery was also used. Hippocrates used a red-hot iron and placed it in the anus to burn off the hemorrhoid, which would have been very painful for the patient. A thirteenth-century image shows a surgeon cutting off a hemorrhoid, and a fourteenth-century illustration shows a surgeon placing a hot poker in the anus to cauterize the pile. Today electro cautery is used, with anesthesia to take away pain, and it is generally effective.

Up until the mid-nineteenth century, many physicians prescribed bloodletting as a treatment, which was useless and could have caused weakness in the patient. On the other hand, American physician Elijah B. Hammack (1826–1888) destroyed piles by nitric acid painted on the protrusion or removed them with scissors, which would have been painful. By the late nineteenth century, numerous products were advertised to reduce hemorrhoids.

Colorectal Cancer

Since colorectal cancer was generally incurable until the late twentieth century, regular physicians competed with folk medicine, patent medicines, and alternative systems that claimed cures. Healers over the centuries have tried many remedies. For example, in the second century Galen suggested burning crabs and smearing the ashes and crab bits onto rectal and other tumors. Liquid tobacco enemas were used in the sixteenth century as tobacco was a cure-all. Throughout the mid-nineteenth century, bleeding, purging, vomiting, and opiates for pain were commonly used but did not cure the condition. By the late nineteenth and into the twenty-first century, hundreds of alternative and folk remedies emerged, and numerous health gurus claimed dubious treatments. A notorious example was American Albert Abrams (1893–1924). In 1916, he proclaimed that a disease radiates certain radio frequencies, and to measure these frequencies and diagnose the disease, he developed the "dynamizer." To cure the malady, he developed the "Oscilloclast" radio frequency device. They were nothing but a jumble of radio tubes and wires and had no effect. It was exposed as fraudulent and it fell out of use in the late 1920s.

Another treatment was the grape diet and enema regime for colon cancer. Its creator advertised that grape juice would eliminate poisonous wastes in the colon and would cure the disease. In 1940, the Federal Trade Commission ordered that such misrepresentations be discontinued. In the 1970s, several cures involving vitamins were touted. American chemist Linus Pauling (1901–1994) claimed that mega doses of vitamin C could prevent and cure colon and other cancers. Research did not support this claim, although high vitamin doses along with conventional therapy in some cases have been shown to reduce the death rate from colon cancer. A synthesized cyanide containing substance found in apricot pits called laetrile (Amygdalin or vitamin B17) was and is marketed by alternative therapists as a cancer cure. However, there is little reliable evidence that it works.

In the late twentieth and early twenty-first centuries, numerous remedies were advertised as cures for cancers. These included, but are not limited to, electromagnetic, light, and energy devices; physical remedies such as acupuncture, colonic irrigations, and spinal manipulation; special diets, fasting, and certain foods; and the elimination of toxic substances. As an example, fear of mercury from amalgam in filled teeth as a cause of colorectal cancer led some alternative practitioners to advise cancer patients to have their teeth pulled, claiming it would cure the disease. Since mercury would have supposedly leaked into the circulatory system, in addition, practitioners recommended chelation therapy. This treatment involves the injection or consumption of chelating agents that chemically bind with metals. Little evidence has been found as to the effectiveness of any of these remedies to cure colorectal or other cancers, but they are still advertised today by alternative healers.

Burns

Burns are tissue injuries caused by excessive heat, chemical, radiation, or electrical contact. They are classified by the depth of tissue damage. Superficial or first-degree burns of the epidermis are red, swollen, and painful, but the burnt area does not blister. Second-degree, or partial-thickness, burns extend into the middle layer of the skin or dermis. They are deep red, form blisters, and are painful to touch, and the burnt area may appear shiny. Third-degree or full-thickness burns involve the top two layers of the skin along with the fat layer, sweat glands, hair follicles, and nerve endings. They result in white, blackened, or charred skin that may be numb and appear leather like. German physician Guilhelmus Fabricus Hildanus (1560–1634), in 1634, was the first to recognize the association of the length of time that heat is in contact with the body with the degree of burn damage.

Burns caused by heat generally result from contact with flames, touching hot objects such as stove top burners, or being assailed with gunpowder or military weapons. If they are caused by hot liquids such as boiling water, coffee, steam, or oil, they are often referred to as scalds. Chemical burns may occur from strong acids or alkalis, such as nitric acid or lye, respectively, or white phosphorus used in munitions, mustard gas, quicklime, or certain petroleum products. Radiation burns most commonly result from prolonged exposure to solar ultraviolet radiation (sunburns) and rarely from x-ray or nuclear disasters. Electric burns from lightning strikes or powerlines frequently cause extensive deep tissue damage. Burns have always frightened people as they have been associated with pain and often death. In the past, death has mostly occurred from infection and loss of body fluids.

In the United States, over a million people each year suffer burn injuries that require medical attention. Approximately 20,000 suffer major burns covering 25 percent of their body surface—burns that exceed 30 percent of the body are potentially fatal. To estimate the body surface area of a burn, Scottish physician Alexander Burns Wallace (1906–1974) in 1951 published the Rule of Nines. This assigns a percent to each major body part that can be divided by, or is, a multiple of nine. (For example, an arm is 9 percent of the body surface, and each side of the arm is 4.5 percent.)

Burns and scalds are most common among males and those under 2 and over 60 years of age. Globally, burns are a serious public health problem with an estimated 265,000 deaths each year from fires alone mostly in developing nations. Burn deaths were reported in the past when mortality statistics were first collected but were not likely accurate. For example, English civil servant John Graunt (1620–1674) reports that in 1632 five people in London died from burns and scalds out of the 9,535 deaths. In the United States, the 1850 mortality statistics reports that over 1,700 died from burns. However, by the early twenty-first century, around 10,000 people die every year from burn-related causes in the United States. About 75–85 percent of these deaths are due to sepsis (blood poisoning). Due to advancements in medicine, 96.7 percent of those treated in burn centers now survive although they often have permanent impairments and scars.

The first use of the term "burn" in English was in the twelfth century and is likely from Old English. "Scald" is from Middle English from the Late Latin *excaldare*, meaning "to wash in warm water." In warfare, using flames and/or hot liquids to burn and scald an enemy has been used for centuries and has shaped the course of history. For example, the infamous weapon of the Byzantine Eastern Christian Empire was "Greek fire"—a burning mixture of resin, pitch, sulfur, and petroleum—that was difficult to extinguish. It was first used when Constantinople was besieged by a caliphate during 673–678. The fire burned the caliphate's ships and crew resulting in a Byzantine victory. To injure and disable opponents during a siege, the simplest and most common method was pouring boiling water and hot sand over attacking soldiers. Hot oil, a scarcer substance, was also used. The French, for example, during the Siege of Orleans (1428–1429), poured scalding oil, hot coals, and quicklime on the English combatants who retreated from the city.

On the other hand, survival from burn injuries increased over the twentieth century due to lessons learned from conflicts. These experiences resulted in the initial medical management of first- and second-degree burns of placing cool water on the burn and covering it with sterile pads. Treatment for serious burns is now accomplished in specialized burn centers with intravenous fluid, antibiotics, excising of dead tissue, grafts from several substances, stem cell therapy, and other measures.

Combat burns may have influenced the outcome of battles. Burns became increasingly common with the use of high explosives in WWI (1914–1918). These burns were partially due to white phosphorous, which can spontaneously ignite and burn flesh. During WWII (1939–1945), Nazi Germany carried out firebombing with incendiary bombs against Britain during the Blitz. The Allies then used these weapons against Germany and Imperial Japan, resulting in victory. These devices turned battlefields into burning infernos, destroyed communities, and burned and demoralized citizens. The United States used flamethrowers and two nuclear weapons against Japan, resulting in many burn injuries, but this led to the war's end. In the early twenty-first century, during conventional warfare, between 5 and 20 percent of combat casualties were from burns; however, due to medical advancements many now survive. But even small burns can incapacitate and strain the resources of medical facilities.

UNUSUAL TREATMENTS

Many folk treatments were developed over the centuries for burns by trial and error. A common one, still used by some today, is coating the burn with grease, butter, or oils, which can slow the release of heat from the skin and cause more damage from the retained heat. Vinegar for centuries has been applied to burns. The acetic acid in vinegar relieves pain, itching, and inflammation. By the early twenty-first century, the medical field rediscovered vinegar soaks for its antibacterial properties. Other folk remedies include putting toothpaste or even cool mud on burns but may contaminate the wound and cause infections.

Several treatments for burns were prescribed by the ancient Egyptians. The *Papyrus Ebers* (c. 1500 BCE) gives many remedies. Some involved animals or their products. For example, healers warmed a frog or the head of an electric eel in oil and rubbed them on the burn. Cat dung and hair balls were also placed on the burn. These remedies likely caused infections. They also gave incantations to the god Horus spoken over milk from a woman who had given birth to a son.

Some ancient Egyptian healers applied a series of remedies to burns. These included crushing goat dung and mixing it with fermenting yeast or adding roasted cow's fat to palm fibers and grain to make a plaster. Another burn covering consisted of onions mixed with red lead, the fruit of the unknown "Am-tree," and pieces of copper. The lead and copper could have helped prevent infections, but goat dung could have caused infections.

Likewise, herbs and plants were used as poultices to treat burns. Ancient Egyptian healers applied honey—which has antimicrobial properties—and the papyrus plant mixed with resin and seeds from the acacia tree that contained tannic acid and helped cool the burn along with aloe vera. Today cooled tea bags, which also contains tannic acid, aloe vera, and honey, are still used as a folk remedy, which may quell the pain and help prevent infection. In addition, in the early twentieth century to dry up serious burns, clinicians sprayed tannic acid on the wound, but found it damaged the liver, so this treatment was discontinued.

In ancient Greece, snails were used to treat burns and skin irritations. Hippocrates (c. 460–370 BCE), for example, prescribed crushed snail shells mixed in oil to treat burns and other skin inflammation. By the late twentieth century, snail slime (what the snail leaves in a trail) in beauty creams were and are still marketed, after snail farmers found their hands remained soft from handling the snails. By 2010, dermatologists used purified snail slime to successfully treat second-degree burns.

In addition to snail slime, Hippocrates prescribed several other concoctions to be placed on burn wounds. He alternated a mixture of melted pig fat, resin, and tar with vinegar. Roman physicians in the first century alternated vinegar with a tanning solution made from oak bark, which may have helped quell infections. Physicians mixed wine and myrrh as a burn lotion, which had bacteriostatic properties. Another remedy prescribed was a mixture of lilies and poisonous Hound's-tongue leaves boiled in oil and wine and placed on the wound. Cimolian chalk (probably talc) was ground with the bark of frankincense and turned into a paste with water. Before it was applied, it was mixed with diluted vinegar. Once the scabs had fallen off, vetch plants with honey, or iris, or turpentine resin were applied, which likely had little effect in preventing scars.

Later, Byzantine physician Paulus Aegineta (c. 625–c. 690) recommended earth mixed with vinegar and water, whole raw eggs, or gall from a bull be placed on burns. Another remedy called for pigeon's dung to be burned with linen and the ashes mixed with oil and applied to the wound. For scalds, before blisters arose, the brine of pickled olives was poured on the burn. This had antiseptic properties, but these other remedies likely caused infections.

During the seventeenth and eighteenth centuries, some British physicians recommended holding a slight burn near a fire, rubbing it with salt, and placing a compress upon it dipped in brandy. If the burn was deeper, they applied olive oil or an egg beaten with an equal quantity of the "sweetest salad oil," or wax melted in oil. Bloodletting and cathartics—to keep the "belly open" and expel bad humors—were also prescribed. Other than brandy that could numb minor burns, these other treatments likely increased the injury, caused more pain, introduced infection, or hastened death.

By the early nineteenth century, Scottish physician Alexander Macaulay (1783–1868) admonished people for placing a burnt part near a fire or placing turpentine on it. He correctly suggested immediately applying a cloth with cold water to the burn with, or without vinegar, but incorrectly suggested pricking blisters with a needle to allow fluid to escape or, for deeper burns, smearing carron oil, made with equal parts of linseed oil and lime water, on the burn, which could have introduced infection and often brought pain. Physicians also recommended cotton, glycerin, starch powder, lard, molasses, or even fruit jellies be placed on burns.

By the early twentieth century, clinicians applied silver salts such as a silver nitrate solution, which reduced infection and mortality from serious burn wounds. In the 1960s, silver sulfadiazine cream—a sulfa antibiotic—was developed and is still a mainstay of topical burn therapy. Silver, unlike other heavy metals, is nontoxic to humans and has antimicrobial properties. By the 1970s, pig and cadaver skin grafts became more widely used to reconstruct serious burns.

Cancer

Cancer, or malignant neoplasm, is a generic term for a large group of diseases that can affect any part of the body and is caused by an uncontrolled division of abnormal cells that often spread (metastasizes) to other sites. Worldwide, the most common cancers are lung, breast, colorectal (discussed in the "Bowel and Rectal Issues" entry), prostate, skin, and stomach. Symptoms are different for each type and include coughing, lumps in the breast, rectal bleeding, difficulty in urination, discolored skin lesions, and pain in the abdomen. Fatigue and loss of weight are common in later stages. Cancers are initiated by hereditary and environmental influences. Risk factors include smoking, heavy alcohol consumption, obesity, and a diet low in fruits and vegetables, in addition to environmental factors such as exposure to certain chemicals, radiation, and microorganisms.

Cancer incidences have increased dramatically since the Industrial Revolution. This is likely due to pollution and change in diet and lifestyles. In addition, before the twentieth century cancer was less common because people generally died in midlife from infectious diseases. By the mid-nineteenth century, cancer was viewed as a woman's disease due to high death rates from cervical, uterine, and breast cancer. Because these often ran in families, clinicians saw this as evidence of the inheritability of the malady. For centuries, cancer was stigmatized as it was considered a death sentence in most cultures, and clinicians believed that intervention might be more harmful than no treatment at all. This fatalistic attitude was accepted by family and physicians until the late 1960s. The diagnosis generally was not revealed to the patient, and cancer was a taboo subject of conversation, leaving children ignorant of a potential inherited health issue. In many developing nations, moreover, cancer in the 2020s is still surrounded with stigma, myths, and taboos.

By the 1990s, worldwide cancer was the second leading cause of death, and by 2020, it resulted in about 21 percent of all deaths. Higher-income regions such as North America, Europe, Australia, and New Zealand saw a plateauing effect or reduction in the mortality from some cancers due to decreased smoking, screening and early detection, and treatment. However, cancer was the leading cause of death in Europe and Canada as opposed to cardiovascular disease in the United States. On the other hand, many developing nations experienced a surge in some cancers. For example, lung cancer has dramatically increased since the 1990s in Asia, in particular China, due to smoking, and it is the leading cause of cancer deaths throughout the region and in the Latin American and the Caribbean region.

Cancer was diagnosed in antiquity and is as old as prehistoric animals. Bone cancer has been found in dinosaurs going back at least 64 million years, and rectal/

prostate cancer has been identified in an Egyptian mummy in around 200–400. In addition, the *Edwin Smith Papyrus* (c. 1500 BCE) described eight cases of tumors of the breast. Greek physician Hippocrates (c. 460–370 BCE), based upon humoral theory, believed that an excess of "black bile" in a particular site would cause cancer. He used the term "carcinos" to describe a tumor and "carcinoma" (*karkinos*) to describe a malignant tumor as these tumors often had creeping projections that resemble crab legs. Roman physician Aulus Celsus (c. 25 BCE–50 CE), in his *De Medicina*, translated this Greek term as "cancer," the Latin word for "crab." The first use of the term "cancer" in English was in the seventeenth century from this Latin word. Although it was generally incurable in the past, management of cancers today include surgery and radiation combinations, chemotherapy, immunotherapy, experimental gene editing and other procedures, and immunizations against viruses that cause neoplasms. Screening tests such as mammograms to detect breast cancer, the Papanicolaou (Pap) test to detect cervical cancer, and colonoscopy to detect colon cancer have saved many lives beginning in the late twentieth century.

Cancer has been depicted in the arts. The earliest evidence is found in two Italian Renaissance paintings that show the artist's models with probable breast cancer. Artists tended to paint objective portraits of paid models who were usually from the lower classes. Michele di Ridolfo or Tosini's (1503–1577) oil, *The Night* (c. 1554), shows the model's left breast smaller than the right with her nipple indented. In another painting, *The Allegory of Fortitude* by Maso da San Friano (1531–1571), the woman sitting on top of a lion has a possible ulcer under the breast where a tumor has broken through the skin. Fiction has also dealt with cancers. For example, the romantic novel and film (1970), *Love Story*, by American writer Erich Segal (1937–2010) is about two college students who meet, fall in love, and get married against his father's wishes. The wife attempts to get pregnant and is diagnosed with leukemia and dies.

UNUSUAL TREATMENTS

Numerous alternative and folk treatments for cancer have been advanced. Many emerged out of the twentieth century. In earlier times, healers attempted to treat various forms of cancer but realized there was no cure for the scourge. For example, the Egyptian *Edwin Smith Papyrus* recommended cauterizing breast cancer tumors in women with a heated instrument called the "fire drill." This was a sharp piece of wood that was heated by rapidly twisting it on a flat piece of wood. This treatment would have been painful and would have only temporarily halted the progression of the tumor. Likewise, the *Papyrus Ebers* recommended surgery to cut out various superficial tumors. For uterine cancer, it advised injecting a concoction of fresh dates and pig brains, which had been left overnight in the dew, into the vagina. Celsus advocated applying boiled cabbage and a salted mixture of honey and egg white to superficial tumors. These were ineffective but usually did not harm the patient.

Greek physician Paulus Aegineta (c. 625–c. 690), based upon humoral theory, recommended placing desiccants and astringents such as sumac, galls, alum, frankincense, and calamine (zinc and ferric oxide) on an open ulcer from a tumor to dry it up. But these did not stop cancerous growths. By the Middle Ages, for skin cancers, a freshly killed rabbit, puppy, kitten, or lamb was held against the tumor. It was hoped that the cancer would feed off the animal rather than the human. Leeches were also placed on tumors to suck out the bad humor. However, these remedies likely led to skin infections or death from over-bleeding.

Scottish physician William Buchan (1729–1805), based upon tradition, administered crushed hemlock leaves and also used them as a poultice along with ground carrots for open tumors and breast and prostate cancer. He bled his patients, gave them mercury pills, and rubbed the tumor with mercury ointment. For breast cancers, he also administered an infusion of the nightshade plant. These recipes likely hastened death due to the poisonous nature of hemlock, mercury, and nightshade. Some physicians also suggested patients eat grapes. Although red grapes have been found to have some anticancer properties, a large amount would need to be ingested to have any effect. Clinicians also advised patients to drink three or four pints of wort—malt infusions—and as a last resort take opium. Possible alcohol in the beer wort and opium likely reduced pain but had no effect on the progress of the disease.

In the twentieth century, almost every decade produced a new alternative or folk cancer treatment that research showed was not safe or effective, so it was rejected by the FDA. Early in the century, various spurious electronic, radio, and light wave treatments were touted to control cancers. As an example, in the 1910s, American physician Albert Abrams (1863–1924) claimed that illnesses radiated certain radio frequencies that he could diagnose with a "dynamizer." If a patient had cancer, or any other illness, Abrams leased them an "Oscilloclast box," which was supposedly tuned to the radio frequency of the disease to destroy it. This bogus contraption was expensive and did nothing for the patient. Using light waves, Indian American Dinshah P. Ghadiali (1873–1966) proposed that color therapy acts to reinforce or interfere with a person's "aura" and could change a disease condition. In 1920, he developed "Spectro-Chrome Therapy" that delivered "tonations" of specific colors on different parts of the body. The light waves came from colored glass placed in a hole in a box through which the light from a light bulb was shown. For most cancers, lemon and indigo were used. In the twenty-first century, this method is still advertised. The device can be made at home and is harmless but not effective in eliminating cancers.

Numerous alternative chemical and folk herbal treatments were also marketed for treating cancers. The FDA after intensive research studies deemed most of them ineffective. For example, American physician and pharmaceutical entrepreneur William Frederick Koch (1885–1967) developed and touted a "synthetic antitoxin" as a cancer cure that his colleagues declared fraudulent in the 1930s. In the 1940s, he developed and marketed "glyoxylide," which the FDA found was nothing but distilled water that Koch injected into cancer patients for a high fee. Around 1950, after the FDA sued Koch, he moved out of the country but sold the substance out of Mexico. Likewise, American naturopath Harry Hoxsey (1901–1974) promoted a treatment for cancer that included an external paste that contained caustic substances such as arsenic sulfide that could sometimes burn the skin. He prescribed internal tonics that contained various herbs including red clover, licorice, burdock root, and cascara that acted as a purgative. Hoxsey also established many clinics that the FDA closed in 1960 as they were considered unsafe and ineffective. This method is still marketed in the early twenty-first century out of Mexico and over the internet.

A cancer treatment fad in the 1950s and 1960s was "Krebiozen." In 1951, two Yugoslavian refugee brothers brought this substance to the United States. They kept the identity of the drug a secret, claiming it was from the blood of Argentinean horses injected with a bacterium. Studies have shown that it only consisted of the amino acid creatine dissolved in mineral oil and was totally ineffective in reducing tumors. A craze in the 1970s and early 1980s was "laetrile," which was the trade name for a chemical in the kernels of apricot pits and other seeds. Promoters called it vitamin B17, claiming that cancer is a vitamin deficiency disease and that the substance could cure cancers. Research showed it had no effect on cancers and harmed the patient with cyanide toxicity.

Canadian naturopath Hulda Clark (1928–2009) claimed that all cancers and many other diseases are caused by parasites, fungus, and other microorganisms. She recommended treating them with black walnut hulls, wormwood, and common cloves to get rid of the parasites. She also had patients annihilate their tumors with an electrical device called the Zapper that she marketed. None of these were effective against cancer. In the 1980s, Polish born Stanislaw R. Burzynski (1943–) developed "antineoplastons," which he claimed were extracted from urine or synthesized and proposed they "normalize" cancer cells that are constantly being produced within the body. He claimed that this substance had cured many people with cancer. However, in 1992, controlled studies concluded that Burzynski's antineoplastons did not normalize tumor cells or cure cancer.

By the late twentieth century into the 2020s, numerous alternative practitioners attempted to cure cancers by various treatments that were not effective but were advertised on the internet. At best, the patient wasted money on these treatments, and at worst, they sometimes died earlier. These "treatments" included coffee and castor oil enemas, "detoxification" cleanses and regimes, drinking urine, hexagonal water, cannabis oil, vegetable juice diets, pancreatic enzymes, extracting teeth with mercury fillings, shark cartilage, massive doses of vitamin C, and rectal ozone gas treatments among numerous others. Even psychological and relaxation techniques such as guided imaging, medication, yoga, massage, aromatherapy, and hypnosis have been tried, but compared to control groups, they also had no effect in reducing tumors. Relaxation techniques, however, are helpful in reducing anxiety and stress.

Cardiovascular (Heart) Diseases

Cardiovascular or heart disease is a catchall phrase for a variety of conditions that affect the heart's structure and function. Numerous heart disorders exist, and a few have been observed since antiquity. These include chest pain from angina and/or heart attacks, strokes formerly called apoplexy, congestive heart failure (CHF) with a symptom of edema formerly called dropsy, and arrhythmias often called palpitations. However, the risk factors that led to these symptoms often were not known until the twentieth century. These include hypertension (high blood pressure), atherosclerosis (fatty deposits in the arteries), arteriosclerosis (stiffening of the arteries), or heart valve disease. In addition, adequate treatment until the mid-twentieth century was lacking. Some congenital heart conditions found at birth are discussed under the "Infant Issues, Illnesses, and Death" entry. Cardiovascular maladies are the leading causes of death worldwide accounting for roughly one-third of all deaths. These diseases can inflict emotional, physical, and financial burdens on the individual and society. In addition, heart disease is highest among those with African ancestry and lowest among those with South Asian ancestry globally.

Coronary Artery Disease (CAD)

A major heart problem is coronary artery disease (CAD) in which atherosclerosis can inhibit the flow of blood (ischemia) and reduce or cut off oxygen to the heart. It can also be found in arteries of other vital organs, such as the brain or kidneys. Risk factors for CAD include high blood pressure, high cholesterol, genetics, obesity, diabetes, and smoking. In addition, a SARS-CoV-2 (COVID-19) virus infection may injure heart and other blood vessels. Obesity, CAD, and diabetes together are sometimes called "metabolic syndrome" with a high risk of early death. By 2020, nearly one-half of all Americans had CAD. The first use of the term "coronary artery disease" was in 1949, and it became a heart condition that included angina and heart attacks.

Angina and Myocardial Infarction (MI)

Angina pectoris occurs from a partial blockage of the coronary arteries that deprive the heart of oxygen (*myocardial ischemia*) and nutrients. Symptoms include pain in the left shoulder, jaw, neck, arms, or back in addition to pressure, squeezing, or burning in the chest. It often occurs with exertion and lasts only a few minutes. Research is mixed as to whether women experience angina more frequently than men.

The term "angina" is derived from the Latin *angina*, meaning "quinsy" or "strangling, choking," and *pectus*, meaning "chest," and was first used in 1768. However, the symptoms had been noted by Greek physician Hippocrates (c. 460–370 BCE) in antiquity. The difference between angina and MI is that angina attacks generally do not permanently damage the heart muscle. Acute management of angina today is usually with sublingual nitroglycerine, and rarely amyl nitrate, both discovered in the mid-nineteenth century. These potent vasodilators widen and relax arteries and veins, lower the blood pressure, and give quick relief from angina pain.

The symptoms of a MI include prolonged chest pressure; shortness of breath; pain radiating into the jaw or shoulders, back, or arms; and sweating and nausea. It is often an acute event, precipitated by rupture of a cholesterol plaque within a coronary artery, which then leads to a clot that blocks blood flow to the heart muscles, which die from a lack of oxygen. This is sometimes followed by collapse from sudden cardiac arrest (SCA). This sudden collapse usually is fatal and is caused by an irregular heart rhythm. Cardiopulmonary resuscitation (CPR) and cardiac defibrillation sometimes can revive the person. Globally, men aged 45 or older and women aged 55 or older are more likely to have a heart attack compared to younger people.

The first-known use of the term "heart attack" was in the early nineteenth century while the medical term "myocardial infarction" was first coined in the late nineteenth century. "Myo" is from the Greek *mus*, meaning "muscle or mouse," and "cardium" is from the Greek *kardia*, meaning "heart." "Infarction" is from the Latin *infarcire*, meaning "to plug up or cram." Hippocrates in the fourth century BCE described a heart attack when he noted that a patient with severe pain in the chest would often "suffocate" and die. However, until the eighteenth century chest pain was not necessarily associated with the heart. To prevent heart attacks today, health professionals recommend exercise, eating a well-balanced diet, and refraining from smoking and heavy drinking. Clinicians in addition often prescribe antihypertensive and cholesterol medications and sometimes low-dose aspirin.

Congestive Heart Failure (CHF) and Edema

Another common heart disorder is heart failure (HF), also called congestive heart failure (CHF). Heart failure is the end stage of most heart diseases and is a major cause of morbidity and mortality. Heart failure occurs when the heart muscle does not pump as much blood as the body needs. Symptoms include breathing difficulties, edema in the limbs and abdomen, fatigue, and irregular heartbeats. More than 5 percent of people aged 60–69 have CHF, and it is equally frequent in men and women.

The term "heart failure" was used by the sixteenth century along with "dropsy." Dropsy is from the Middle English *dropesie*, from the Latin *hydropisis*, from the Greek *hydrōps*, from *hydōr*, meaning "water." The oldest identified case of this malady was in an Egyptian mummy circa 1400 BCE. Hippocrates believed that dropsy and difficulty in breathing (dyspnea) were caused by impurities in the blood. Later Arab physician Ibn al-Nafis (1213–1288) who discovered pulmonary circulation of blood before William Harvey (1578–1657) demonstrated that the heart was a pump and that valve abnormalities caused CHF symptoms. Until the mid-twentieth century, to manage HF/CHF, physicians prescribed digitalis and diuretics. By the 1990s, other drugs had been developed, and in the early twenty-fist century, implantable defibrillators to keep the heart beating, stem cell, and other genetic therapies were in the process of being developed to manage CHF and to prolong life.

Hypertension and Stroke

High blood pressure (HBP) or hypertension may lead to strokes, heart failure, heart disease, dementia, and sometimes death. It is caused by abnormal narrowing of the arteries. However, most people are unaware that they have HBP as it is generally asymptomatic, which is why it is known as the "silent killer." In the United

States, hypertension is defined as a systolic blood pressure above 120 (number on top that is the maximum pressure the heart exerts when beating) and a diastolic pressure of over 80 (the number on bottom or pressure between beats) in millimeters of mercury. In other English-speaking nations, including Australia, Canada, and the United Kingdom (UK), HBP is 140/90 mm Hg or higher. About 40 percent of adults in the United States and about 30 percent in other nations have HBP. Hypertension was not taken seriously until after longitudinal studies, such as the Framingham Heart Study (1948–), showed that it was a major factor in cardiovascular diseases. However, since antiquity clinicians have noted that a strong pulse was often found prior to a stroke. By the 1960s, various drugs were developed to reduce HBP.

A stroke, cerebral vascular accident (CVA), or apoplexy occurs when blood is prevented from going into the brain. Signs of a stroke include sudden severe headache, numbness or weakness on one side of the body, confusion, trouble speaking or understanding speech, dizziness, and loss of balance and collapse. Globally, around one in four adults will have a CVA in their lifetime that can result in permanent disability. About 90 percent of strokes are caused by a blood clot that prevents blood from reaching the brain. In some cases, the clot forms in the vessel, and in other cases, it can travel from the heart (embolism). If a blood vessel ruptures in the brain, it is a hemorrhagic stroke. In a "mini stroke," or transient ischemic attack (TIA), blood flow to the brain is interrupted for a brief period. The *Papyrus Ebers* notes an association with "hard pulse disease" and heart attacks, dropsy, and apoplexy. In 1935, the term "cerebral vascular accident" was first used. Management of stroke today is by anticoagulation drugs, surgery, and physical therapy.

Arrhythmia and Palpitations

Cardiac arrhythmias or palpitations may feel like a skipped beat, fluttering, rapid heart rate, or pounding in the chest sometimes accompanied with fatigue, dizziness, or lightheadedness. Normally electrical impulses originating in the right atrium pass through the rest of the heart to control the contractions and the pumping of blood throughout the body. Electrolytes such as sodium, magnesium, and potassium help trigger and regulate these impulses. However, if there is an imbalance in electrolytes, heart damage, hypertension, or other factors, arrhythmias may occur.

The most common cardiac arrhythmia is atrial fibrillation (AF), which can last a few minutes to hours or longer. During an attack, the atria beat irregularly and are unable to adequately pump blood. This can cause the pooling of blood and potential clot formation. About 7 percent of people over age 65 have the condition, and it is more common among men than women, but women have a greater incidence of stroke-related deaths. The second most common arrhythmia is atrial flutter in which the rhythm in the atria is more organized and less chaotic but very fast. The most dangerous cardiac arrhythmia is ventricular fibrillation, which is often triggered by a heart attack. In this arrhythmia, the ventricles quiver, do not pump blood, and can lead to SCA and death. Today to manage these conditions, anticoagulants, antiarrhythmic, and beta-blocker drugs are prescribed. In some cases, surgery is done to stop erratic electrical signals in the atria, or electric devices are implanted to bring the heart into normal sinus rhythm.

The first use of the term "arrhythmia" in English was in the late nineteenth century. It is from the Greek *arrhythmia*, meaning "lack of rhythm." Palpitations have been described since antiquity, but their cause was not known. In the nineteenth century, physicians, such as American Henry Hartshorne (1823–1897), believed that palpitations were primarily found in anemic people and hysterical females, in addition to heavy alcohol, coffee, and tobacco consumption—which have been shown to be factors—but also due to "excessive venery or self-abuse [masturbation]."

Heart disease and stroke have been found as a plot in fiction throughout the centuries. William Shakespeare's (1564–1616) *Henry IV* (1598), for example, has an obese comic character Falstaff who is a companion to Prince Hal, the future King Henry V of England. Falstaff discussed the king's apoplexy (stroke) of which the king eventually dies but not before Hal attempts to become king. American novelist John Steinbeck's (1902–1968) *East of Eden* (1953) is loosely based on the biblical story of Cain and Abel. The father of twin sons upon hearing of the death of one in WWI dies of a stroke. This novel was made into a film in 1955. American playwright Tennessee Williams' (1911–1983) *Summer and Smoke* (1948) describes a potential spiritual/sexual romance between a young uptight spinster who has heart palpitations and her young wild physician neighbor who has come home for a summer vacation. The following summer he tells her he plans to marry someone else, and she gravitates to sleeping with strangers while becoming addicted to prescription drugs for her palpitations.

UNUSUAL TREATMENTS

Over the centuries, numerous treatments and remedies for heart disease have evolved. Many of the folk and alternative remedies were based upon herbs, diet, and meditation.

Coronary Artery Disease (CAD)

In ancient Egypt, the *Papyrus Ebers* (c. 1500) recommended that for general heart issues the patient should eat fat, honey, and wax, which likely caused more wax buildup in the arteries. Garlic has also been used for years to battle heart problems by folk and alternative practitioners, and some research has shown that garlic extract may help prevent plaque buildup in the arteries. Some other substances for which research is mixed or has shown some effect in reducing inflammation and in turn atherosclerosis and CAD include ginger, curcumin, capsaicin in peppers, ginger, alfalfa, and hibiscus tea.

During the early twentieth century, alternative medicine practitioners and authors such as Upton Sinclair (1878–1968) in his *The Fasting Cure* (1911) promoted "cleansing the body" through fasting to reduce heart disease and other illnesses. Research is mixed as to the effectiveness of fasting. Another alternative remedy is chelation therapy designed in the 1930s to remove heavy metals from the body. Supposedly chelating drugs bind to calcium in the arteries and remove it. Some studies have shown that the remedy in some cases may reduce atherosclerosis, but it can lead to low calcium levels, kidney damage, and death.

By the twenty-first century, other alternative remedies were advertised. Crystal therapy became popular. Proponents believe that "vibrations" from crystals correct "imbalances" in the body. The primary crystal for heart disease is rose crystal. The person carries it, uses it in meditation, or places it over the heart while resting. No evidence exists about cures from crystals unless due to placebo effect, likewise, for color or color light therapy. Advocates claim that different colored light balance the energy lacking in the person. Some other twenty-first-century alternative and folk cures that have modest scientific support include eating daily 100 mg of chocolate, listening to classical music, and meditation.

Angina and Myocardial Infarction (MI)

For pain or "fever in the heart," likely angina, the *Papyrus Ebers* recommends milk, honey, and water, but if this did not work, the healer administered onions, sweet beer, mustard seeds, and date meal. Mustard seeds may help reduce cholesterol and widen the arteries, but the other ingredients were ineffective although harmless to the patient. Greek physician Hippocrates in the fourth century BCE bled from the elbow vein to rebalance the humors for angina pectoris or a myocardial infarction. Some studies report that venesection may reduce hypertension, which in turn could have helped to reduce chest pain.

In ancient Rome, Pliny the Elder (c. 24–79) administered juniper berries in wine for angina and heart pain. Some research suggests that these berries contain anti-inflammatory substances that may reduce total cholesterol. In addition, moderate alcohol consumption is correlated with a reduced risk of atherosclerosis, so these remedies may have been helpful. Into the mid-nineteenth century, physicians still used leeches and venesection to treat angina and MI to "bleed" impurities out of the body. In the late nineteenth century, clinicians also used chloroform, which would have reduced blood pressure and pain; however, it also led to sudden cardiac death so was soon abandoned as a treatment. Scottish physician Alexander Macaulay (1783–1868) besides bloodletting also applied blisters to the chest, which would have been painful and but not effective.

A folk remedy from nineteenth-century Austria is walking barefoot in cold wet grass from the morning dew or soaking the feet in cold water. It supposedly increases blood circulation to prevent angina and heart attacks but likely had little effect. In the 1970s, a vitamin E craze for preventing angina and heart attacks was advertised in health magazines. However, research in 2004 showed that it was not effective in preventing heart issues and increased the risk of dying, in particular, from cerebral hemorrhage, as the vitamin is an anticoagulant and increases the risk of bleeding.

Congestive Heart Failure (CHF) and Edema

The *Papyrus Ebers* gives a recipe to relieve edema in the limbs from heart failure. Powdered dates were boiled in about five quarts of water and consumed at body temperature. This emetic caused vomiting but likely did not decrease the edema. From Roman antiquity, the hawthorn plant has been used to treat CHF. Some studies have shown that chemicals in the plant dilate blood vessels and improve the strength of heart contractions and may have been effective in reducing edema. Likewise, the squill, or sea onion, bulb, which acts as a diuretic and slows and strengthens heart contractions, helped keep fluid retention to a minimum.

Noted Roman naturalist Celsus (c. 25 BCE–c. 50 CE), for fluid in the abdomen (ascites), recommended inserting a copper or lead tube into the abdomen to drain fluid, which was a temporary fix and likely led to extreme pain, infection, and death. In the seventeenth century, English physician, Jonathan Goddard (1617–1675), developed a "cure-all" remedy that included boiling together dried viper, hartshorn (ammonia), and the skull of a recently hung person. His concoction had little effect on any disease and was mostly harmless. In 1775, British pharmacologist William Withering (1741–1799) first described the use of foxglove (*Digitalis purpurea*) leaves for the treatment of dropsy, which he obtained from a traveling peasant woman. It did reduce edema and caused the heart to pump more efficiently although ingesting too much could be lethal. Macaulay, in the nineteenth century, besides using foxglove, administered several small punctures in the edematous skin to allow the fluid to run out. However, the wounds could become infected.

In the mid-nineteenth century, physicians advised bloodletting for heart failure to reduce blood volume, thereby reducing swelling. Although it was often successful, too much bleeding could lead to death. Another treatment in the nineteenth century was the Southey tube, a three-inch large needle with perforated sides. This was placed in edematous tissue, attached to a rubber tube, and successfully drained off the fluid.

Hypertension and Stroke

Clinicians in the past advanced some treatments for "hard pulse disease," which was associated with heart attacks and stroke. Bloodletting was the major treatment for this from antiquity into the late nineteenth century to prevent a stroke. It was also used for cerebral hemorrhage, in particular subarachnoid hemorrhage accompanied by severe hypertension, to rapidly lower blood pressure. Physicians also gave patients malaria and typhoid bacilli to cause a fever to prevent a stroke. However, this was ineffective and often killed the patient.

By the twenty-first century, alternative remedies such as crystal therapy were proposed for HBP which had no scientific support. Proponents suggested amethyst had soothing and healing effects on the arteries and blood vessels and stabilized blood pressure. Aromatherapy using short-term exposure to orange essential oil reduces blood pressure, but the research is mixed. On the other hand, soothing classical music has been found to reduce blood pressure.

With regard to strokes, the *Papyrus Ebers*, for example, gives several recipes for poultices that include tar, mustard oil, and frankincense, which were placed on the limbs to treat hemiplegia (paralysis on one side) from a stroke. These did not cure the condition but were generally harmless to the patient. Buchan in the late eighteenth century in his *Domestic Medicine* (1774) recommended that during a stroke the patient's head should be raised high and his "feet hung down" to reduce circulation toward the head. Then he bled and gave the patient an enema to "reduce congestion" based on humoral theory. He also placed blistering plasters between the shoulders and to the calves of the legs and advocated electrical shock to the "part affected." Bleeding likely lowered blood pressure, but the other treatments only resulted in pain. In the late nineteenth and early twentieth centuries, clinicians placed ice packs on the head during and after a stroke, which research in the twenty-first century suggests might help prevent some damage to the brain.

By the late twentieth century, several alternative treatments for stroke rehabilitation became popular. Mirror therapy, in which a mirror is placed over the affected hand while the other hand exercises and is reflected in the mirror, is thought to rewire neurons. In the 2010s, deep "brain stimulation" in which electrodes are implanted into the brain after a stroke has been used. Although generally safe for the patient, the research is mixed as to the effectiveness of these remedies.

Arrhythmia and Palpitations

Various remedies over the centuries have been tried for cardiac arrhythmias or palpitations. These include numerous folk treatments that were likely found by trial and error that may, or may not, work. Most of them do not harm the sufferer. Some popular ones today include holding one's breath and bearing down as if having a bowel movement (Valsalva maneuver), coughing, placing ice or a cold, damp towel on the face for a few seconds, plunging the face into cold water, chanting, meditation, or taking a cold shower. Sometimes consuming magnesium, potassium, and calcium will help halt the attack.

Various herbs have been used to stop palpitations since antiquity although most were not likely effective. The *Papyrus Ebers*, for example, recommended ingesting a mixture of figs, berries, and goose grease. The ancient Romans into the Middle Ages used hawthorn for irregular heartbeat. Studies of this herb's effectiveness in treating arrhythmia are mixed. This is also the case for various herbal mixtures with regards to traditional Chinese medicine and moxibustion treatment in which the herb mugwort is burned on the skin on the "energy meridians."

Childbirth: Maternal Complications

Childbirth throughout the late 1930s was an extremely dangerous time in a woman's life. Serious complications could occur during the three stages of labor and the postpartum period, which is up to six weeks after birth. (The stages of normal labor are discussed in the entry, "Childbirth and the Normal Postpartum Period.") Women often feared childbirth as death was possible. In the sixteenth century, women were told to make wills before their delivery day. From 1800 through the late 1930, maternal mortality remained on a high plateau in Western nations with a death rate of around 400–500 per 100,000 live births. However, maternal deaths dropped precipitously in the post–WWII era in the United States, Canada, Britain,

Australia, New Zealand, and northern European nations due to the adoption of rigorous antisepsis, medical interventions, better trained health-care professionals, antibiotics, and pre- and post-natal care.

On the other hand, maternal mortality began to rise in the United States in the late twentieth century. Due to this trend, in 1986 the Center for Disease Control and Prevention initiated a national surveillance of pregnancy-related deaths. Maternal deaths continued to rise and climbed from 7.2 in 1987 to 19 per 100,000 live births by 2020. Conversely, a decline was seen in other industrialized nations resulting in the United States having the highest maternal deaths of all developed nations. In addition, the risk was three to four times higher for Black compared to white women, irrespective of income or education in the United States. In contrast, Canada, which is also a nation of various ethnic groups and immigrants, had a stable maternal death ratio of around 10:100,000.

This upsurge in maternal deaths in the United States is thought to be caused by numerous factors. A growing number of pregnant women have chronic health conditions such as hypertension, diabetes, obesity, and chronic heart disease, and an increasing number of older women are getting pregnant who are more prone to complications. Another factor is the increased use of unnecessary medical interventions in low-risk pregnancies such as C-sections (Caesareans) for the convenience of the physician or mother. In addition, many states enacted laws that decreased access to family planning and reproductive health services. The World Health Organization (WHO) also implies other factors that include a lack of a midwife-oriented care system and the absence of a universal standard of maternity and delivery care throughout the nation. By 2020, around 14.6 percent of maternal deaths were caused by embolisms—blood clots or amniotic fluid that traveled to the lungs or brain. Infection, hemorrhaging, and pregnancy-induced high blood pressure were other leading causes of maternal mortality and morbidity (illness).

"Childbed" or Puerperal Fever

Before the mid-1930s, about 40 percent of all maternal deaths were from infections caused by unsanitary labor and delivery practices and unsafe induced abortions. In the United States by the late 2010s, 12.5 percent of all maternal deaths were caused by infections. It affects women usually within the first three days after childbirth, causing severe abdominal pain, fever, prostration, and death in about 80 percent of cases. Childbed or puerperal fever—usually caused by the *beta-hemolytic streptococcus* bacterium—generally starts in the uterus, but the bacteria can spread to the circulatory system and tissues (sepsis), leading to the breakdown of organs. The term "puerperal fever" was first used in English in the eighteenth century from the Latin *puerperal*, meaning "woman in childbirth." The condition had been recognized from antiquity, but its cause was unknown. For example, the *Hippocratic Corpus*, the writings of Greek physician Hippocrates (c. 460–370 BCE) and his later followers, believed that women in childbirth were just prone to fevers. In the eighteenth and nineteenth centuries, some thought it was an inflammation caused by a contagion or putrid air. However, in 1847,

Hungarian physician Ignaz Semmelweis (1818–1865) reduced the rate of the fever in his obstetric ward by ordering physicians—many of whom had just come from the autopsy room—to wash their hands after noticing that midwives had a lower rate of infection. But washing hands and the germ theory of infection was rejected by many physicians until the late nineteenth century. Antibiotics are used today to treat the condition.

Postpartum Hemorrhage

Another major complication of childbirth is postpartum hemorrhage. By the late 2010s, it accounted for about 11.0 percent of all maternal deaths in the United States. Certain medical conditions and treatments increase the risk of developing severe bleeding including the placenta placed over the opening of the uterus, placenta material remaining inside the uterus, multiple births, pregnancy-induced hypertension, obesity, and the use of forceps—or in recent years, vacuum-assisted delivery. The term "hemorrhage" is from the Latin *haemorrhagia*, meaning "violent bleeding," and was first used in the seventeenth century. In the past, it was noticed that if the womb contracted, "flooding," or heavy bleeding, would stop or slow down. Before the mid-1800s, there was little a midwife or physician could do, and many women simply bled to death. Management of hemorrhage today is by medication such as ergotoxin developed from ergot—a fungus found on rye—in 1906, abdominal uterine massage, surgery to remove retained placenta, tying off blood vessels, blood transfusions, or even a hysterectomy to remove the uterus.

Eclampsia or Pregnancy-Related Hypertension

A serious pregnancy disorder is preeclampsia—pregnancy-related hypertension—which is most common during the second half of pregnancy. It can also develop in the postpartum period and occurs in about 3–5 percent of pregnant women and can be frightening to the woman. Symptoms include a sudden rise in blood pressure, excessive weight gain and edema, headaches, abdominal pain, shortness of breath nausea and vomiting, confusion, and blurred vision. It can lead to fatal strokes or convulsions (eclampsia). The term "eclampsia" is derived from the Latin meaning "sudden development" or "violent onset" and was first used in the late nineteenth century. American physician Elijah B. Hammack (1826–1888) called the condition puerperal, or childbirth convulsions. He believed it was caused by irritation of the spinal cord from undigested food in the stomach. Treatment by the mid-twentieth century included newly developed antihypertensive drugs and magnesium sulfate to prevent seizures.

Obstructive and Slow Labor

Another complication is obstructed labor in which despite strong uterine contractions the fetus does not move into the birth canal. This can be caused by a large or abnormally positioned fetus such as the buttock facing the birth canal (breech birth) or a transverse lying position, an unusually large head, or a woman with a narrow of abnormal pelvis or birth canal. The term "obstructed labor" was first used in the sixteenth century. It is derived from the Latin *obstructus*, meaning "to obstruct," and from the Latin *labor*, meaning "toil." This condition increases the risk of uterine rupture, an obstetric fistula (hole) between the vagina and the bladder wall or the rectum along with the complications already discussed. If the fetus cannot be turned around by manipulation on the abdomen called External Cephalic Eversion, vacuum, or forceps delivery, or even a C-section is now used. Similarly, if a woman has been in labor for 18–24 hours, it is considered a difficult or slow labor. Lying flat on the back with knees bent or on stirrups, epidural injections to dull pain, and lack of moving around can contribute to slow deliveries. Slow deliveries can also be caused by sluggish cervical dilations, weak contractions, or even worry or fear. Pitocin (the hormone oxytocin) for stronger contractions may be given, or if the infant is already in the birth canal, previous mentioned interventions can be used.

Stuck or Retained Placenta

During the third stage of labor, the placenta is delivered. If it is not expelled within about 30 minutes of birth, it is considered a retained placenta. After the placenta has been delivered, it is checked by the health-care professional to make sure all of it was removed as even a small piece remaining in the uterus can cause severe hemorrhage. Swiss physician Jacob Rueff (1500–1558), in his *The Expert Midwife* (1636), noted that a stuck placenta often led to stroke, convulsions, and eventually death. Today ergometrine (derived from ergot) and oxytocin, which cause the uterus to contract, are used to push out the placenta, but sometimes it needs to be removed surgically.

Childbirth complications and death have been depicted in the arts for centuries. For example, a woodcut, by German surgeon Johann Schultes (1595–1645), shows doctors performing a C-section in the mid-seventeenth century. In the nineteenth century, Charles Dickens' novel *Oliver Twist* (1839) mentions that Oliver's mother died giving birth to him in a workhouse, and in *Snow White*, a Grimm Brothers' fairy tale, Snow White's mother also dies in childbirth. In the 2010s' British TV series *Downtown Abbey*, a main character dies from eclampsia, but her child lives.

UNUSUAL TREATMENTS

Since childbirth was considered so life-threatening, midwives said prayers and incantations for a safe delivery. Many unsuccessful remedies and procedures were tried over the ages for birth-related complications based upon trial and error and the prevailing concepts of health and medicine.

"Childbed" or Puerperal Fever

Childbed fever was treated like other inflammatory disorders in the eighteenth century. Physicians bled the new mother from a vein or with leeches at the first sign of an infection, which they claimed often helped. Scottish physician William

Buchan (1729–1805) gave the new mother much to drink including gruel or barley water mixed with potassium nitrate. He also administered warm water enemas and placed bladders of warm milk or water on her abdomen. Hammack in the nineteenth century placed flannel cloths soaked with turpentine on the abdomen and administered a cathartic such as calomel, which contained mercury. He also favored brandy toddies or a "good port wine" along with enough morphine or opium to keep the woman sedated for several days. These palliative treatments likely had little effect on the infection.

Postpartum Hemorrhage

Since excessive bleeding could quickly lead to death, various remedies were attempted over the centuries. Some included stuffing the uterus with cloth or various wax and herbal mixtures, which usually were not very effective. Rueff in the sixteenth century recommended that the midwife should give the woman roasted or fried hen or capon meat, have her abstain from fluids, and rest in bed with her head down. He also ordered her to consume confections and toast seeped in wine. By the late sixteenth century, ergot was used to help slow postpartum bleeding, which may have worked in some cases.

To control hemorrhage, Buchan in the seventeenth century applied wine vinegar on a cloth to the belly. He also administered a mixture of "syrup of poppies" (an opiate) and toxic sulfuric acid or a mixture of crab claws, saffron power, sulfuric acid, and potassium nitrate every few hours, which likely had minimal effect. Hammack in the nineteenth century advised the midwife or general practitioner to grab the uterus through the vagina and squeeze it to make it contract as fast as possible, which may have helped. To induce contractions, he also recommended pouring a pitcher of ice water on the abdomen while grabbing the placenta through the vagina and letting the contractions expel it along with the hand. Also, he suggested putting a small piece of ice into the vagina or uterus or injecting ice water into the rectum to start contractions and even ergot. Some of these methods may have helped slow bleeding, but others could have led to infections.

Eclampsia or Pregnancy-Related Hypertension

In the nineteenth century, some physicians bled the woman in labor if she had a bounding pulse to prevent convulsions as it tended to foreshadow seizures. Others recommended an enema, of castor oil, molasses, and warm water. Physicians also gave Dover's powder (ipecac and opium), an emetic, along with brandy and quinine, but these were not very effective. However, physicians also administered potassium bromide, which does help prevent seizures.

Obstructive and Slow Labor

A major cause of death leading to infection and hemorrhage is obstructive labor. If the infant could not be turned around with manipulation on the abdomen, European physicians from the fourteenth century such as John Arderne (1307–1392)

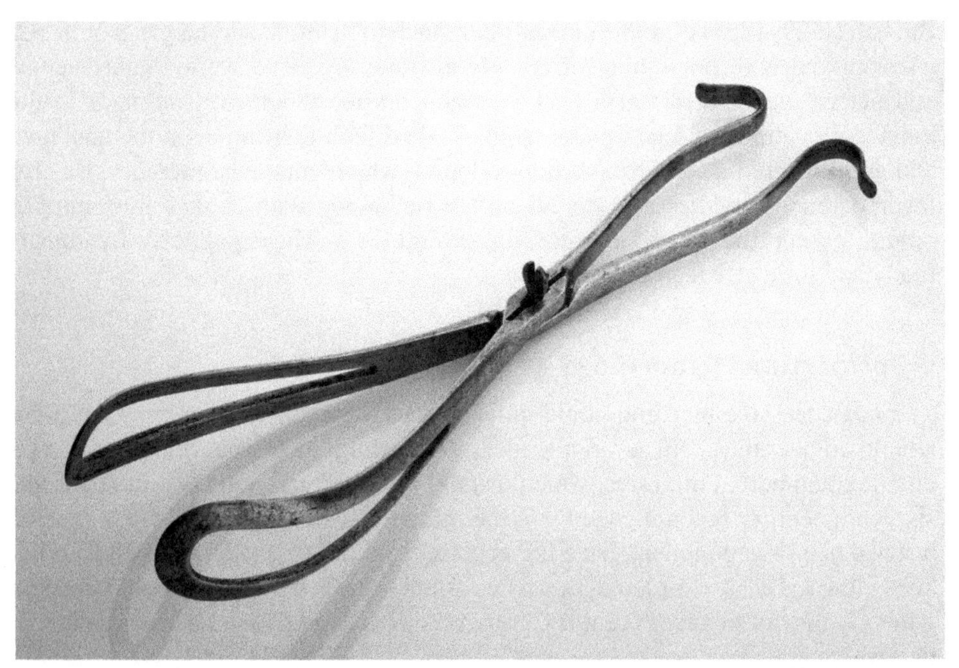

These obstetrical forceps in the late eighteenth century were used for difficult births to pull out the infant. However, forceps sometimes caused brain damage to the infant and tissue damage to the mother and often introduced infections. (Science Museum, London. Attribution 4.0 International (CC BY 4.0))

and later Rueff advised the midwife to reach into the vagina with greased hands, seize and push the child backward, and turn its head toward the opening. If the mother was dying a choice had to be made between saving the life of the mother or the child. Rueff recommended a few tools including a hook called a "crotchet" be used to pull out stuck and/or dead babies to save the mother's life. When forceps were developed by a man-midwife—a barber/surgeon—in the sixteenth century but kept secret until the seventeenth century, the infant was pulled out of the vagina. Forceps have their own issues as they can cause bruising and brain damage.

However, in some Catholic nations, theologians deemed the life of the child was more important than the life of the mother as the living infant could be baptized. Therefore, the C-section, which had been used from antiquity for women who were dying or had died, was performed. The procedure was almost certain death for women until the development of antibiotics and the use of blood transfusions in the 1940s. By the late eighteenth century, man-midwives and general practitioners usually carried destructive instruments and were often asked to remove dead or obstructed fetuses.

In situations when performing a C-section would be a death sentence for the woman, a surgical procedure, symphysiotomy, to widen the pelvis was tried. First used by the late sixteenth century, the practitioner cut the cartilage between the two pelvic bones with a knife. Since this could take time, in the late eighteenth century two physicians developed a hand chain saw (the osteotome) to cut the cartilage for quickly delivering an entrapped infant, but symphysiotomy often caused chronic pain, impaired mobility, and incontinence for these woman. It was used in

Ireland from the 1940s into the 1980s for religious reasons long after it was discontinued in other developed nations. However, it is still performed in some countries in situations where a C-section is not feasible.

Various techniques have been used for slow labor. Midwives for centuries have given women ergot for stronger contractions. However, it often caused stillbirths and could rupture the uterus. Rueff remarked that some midwives hung a "Jasperstone" or *lapis Aquilinus* stone on the left hip to hasten labor. Buchan (1729–1805) in the eighteenth century, for a difficult and slow labor, bled the woman. He also administered an enema, had the patient sit over steam, and rubbed the birth canal with fresh butter. If the woman was exhausted, he gave her wine. However, if these were not effective, he recommended that a man-midwife be called to extract fetus, dead or alive, in an attempt to save her life, which likely induced an infection.

Stuck or Retained Placenta

If the placenta was not delivered in a timely manner, Rueff advised midwives to give the woman meat and drink, and broths made of capons or hens and seasoned with cinnamon and saffron to strengthen her along with herbs. He also advised midwives to fumigate the womb with myrrh, cinnamon, and castoreum (dried beaver testicles). Rueff mentions other concoctions to put into the womb and pepper into her nostrils to make her sneeze while holding her breath. If these did not deliver the placenta, he advised the midwife to grab hold of it and pull it out with her hand. If nothing worked, the midwife called a man-midwife. However, the mother likely died anyway of bleeding or infection.

Childbirth and the Normal Postpartum Period

Childbirth involves three states of labor. During the first and longest stage, contractions start, the cervix softens and dilates, and the amniotic sack usually ruptures. The cervix dilates to 6 cm during early labor and from 6 to 10 cm during active labor, which often lasts 4–8 hours or more. The second stage of labor is when the mother pushes the infant out through the birth canal (vagina) into the world; this can take a few minutes to several hours. Most women now make a "birth plan" prior to delivery with various options such as delayed cord clamping and cutting and skin-to-skin time between mother and baby before other procedures such as an antibiotic being placed in its eyes, a health assessment, and washing. The third stage of labor ends with the delivery of the afterbirth (placenta). The woman now enters the postpartum period—the first six weeks after delivery—when hormone levels and size of uterus returns to a nonpregnant state.

In terms of birthing positions, women in Western cultures since antiquity have used birthing chairs or stools to deliver their infants. These were generally straight-backed, horse-shoe-shaped chairs or stools with sometimes a curtain surrounding a part of it. Through the opening, the midwife was able to catch the child. In numerous cultures, kneeling, squatting, standing, being on hands and knees, and side lying have also been common birth positions. However, the Western birthing position began to change in the early eighteenth century.

Some scholars suggest that King Louis XIV (1638–1715) of France wanted to see his many children born and insisted his women lie on their backs on a table and put their legs up in stirrups. This lithotomy position was the same as for removal of bladder stones (see "Urinary Tract Issues" entry), and the position was fostered by the same French physician for both situations. Lying down along with semi-reclining has been the most common practice in the United States and many Western cultures for hospital deliveries from the late nineteenth into the twenty-first century. However, Scottish physician Alexander Macaulay (1783–1868) did recommend that the mother should lie on her left side during delivery but encouraged her to take any position during labor.

From at least Greek and Roman ancient times through the mid-twentieth century in Western cultures, for several weeks after delivery the mother and child were kept at home. In antiquity, they were considered "polluted" until they went through a purification ritual, and by the Middle Ages, women were "churched" after about 10 days to counteract the original sin that led to childbirth. Later the term "lying in" or "confinement" was used for the period of labor, delivery, and exclusion from society for several weeks. The term "lying in" was first used in the fifteenth century and is from the late Middle English word *lyynge in*. "Confinement" is borrowed from Middle French, from *confiner*, "to confine," in the sixteenth century. This practice has been found in many cultures and is still found today in some East and South Asian and Latin American traditions. However, poor women in many cultures often did not have the luxury of lying in.

During confinement, women in the past were expected to "bleed away" any remains of their pregnancy. During much of the twentieth century, women often rested in the hospital and then at home for a couple of weeks in bed after delivery. However, by the early twenty-first century women were now encouraged to be active and do light exercise a few days after an uncomplicated vaginal birth as it helped prevent complications such as blood clots, muscle weakness, or even depression.

Childbirth in all cultures has been considered a natural part of human existence, and deliveries were at home usually with the help of a midwife or female family members. The term "midwife" from the Old English meant "with woman." By the late 2010s, in European nations, 75 percent of deliveries were by midwives while in the United States it was only around 10 percent, and about 98 percent of births were in hospitals in most developed nations. In countries with the lowest maternal and child mortality rates, trained midwives are the primary providers for pregnant women.

However, in Western nations beginning in the seventeenth century, interprofessional rivalry between surgeons, physicians, and midwives and the professionalization of medicine led to changes in long-standing traditions. From at least this period, during difficult deliveries, female midwives usually asked for help from male barber/surgeons. "Man-midwives" or "accoucheurs" also arose in the seventeenth century. Over time, these men evolved into physicians and surgeons who based their practices on more scientific information and gained power.

Consequently, by the late nineteenth century, birth had changed from being a normal, home-based event to becoming an illness-oriented, hospital-based procedure.

Although lying-in hospitals for impoverished women had previously been established in several nations, in the United States private "lying-in hospitals" and maternity wards were developed for higher socioeconomical women, which were staffed by male physicians who took control of all health care. As a result, midwives, in the United States, were pushed out and rarely allowed to practice other than in remote rural areas. On the other hand, midwives, in other developed nations, such as the United Kingdom, maintained their power, and the mostly male physicians, in particular obstetricians, took on high-risk pregnancies or childbirth emergencies.

By the 1970s, women had become dissatisfied with the hospital birthing experience, and as part of the feminist movement, many American women demanded more say over their birthing procedures and a desire for more natural childbirth. This led to hospitals establishing homelike "rooming in" maternity units with physicians allowing more flexibility in labor and delivery. More importantly, birthing centers or home deliveries with midwives emerged that fostered delivery in any position the woman desired without medical interventions for low-risk patients. Other outcomes of this movement was the doula or trained birth attendant to offer physical, emotional, and psychological support and the slow emergence of certified nurse midwives in hospitals to deliver infants. Most importantly, research has shown little difference in maternal mortality or morbidity (sickness) between in or out of hospital births for low-risk births attended by trained midwives.

The birth and the postpartum period have been depicted in the arts from antiquity until the present times. For example, a tomb of a Roman midwife has a bas relief of a child being delivered. Many childbirth paintings exist with Christian religious themes. These include numerous "birth of Jesus" art works where Mary is often shown during her confinement. In contemporary times, Canadian oil artist Amanda Greavette paints life-size portraits of labor, birth, and postpartum as part of her *The Birth Project* series. Childbirth has even been a central theme in the popular 2010s British TV series, *Call the Midwife*, which details home birthing attended by midwives in 1960s London.

UNUSUAL TREATMENTS

Many folk and other remedies were developed over the centuries to hasten labor and delivery and for safe childbirth. For example, The *Papyrus Ebers* (c. 1500 BCE) lists several recipes to help a woman have a faster delivery. One includes having the woman rub peppermint on her bare buttocks. Another involves plastering a mixture of sea salt, grains of wheat, and a female reed (unknown plant) on her abdomen or even drinking a mixture of sea salt and honey that had been strained. In Roman times, some believed if a torpedo electric fish was brought into the laboring room, the woman would have an easy speedy delivery.

In the sixteenth century, Swiss physician Jacob Rueff (1500–1558) in his book *The Expert Midwife* (1637 English translation) suggested that as labor would begin the midwife should lubricate her hands and the vagina with oil of lily and almonds and chicken grease to make birth easier, especially for fat women, and egg white could be added for a first birth. If the birth canal was tight, the midwife enlarged it by stretching it out with her fingers. These procedures likely introduced bacteria and caused infections. Midwives also gave the woman an enema, a practice that continued into the late twentieth century.

For most women from antiquity through the late nineteenth century, it was assumed that the mother would suckle her own child. Women of the middling and upper class, however, could hire a wet nurse who had recently had a child of her own and was still lactating. For nursing mothers, to encourage her milk "to let down" for suckling, the *Papyrus Ebers* gives milk-stimulating recipes. Ground-up Nile swordfish bones were heated in oil and rubbed on the mother's back, or the mother was told to eat fragrant bread made from sour durra, a wheat mixture, and the poppy plant while sitting cross-legged. From the Middle Ages into the late twentieth century, some midwives or physicians recommended a woman should drink beer or stout to help let down her milk as it would relax her.

In the sixteenth century, Rueff suggested that every day for one hour before the child first suckles it should be placed on the left side of its mother so she could attract all diseases from the infant as this would protect the child from epilepsy and leprosy throughout life. At the same time, red coral flowers mixed with sugar and fresh butter were placed in the child's mouth to prevent these diseases. American physician Elijah B. Hammack (1826–1888) in the nineteenth century recommended that a young puppy should suckle from the mother's breast if the baby was not consuming enough milk as he believed swollen breasts would lead to complications. If a mother could not nurse for health and other reasons, a binding was placed around her breasts to stop the flow of milk.

Nourishment for the new mother was important. Scottish physician William Buchan (1729–1805) in the eighteenth century recommended that after delivery she should only be given tea and toast. However, family and friends often administered special foods, such as a caudle—a thick hot drink made with eggs and milk and sweetened with sugar or honey—to strengthen her into Victorian times. "Taking caudle" was also a euphemism for postpartum social visits. Macaulay in the nineteenth century suggested that after delivery a cordial of wine mixed with water or some laudanum (alcohol mixed with opium and by the 1840s morphine) should be given to the new mother to build up her strength and take away pain.

From the late nineteenth through the early twenty-first century, some procedures originally developed for complicated deliveries became common for routine low-risk vaginal births sometimes leading to difficulties. Physicians increasingly performed caesarian deliveries, C-sections, for the convenience of the physician, or even the patient, which could result in infection and blood clots; around 30 percent of women in the United States had C-sections into the late 2010s. Women were often confined flat in bed and told not to move during labor, which often prolonged the process without the benefit of gravity and caused more pain. Some physicians insisted any other position for delivery caused more vaginal tearing, so they used episiotomies (cutting into the perineum to allow more room for the emerging infant); this practice is now under 20 percent as studies showed more damage was done to the vaginal floor from episiotomies than from normal tearing. Physicians also used interventions such as forceps and—in the late twentieth century—a vacuum apparatus to extract the infant from the birth canal, which sometimes caused brain damage to the infant and tissue damage to the mother. Their use has decreased.

Other procedures routine throughout most of the twentieth century also decreased over the century. So-called twilight sleep had been introduced by 1914. This was a combination of morphine and scopolamine that provided pain relief and wiped out any memory of childbirth. It was initially welcomed as a sign of medical progress. However, women were strapped to the bed as they often thrashed around irrationally from this medication. The experience sometimes caused psychological trauma, and by the 1970s, it was rarely used. During the first stage of labor, nurses routinely gave women an enema, and the vaginal area was shaved for "hygienic reasons." However, shaving was more likely to cause infections, and the infant encountering fecal material helped to establish the immune response, so these have been largely discontinued. Until the late twentieth century, spouses, partners, and family members were generally not allowed in the labor or delivery rooms. As soon as the baby was born, it was whisked off to a separate nursery and only brought to the mother a few times a day.

Some new practices had emerged by the early twenty-first century. In the United States, women were often required to see a rotating team of doctors within an obstetric practice during prenatal visits but had no idea who might be present at their labor and delivery. In some large cities by the 2010s, it was often a strange physician who attended them in the hospital. These situations often caused anxiety and fear. On the other hand, virtual reality headsets that provide relaxing scenes and messages have been reported to reduce pain during contractions. Some folk practices have also developed. For example, eating the placenta had become fashionable among a few women who claimed it prevented postpartum depression, improved milk supply, and provided some nutrients. However, no research has supported that eating the placenta provides any health benefits, and pathogenic microorganisms have been found in products made from this organ.

Childhood Diseases

Diseases of children such as diphtheria, whooping cough, measles, mumps, and chickenpox have been known for centuries. They often came in epidemic waves and were viewed as a normal part of childhood throughout the mid-twentieth century. By the 1980s, these diseases had almost disappeared from most developed nations as they were prevented by vaccinations in infancy. However, at the end of the twentieth century, an anti-vaccination movement arose in North America, Australia, and a few European nations. Parents refused to have their children vaccinated based upon a fraudulent medical report that claimed vaccinations caused autism. Therefore, these diseases sporadically emerged again in the early twenty-first century in some developed countries. They were still endemic (generally found) in poorer developing nations that did not have adequate vaccination programs and were spread by tourists going to, or visitors from, these countries.

Although the cause of childhood diseases was not known until the late 1800s, since ancient times people had noted that they spread quickly from one person to another particularly in crowded conditions. They also noticed that after children had recovered from the malady, they generally did not get it again. But if adults had not been infected during childhood, they often became victims of the infection during an epidemic and were more likely to suffer serious consequences that caused fear among these adults. From antiquity, depending upon the medical philosophy of the time, these childhood illnesses were thought to be caused by an imbalance of the body humors, a contagion, bad air, putrid blood, or evil spirits. By the 1930s, science had shown they were caused by specific bacteria or viruses and were spread by person-to-person contact, generally from infected respiratory droplets being projected into the air or upon objects.

Diphtheria

One of the most studied childhood diseases is diphtheria, "throat distemper," or "bladders in the windpipe." It is caused by the *Corynebacterium diphtheria,* or Klebs-Löffler bacillus, which was first identified in Germany in 1884. In its early stages, the symptoms of diphtheria are like a strep sore throat (discussed in the "Sore Throats" entry). Besides a sore throat, symptoms include fever, swollen neck glands, fatigue, and difficulty in swallowing. The diphtheria bacillus produces a toxin that can produce a thick putrid-smelling grayish white pseudomembrane that often looks like leather. In 1826 a French physician officially named the disease *diphtérite,* which is based on the Greek word for leather. This pseudomembrane can cover the tonsils, the back of the throat, nasal passages, and sometimes the larynx. If it grows over the larynx, a child, and especially an infant, can suffocate in as little as three days. In milder cases, only a slight membrane is seen. The disease can cause permanent heart, kidney, and brain damage, as well as paralysis. A mild form of the disease is found on the skin and results in sores particularly in tropical regions.

Diphtheria was described in antiquity and the Middle Ages and was thought to be caused by an "imbalance of humors" or bad winter air. Diphtheria symptoms were reported in the early 1600s in Europe during a deadly epidemic called the "strangulation disease" in Spain and "the gullet disease" in Italy. It primarily affected infants and young children and frightened parents.

The first recorded outbreak of this deadly disease in North America erupted in New England during 1735–1740 and killed 22 of every 1,000 people. Of these deaths, more than 75 percent were children. Official records noted that it wiped out whole household, and in one community, half of all children died from the illness. People theorized that the disease was connected to a dramatic increase in black caterpillars during the summer of 1735. The epidemic caused social disruption and led one preacher to admonish people who did not help their sick neighbors out of fear of catching it. He also suggested the disease resulted from "impious" and sinful behavior. Even in the late nineteenth century, some physicians still believed it was caused by an imbalance in the humors. In the early twentieth century, diphtheria was one of the major causes of infant and child mortality and worldwide killed up to one-third of all children who contracted it.

By the mid-1920s, a diphtheria antitoxin had been developed, and by the 1940s, the DPT (diphtheria, pertussis, tetanus) vaccine to immunize against this disease along with pertussis and tetanus—discussed in a separate entry—had evolved. The subsequent widespread vaccination campaigns led to a dramatic decrease in diphtheria worldwide in the 1990s.

The World Health Organization (WHO) reported about 5,000 cases in the late 2010s worldwide compared to 100,000 cases in the late 1970s. Today—even with treatment—up to 10 percent can die. A curious story concerning the diphtheria antitoxin occurred in 1925. During the winter, an epidemic broke out in Nome, Alaska. The nearest supply of diphtheria antitoxin was over 500 miles away in Anchorage. However, a relay of sled dog teams rushed to deliver the medication as it was the only way to get to the village. This is allegedly the origins of the modern-day Iditarod race that commemorates this event that saved hundreds of lives.

Pertussis or Whooping Cough

Another dreaded disease of infants and young children was pertussis (whooping, hooping, chin or kink cough) caused by the *Bordetella pertussis* bacterium identified in 1906 by a Belgium researcher. The first symptoms of pertussis are a runny nose, fever, and mild bronchitis. The cough in a couple of weeks develops into an intermittent spasmodic cough that can last up to several months. During a coughing spasm when the child inhales, a high-pitched "whoop" sound occurs, thereby giving the disease its common name. This violent coughing in infants and young children can cause vomiting, seizures, brain damage, hernias, and death from suffocation. Coughing can lead to broken ribs and hernia in teenagers and adults. Prior to antibiotics, pneumonia was a most serious complication that often led to death.

Outbreaks of whooping cough were first described in England in the late twelfth century, but increased deaths from the disease were not reported until the 1700s. Throughout the late 1800s, the disease was thought to be caused by a hereditary predisposition, a spasmodic disorder, too many bad humors in the stomach and guts, worms, or catching cold. During epidemic cycles of three to five years, it sometimes killed all the children in a family. In London in the early 1840s, the illness was more fatal for those under five years of age than smallpox, measles, or scarlet fever. Roughly 1 in 30 young children died from the disease. The highest mortality was found in infants under a year old, the poor, and those who lived in rural areas.

In the early twentieth century, pertussis was one of the most common childhood diseases and a major cause of childhood mortality worldwide. Like other childhood diseases, adults were susceptible if they had not been exposed as children. After 1943, with the development of the DPT vaccine, a dramatic decline in the illness was found in higher-income nations. However, around 2012 a rise in incidence occurred in Australia, North America, the United Kingdom, and Europe due to unvaccinated children. Globally most cases are found in developing nations, especially in the WHO African region where thousands of babies still die each year. WHO reported a decrease from 2 million cases in 1980 to about 144,000 worldwide in 2017.

Measles and German Measles

Another highly contagious disease, measles, is caused by the *rubeola* virus, which was first identified in the United States in 1954. Physicians for centuries had observed that the early symptoms of both measles and smallpox—discussed in a separate entry—are similar. The onset of symptoms for both diseases includes red body rash, prickly feeling in the skin, and fever. However, additional signs of measles are nausea, red eyes, sore throat, coughing, and sensitivity to light. A rash starts on the face and spreads down over the rest of the body. The flat red blotches sometimes have a small bump in the middle and often flow into one another.

Measles unlike many diseases was not described by ancient Greek or Roman physicians. In the tenth and eleventh centuries, Middle Eastern physicians Rhaszes (c. 865–925) and Avicenna (c. 980–1037), respectively, provided the first descriptions of a measles-like disease. Some recent genetic research suggests that measles may have first evolved in the eleventh or twelfth century from an animal disease resulting in deadly outbreaks for the next two centuries throughout Europe. The noted English physician Sydenham (1624–1689) in the 1600s believed the disease was from miasmas (vapors) arising from the ground, but in 1757, a Scottish physician demonstrated that measles was caused by an infectious agent in the blood of patients. Throughout the mid-1800s, many physicians still accepted the humoral theory and defined measles as an acute hot disease as it produced fever.

In the pre-vaccination era, about 90 percent of all children got measles by the time they were 15 years of age, and most children survived the infection. But the measles virus silently wiped out the immune system's memory of past infections, which left people vulnerable to infections they previously fought off. Complications of blindness, brain damage, and deafness could occur in up to 30 percent of cases but were more common in undernourished children. Of children and youth with complications, up to 20 percent died.

In the United States from 1912 to 1916, 26 deaths for every 1,000 measles cases were found. In 1963, a rubella vaccine was developed; and by the 1980s, the MMR (measles/mumps/rubella) vaccine was widely used. WHO reported that between the years 2000 and 2016, vaccinations resulted in an 84 percent drop in measles deaths worldwide. But in the late 2010s, about 20 million children per year still contracted the disease primarily in developing regions of the world. In addition, a measles epidemic arose in Europe and North America among children and teenagers who had not been vaccinated due to vaccination-hesitant parents.

A somewhat similar but milder childhood disease is *rubella* (three-day or German measles). It was first identified by German physicians in 1814. Compared to *rubeola*, the rash is less bright and only lasts for about three days. Throughout the late 1800s, some physicians thought it was a hybrid of scarlet fever and measles. Rubella typically occurs between ages five and nine, and few die from the disease. It was only in the 1940s that researchers realized that if a pregnant woman was infected early in her pregnancy with rubella, the infection could cause miscarriage and Congenital Rubella Syndrome (CRS) birth defects. These issues included cataracts, heart defects, hearing loss, and mental retardation, which brought fear to pregnant women who were advised to avoid contact with sick children.

Between 1962 and 1965, a pandemic of rubella started in Europe and spread to North America, resulting in CRS birth defects in 1–2 per 1,000 live births. As a result, these nations began a concerted effort to vaccinate not only infants but also young women when a vaccine was developed in 1969. Due to mass vaccination programs, in 2015 the Americas became the first WHO region to officially eradicate the disease although some cases were still being imported from developing nations, where thousands of infants are born each year with CRS.

These diseases influenced the course of history. Early records in London showed that in the late 1600s, 295 died of measles out of 12,000 burials. However, among populations that had not been previously exposed to the disease, about 30 percent died. Measles was one of the diseases that killed millions of indigenous peoples in the Americas and Pacific islands after 1492 when European explorers, clergy, and colonizers brought this and other diseases to these regions. With the native population experiencing much disease and weakness, the Europeans easily defeated and dominated them.

Mumps and Chicken Pox

Some other childhood diseases were, also, thought to result in a few problems until the late 1900s. Mumps, for example, caused by the *paramyxo* virus, generally lasts only a few days. The salivary glands in front of the ears swell giving the child a swollen neck, which becomes hot and painful to the touch. Swallowing is sometimes affected. There is little or no fever, but general malaise is found. It was described in Greece around 500 BCE by Hippocrates—the father of Western medicine. Symptoms occur in 31–65 percent of individuals infected with the disease. In rare cases, it can infect the testes which can cause sterility, and if it infects the brain, deafness, paralysis, and seizures may occur. Pregnant women who get mumps have a higher rate of miscarriage. A vaccine against mumps was developed in 1967.

Another mild childhood disease is chicken pox, which is caused by the varicella-zoster virus. It causes an itchy red rash with small, fluid-filled blisters. It is an ancient disease going back to hundreds of thousands of years. Before a vaccine was developed in 1995, about 95 percent of children caught the disease between the ages of 6 and 10. It has a mortality rate of less than 1 percent. For centuries, chicken pox was sometimes thought to be a mild case of smallpox until 1767 when an English physician demonstrated it was a different disease. Until the early twentieth century, it was not known that the virus migrated to the dorsal root ganglion near the spine, which later in life could cause shingles or zoster and which is discussed under the "Skin Conditions" entry. Shingles was described in antiquity. In the early twenty-first century, an immunization against shingles was developed. German measles was generally untreated as it only lasted a short time. Today the MMRV (measles, mumps, rubella, and varicella) vaccine helps prevent these childhood diseases.

UNUSUAL TREATMENTS

Until the twentieth century, few standard treatments for childhood diseases were found. Most were ineffective and sometimes they were deadly.

Diphtheria

To treat diphtheria many remedies were tried. By the nineteenth century, physicians recommend that good brandy and wine should be freely given to a child with diphtheria as it was thought alcohol would prevent the formation of the membrane. To balance the humors, they ordered a purge in the first three days of the disease. In addition, physicians placed leeches and blistering agents such as cantharis (Spanish fly) on the swollen neck to draw out the "inflammation" from the body and blood. This only caused pain to the child.

If a pseudomembrane developed—to allow the child to breathe—some physicians cauterized with a hot iron, which would cause bleeding and was not effective in removing the membrane. By the 1800s, doctors coated the membrane with various caustic solutions to destroy it. First, they applied a solution of silver nitrate, which was considered an all-purpose disinfectant. When this was not effective, they applied a mixture of hydrochloric acid (the strong acid found in the stomach) mixed with honey to the membrane. When this did not work, sulfuric acid (now found in automobile batteries) was then swabbed on the membrane. The physician needed to avoid touching healthy areas with these mixtures as they could cause permanent damage to the vocal cords and throat; they rarely removed the membrane. However, if a child was hospitalized, it was documented that 30 percent survived if a tracheotomy (cutting into the trachea to allow breathing) was performed.

Pertussis or Whooping Cough

In the 1600s, families often used home remedies to treat whooping cough and other illnesses. To calm the cough, both children and adults smoked rosemary herb in a pipe, which likely made coughing worse. Another home remedy was a drink made with opium and herbs. Due to the opium, it may have calmed coughs in some cases. In simple cases of whooping cough, physicians attempted to prevent bronchitis or pneumonia, which often led to death. Some physicians thought it was best to let the disease run its course, but others, based upon the humor theory, treated it with various remedies to prevent the deadly complications. Throughout the mid-1800s, physicians placed blistering agents on the back of the neck to "draw out the poisons." They also bled children by placing leeches on their necks as it was believed that if not enough blood were taken the child would die of suffocation. These could cause infection.

In the early 1800s, some physicians claimed that a dilute nitric acid (a caustic substance) solution would "cure" the disease and reduce coughing fits as the mixture was believed to be an antispasmoic. The nitrogen, in the acid, supposedly removed a substance called "fibrin," which was thought to be the poison causing the affliction. Physicians also gave the child juice from the deadly hemlock plant mixed with water and sugar as a cure. Throughout the late 1800s, many physicians sprayed belladonna from the poisonous nightshade plant mixed with water and other herbs into the throat to relieve the coughing and to dry up secretions. These could have caused poisoning. In the mid-1840s, they sometimes gave children ether or chloroform—used to anesthetize during surgery—by the spoonful to stop the spasmodic coughing. This may have worked temporarily as the child was likely rendered unconscious. Calomel (mercury compound) was also given to reduce inflammations, but if the gums became red (now known as a sign of mercury poisoning), it was stopped. Most of these treatments likely made the condition worse by poisoning or killing the child. Without any treatment in mild cases, the disease tended to run its course in about 100 days.

Measles and German Measles

Since the Middle Ages, physicians believed imbalance in the humors heated the blood; therefore, the first treatment for measles was to extinguish this heat and to prevent further rashes from breaking out. They did not allow patients to consume hot drinks and food, allowed only cold foods, and kept the patient in a cool room. Acid fruits such as pomegranates and rhubarb only were allowed as clinicians believed they extinguished heat. If a child was over 14, the physician bled from a vein but used cupping glasses to draw blood out of children and infants. Physicians also recommended purging, vomiting, and bleeding to get rid of the supposed poisons in the body. Since measles caused a few children to become blind, doctors into the late twentieth century recommended that those with measles be kept in a darkened room to prevent eye damage. However, darkened rooms did not prevent eye problems.

Mumps and Chicken Pox

Since other childhood diseases, including mumps and chickenpox, were not considered serious diseases until recently, they were given minimal treatment. In ancient Egypt, for example, healers treated mumps with a black salve slathered on the swollen neck of the child made from bitumen (a natural tar), as it was thought to draw out the inflammation. Families, as a home remedy, used a similar salve into the late twentieth century in many Western nations. After the salve was put on the neck, it was bound with a white cloth and knotted on the top of the head. However, the treatment had little effect on reducing the swelling. In the nineteenth century, doctors treated mumps with a laxative on the second day of the illness to eliminate the "bad humors." They treated chicken pox (*varicella*) with wine, brandy, and quinine. Since quinine is specific for malarial infections, it would have had no effect on the illness. The alcohol likely made the child sleepy and less likely to scratch the blisters.

Cholera

Asiatic cholera is an acute diarrheal disease, and most people who become infected only have mild diarrhea or no symptoms at all. However, about 5–10 percent of cases exhibit a severe form with profuse watery diarrhea—described as "rice-water stools"—vomiting, severe abdominal cramps, rapid heart rate, loss of skin elasticity, and muscle cramps. It generally lasts a few days. Death can occur within 24 hours from dehydration. Cholera is caused by the *Vibrio cholerae* bacterium that produces a toxin that causes the symptoms. It was first identified by Italian scientist Filippo Pacini (1812–1883) in 1854 a year before British physician John Snow (1813–1858) demonstrated that the disease was carried in the water supply contaminated with raw sewage. These discoveries, however, were not readily accepted until several decades and millions of deaths later.

The mortality for severe cases of cholera, without adequate treatment, is from 30 to 80 percent. Between 1848 and 1849, it claimed over 50,000 sufferers in Britain. The first two outbreaks in the United States resulted in around 150,000 deaths, and it is estimated that in the mid-nineteenth century cholera killed millions in China, India, and Russia. The disease affects all ages but tends to be more lethal in young children and the malnourished.

The term "cholera" is from the late Middle English and Latin, meaning "vomiting and diarrhea"; it was adopted from *choler*, the Greek term for "bile." It was first used in the early nineteenth century when the disease became a pandemic. It was first clinically described in 1563 by a Portuguese physician based upon colonial records from South Asia. These records documented many outbreaks between 1503 and 1817.

When cholera first appeared in Europe, health professionals noticed a link between cholera and filthy overcrowded urban slums. The disease was blamed on foul air rising from sewers and rotting trash. In 1840, when gummed postage stamps were introduced in Britain, licking the back of stamps was thought to be a cause. Very conservative Christians, particularly in the United States, believed cholera was God's punishment for sinful and immoral behavior. If respectable people caught the disease, it was assumed they must have sinned in some way, leading families to sometimes hide cases of cholera in their households.

Cholera had been a local disease around the Bay of Bengal in India and in some areas of Asia until the early 1800s when worldwide travel and trade brought this deadly scourge to the West in a series of pandemics. The disease was so feared that when it first arrived in Europe and North America in 1832 people fled in panic from the cities to the supposedly healthier countryside. In the United States, it fanned anti-immigrant feelings as Irish immigrants were thought to have brought the disease to North America. Outbreaks in the later twentieth and early twenty-first centuries occurred after war, civil unrest, or natural disasters when water and food supplies became contaminated with *V. cholerae*.

By the early 1900s, cholera had disappeared in regions where better sanitation and clean water systems had been implemented. Antiserums and immunizations, developed in the 1930s, were used extensively by the United States' and other developed nations' militaries during WWII. Treatment for cholera was based upon rehydration with oral and intravenous solutions. Antibiotic therapy was added around 1965. Today antibiotics are generally used only for severe cases, and all victims receive aggressive oral and often intravenous rehydration. Curiously in the 1830s, one physician suggested treating cholera with an infusion of a saline solution into the victim's veins, but he was ignored by the medical establishment for decades.

Cholera outbreaks shaped the history of public health and led to public health reform. Since it was believed to be caused by filth, local agencies organized massive cleaning initiatives. They swept and washed the streets and eliminated accumulated trash. Houses with cholera victims were fumigated. Due to the pandemics, the first International Sanitary Conferences was held in 1851 to standardize international quarantine regulations. Vessels were quarantined in Britain, the United States, and other nations for weeks if cholera had broken out. Ships were not allowed to dock until physicians inspected everyone, isolated sick patients, and disinfected the ship. By 1900, many large cities had developed sanitation and clean water supply facilities, thus reducing many waterborne diseases.

UNUSUAL TREATMENTS

Since cholera was a new disease in the early nineteenth century from a Western perspective, treatments included the standard regimes for diarrhea as well as by trial and error. However, in the Bay of Bengal region of India, Hindu Ayurvedic practitioners for centuries had treated cholera with mild purgatives. Later they also prescribed alcohol, an idea borrowed from Western practitioners. In the West, opium was the first line of treatment and helped to stop the diarrhea. However, many physicians recommended the standard harsh remedies of vomiting, purgatives, and bloodletting as they believed a morbid poison existed in the blood that needed to be eliminated. Bloodletting was accomplished in several ways for cholera. In one method American physician, Charles B. Coventry (1801–1875), had the patient be put into a tub of hot salty water. Two large men then held the patient down while the physician opened a large vein in the instep of each foot, allowing the blood to flow into the water for 12–20 minutes. In the last stage of the disease, bleeding with wet cupping (scratching the surface and placing a heated cup over it to draw blood) and placing leeches on the abdomen were tried. In opposition to the hot bath regime, British physician Edward Waring (1819–1891) suggested in 1865 an ice-cold bath along with arsenic. Others suggested ice bags on the spine. All these methods likely hastened death.

In the early 1830s, experiments with clysters (enemas) were tried with different substances. Some prescribed an enema of a pint of "chicken tea," with a tablespoon of table salt. To retain the mixture for a few minutes, pressure was put on the anus, which in some cases may have helped to hydrate the victim. Folklore also suggested that the anus was plugged with beeswax or a cloth to prevent the diarrhea. Experiments with coffee and tobacco smoke enemas were tried with little success.

Different pungent substances were rubbed or placed on the body at the same time the patient received medication. For example, some physicians administered three drops of the spirits of camphor (think chest vapor rubs) every 15 minutes in a tablespoon of water. At the same time, the limbs and body were rubbed with camphor oil to reduce muscle cramps, and the oil was also sprinkled on bed clothes. Some also thought that rubbing limbs with brandy and red pepper and placing a poultice (a hot medicated material) over the stomach was helpful.

Numerous concoctions were administered, which likely hastened death. For example, some clinicians gave drops of sulfuric acid (used in automobile batteries) diluted in water. Others gave a mixture of chloroform and calomel—which contained the poisonous metal mercury. In the 1860s, American physician Elijah B. Hammack (1826–1888) treated the disease with a mixture of chalk, an opiate, and brandy. Some suggested that a strong alcohol solution of camphor given at the onset of the disease saved lives. In Britain, physicians claimed a mixture of calomel and opium resulted in fewer deaths compared to other treatments. Other

German painter Alfred Rethel's (1816–1859) *Death the Strangler* represents the first outbreak and deaths from cholera that occurred in Paris at a masked ball in 1831. (The Cleveland Museum of Art, Gift of Robert Hays Gries 1939.620)

practitioners found that a tablespoon of a mixture of chloroform, oil of camphor, and laudanum (alcoholic solution containing morphine) followed by a piece of ice every few minutes was successful along with a tablespoon of brandy every hour or so. Mercury may have killed the bacteria but could have also poisoned the patient. The other treatments were not likely successful.

Proprietary treatments and folk cures also evolved for curing cholera such as color therapy. For example, in the American post–Civil War years, a former Union General claimed to cure most diseases with light shining though blue glass. Another system claimed that blue-tinted water would cure cholera. By the early twentieth century, therapies were advertised that advocated different colors or light waves to be effective in curing various illnesses. One system, the Spectro-Chrome light box, suggested a green and then indigo light could cure cholera. To achieve this, colored glass was placed over a 1000-watt lightbulb and shown on the patient. The device was marketed into the 1950s. Colored light therapy is still used today in some alternative medicine systems but is not effective for cholera.

Common Cold and Other Nasal Issues

Several conditions are found in the nasal and sinus cavities that result from inflammation or irritation of the mucous lining. These include the common cold, sinusitis, allergies, and nasal polyps, which often exhibit similar symptoms. The common cold, or "cold," is a viral infection primarily of the nose. Inflammation of the nasal passages (rhinitis) is also caused by bacteria, irritants, or allergens. A cold causes sneezing, a runny discharge from the nose, nasal congestion, and malaise. Allergies cause similar symptoms. Some cold sufferers experience chills, sore throat, a brief fever, and cough. A cold is generally mild and self-limiting, and most symptoms improve within a week or so. Acute sinusitis, besides cold symptoms, is also accompanied by facial pressure and pain and can last several weeks. Chronic sinus inflammation and polyps can cause long-lasting congestion along with facial pressure, thick discharge, and discomfort. A nosebleed is bleeding from the nose.

About 200 viruses cause the common cold with the rhinovirus being responsible for over 50 percent. British researcher David Tyrrell (1925–2005) first identified this virus in the 1950s; the coronavirus causes about 30 percent. These viruses are transmitted mostly through hand-to-hand contact rather than droplets in the air. Although colds do not lead to death, they are responsible for much discomfort, lost work, and higher medical costs. People find them a nuisance and inconvenient. Insufficient sleep, stress, and malnutrition ("stuff a cold, starve a fever") have been associated with a greater risk of developing a cold. In higher-income nations, children suffer five or more colds and adults two or three colds a year. Colds occur all year long but are more common in the winter months while allergies are more common in warmer months.

The cold is the most common human illness and has affected people for millennia. It was mentioned in Egyptian papyri, Classical Greek, Roman, and Asian writings. The coronavirus, for example, is thought to have been transmitted to humans from camels around 10,000 years ago. The name "cold" came into use in the sixteenth century, due to the similarity between its symptoms and those of exposure to cold weather. Since chills often occur at the onset, the term may have also derived from this symptom. For centuries, colds were thought to arise from going out in cold or wet weather, into damp night air without warm clothing, or being in a cold draft. Some research has suggested that the rhinovirus replicates better in cooler nasal conditions supporting some of the "old wives' tales."

In earlier centuries, allergic rhinitis and sinusitis were likely confused with the common cold. In the 2010s, for example, about 7 percent of the U.S. population was afflicted with allergic rhinitis or "hay fever," and around 12 percent suffered from sinus infections with twice as many women compared to men having the condition. Chronic sinusitis can be dangerous as the sinuses are near the eyes and brain to which pathogens can spread, causing meningitis and death (this was more likely before the development of antibiotics). Chronic rhinosinusitis can also lead to nasal and sinus polyps. These pendulous growths primarily occur in middle-aged adults. In rare cases, polyps can obstruct breathing. In the nineteenth century, some physicians began to focus on nasal issues as they were the most common illnesses among patients. Various instruments and procedures were developed, and by the early twentieth century, the ear, nose, throat (ENT) medical specialty arose. Common folk remedies for colds and congestion include inhaling steam or eating chicken soup, which has been found to have antiviral properties. To manage a common cold or allergies, physicians recommended antihistamines to relieve symptoms of rhinitis. Steroids decreased the inflammation of the nasal passages and decreased polyps.

Nosebleeds (Epistaxis)

Anterior bleeding from the nostrils is more common than posterior bleeding, which is usually more serious. Nosebleeds affect about 60 percent of people over a lifetime, mostly the young and very old; death is rare. Bleeding can be caused by trauma, dry air, high altitude, anticoagulants, nasal tumors, high blood pressure, and blood and genetic disorders. The term "nosebleed" was first used in the nineteenth century. Ancient Egyptian papyri suggest that during a girl's first menstruation a nosebleed meant she was fertile and healthy. The Greek physician Hippocrates (460–c. 370 BCE) also held this belief and thought that blood from the nose and vagina were basically the same. Middle Ages physicians believed that nosebleeds helped to keep the body humors in balance and were a natural means of eliminating various diseases. The treatment for nosebleeds from ancient times through today includes bending the head forward and pinching the nose, applying cold compresses to the nose, or in serious cases plugging the nose with material. Although antibiotics were used for nasal congestion during the mid-through twentieth century, they were not effective and caused many bacteria to become drug resistant.

Nasal issues have appeared as part of literary and film plots. One example is Stephen king's science fiction horror novel, *Firestarter* (1980), made into a film (1984) of the same name. A man who could control people's minds finds that when he is exerting this control called "the push," his nose bleeds. In American Science fiction writer Alan Edward Nourse's (1928–1992) short story "The Coffin Cure" (2008), a group of scientists find a cure for the common cold. One takes credit for the discovery, but he finds that the world is too malodorous as does everyone else who has taken the anticold vaccine. So, he isolates a new virus to bring the cold back. He then is unable to smell unpleasant smells anymore, but the new virus has created a permanent cold.

UNUSUAL TREATMENTS

Numerous home remedies for nasal conditions have been used since antiquity. Depending upon the culture, sufferers have eaten or chewed hot or spicy items such as garlic, ginger, raw onions, horse radish, turmeric, and cayenne pepper.

Herbs and spices, including peppermint, chamomile, eucalyptus, cinnamon bark, tea tree oil, and Echinacea, have been used to make hot steamy tea often with added honey, lemon juice, or wine. Distilled spirits, such as brandy or whiskey, were added to these hot drinks by at least the eighteenth century for a "hot toddy" to treat a cold, which may have made the person feel better.

However, other old remedies are not common today. For example, in ancient Egypt magic spells and incantations to Ra and Thoth were said for rhinitis. The *Papyrus Ebers* suggests rubbing peppermint on the nose or pouring date juice into the nares for runny noses and congestion. Another cure along with the proper incantations included mixing "milk of a woman who has borne a son" with fragrant bread and placing it in the nostrils, which was unlikely to be effective.

From the classical Greek and Roman era through the eighteenth century, bleeding from a vein, leeching, and cupping (placing a hot cup on the skin after scratches had been made to draw out the blood) were sometimes performed to balance the humors and relieve rhinitis. In the fourteenth century, English physician John Arderne (1307–1392) devised various concoctions to cure runny noses and congestion. For example, he added absinthe, white horehound, and ground ivy to white wine. This was boiled, and the patient inhaled the steam. He made a pill of licorice and seeds of citron melons combined with gum acacia that the patient placed under the tongue. Physicians in the nineteenth century, such as American Elijah B. Hammack (1826–1888), advised patients to take a cold bath at night and place a hot brick in bed at their feet to cure a cold. These all may have helped the patient feel better, but did not cure the cold and were harmless.

In the late nineteenth and early twentieth centuries, a patent medicine cure for a cold was advertised. The sufferer inhaled the smoke from a "carbolic smoke ball"—a hollow rubber ball filled with carbolic acid and fitted with a tube for the nose. The disinfectant would cause the nose to run and supposedly flush out the infection but could have damaged the nasal tissue. From the 1930s through 1950s, Benzedrine (an amphetamine) cylindrical inhalers were used to clear up nasal congestion. The device reduced congestion but also caused drug dependence, so it was removed from the market. Home remedies of the late twentieth and early twenty-first centuries include sucking on zinc lozenges, mild exercise, mega dosing of vitamin C, or listening to jazz, which are thought to increase immune function. Wearing cold wet socks to bed supposedly increases circulation to the feet and relieves head congestion. Research on the effectiveness of these methods is inconclusive.

In terms of nasal polyps, the ancient Egyptians pulled them out through the nose with a hook as did the twelfth-century Persian physician Avicenna (c. 980–1037) who then cut through the attachment with scissors and washed the nasal cavity with vinegar. Some current home remedies for nasal polyps include applying tea tree oil in nasal passages or ingesting magnesium tablets, two tablespoons of apple cider vinegar in water, probiotics, hot spices mentioned above for colds, and the core of a pineapple to reduce inflammation. However, mixed results have been found for the effectiveness of these remedies.

Nosebleeds (Epistaxis)

For this issue, various substances throughout history have been stuffed up the nose to stop the blood flow. For example, Roman scholar Pliny the Elder (23–79 CE) recommended plugging the nose with pounded chives, mint, and nutgalls from oak trees, which may have contained tannic acid that stopped bleeding. He also recommended stuffing the nares with burnt ashes of tadpoles, which could have caused clotting.

From the Middle Ages through the seventeenth century, "usnea," or beard lichen, which grew on improperly buried corpses' heads, was placed in the nostrils. During the sixteenth and seventeenth centuries, blood from the patient was dried, powdered, and stuffed up the nose to stop bleeding, which could have helped. In addition, from the Middle Ages through the early nineteenth century, mumia (a powder made from Egyptian mummies) was placed in the nose. William Buchan (1729–1805) in the eighteenth century suggested soaking the genitalia in cold water would stop the bleeding. In the early nineteenth century, balloons made of animal intestines and later of rubber were stuffed into the nostrils to stop serious bleeding after bloodletting went out of fashion, which likely resulted in infections.

Coughs, Bronchitis, and Laryngitis

Coughs, bronchitis, and laryngitis are maladies of the throat, bronchial of the lungs, and the larynx. Coughing is a reflex found in many animals and keeps the throat and airways clear and can be either acute or chronic. Acute coughs are usually associated with a cold, influenza, certain childhood diseases, and bronchitis. Chronic coughs lasting longer than two to three weeks are the result of allergies, chronic bronchitis, COPD (chronic obstructive pulmonary disease), acid reflux, postnasal drip, air pollution, smoking, and even some medications. Coughing is also a symptom of tuberculosis, pneumonia, and certain cancers discussed in other entries.

Coughs can be nonproductive (dry) or productive when sputum and mucus is coughed up. From ancient times, various types of coughs have been recognized, which are still known today. These include a dry tickling cough, wet chest cough from the lungs with lots of mucus, a cough with yellow gray phlegm associated with cold-like symptoms, and whooping cough. A productive cough is the most common symptom of both acute and chronic bronchitis.

Acute bronchitis is short-term inflammation of the bronchi of the lungs, while chronic bronchitis (CB) lasts for at least three months and is a common cause of a "smoker's cough." In the United States, for example, in the 2010s, about 2.0 percent of males and 4.4 percent of females suffered from CB, which affects the economics of modern societies because it increases health-care costs as sufferers are more likely to be hospitalized, visit emergency department/urgent care facilities, and use expensive prescription medications. Laryngitis or hoarseness can result from a viral infection, allergens, and voice strain and is generally temporary. Chronic laryngitis is caused by long-term exposure to irritants. In a few cases, chronic hoarseness can signal a more serious medical issue such as cancer or growths on the vocal chords.

Coughing, bronchitis, and laryngitis are likely as old as humanity. Names for these conditions in English go back at least several hundred years. The term "cough" was first used in the fourteenth century and is from the Middle English *cough*, as is the term "hoarseness," which is derived from an Old Norse word meaning "dried out" or "rough." Bronchitis and laryngitis are from New Latin, from the Greek, and were first used in the early nineteenth century. The suffix *-itis* means "inflammation" of the structure. Coughing and bronchitis throughout the eighteenth century were thought to be caused by cold drafts, being outdoors in cold damp weather, or by an imbalance of the body humors.

Many modern treatments for these conditions are based upon remedies used for centuries. Some of these are somewhat effective including a spoonful of honey to quell a cough; breathing in steam and consuming hot teas made with various aromatic herbs to loosen bronchial congestion; and ingesting chicken soup that has a mild anti-inflammatory effect. A major remedy for centuries in many cultures to suppress a cough has been opium or one of its derivatives, such as codeine, developed in 1832. Due to a rising opiate abuse problem in the late twentieth and early twenty-first centuries, a prescription for codeine cough syrup was mandated in several nations.

References to coughing have been found in the arts. The first time it was used as a term was in Chaucer's (1343–1400) *Canterbury Tales*. In the "Millers Prologue," a character coughs as a signal that he is outside before he knocks on the door. Throughout Italian musician Giacomo Puccini's (1858–1924) most famous and performed opera *La bohème* (c. 1894), coughing is heard from Mimi a seamstress who has TB and dies.

UNUSUAL TREATMENTS

Healers in the past concocted numerous remedies over the millennia for coughs by trial and error. Recent research has suggested that some may have been effective. For example, ginger has been used for centuries in several cultures to ease a dry cough and has been found to have mild anti-inflammatory properties. Likewise, the herb thyme, which has antioxidant properties, has been a common remedy for cough and bronchitis for centuries. As an expectorant, in the fourteenth century, hyssop, mint, maiden hair, horehound, cinnamon, licorice, and other herbs were boiled and added to honey. For a dry cough, healers suggested sugar mixed with violets, figs, horehound, and opium poppies. The opium likely quelled the cough, and the other substances soothed the throat.

The *Papyrus Ebers* (c. 1500 BCE) illustrates that raw garlic was routinely given to those with coughs. A remedy used by both Egyptian and Greek healers for wet coughs and bronchitis was a "plaster" (crushed aromatic substances wrapped in a cloth) that was placed on the affected area such as the chest. The plaster would heat the area and was thought to draw out "poisons." The vapors helped to break up phlegm and mucus, making it easier to expel. By the late 1500s, the use of mustard plasters was common in northern Europe, and by the mid-eighteenth century, turpentine was used in the plaster. A similar remedy for congestion relief was the menthol rub in which mint extracts were mixed with oil and beeswax and smeared on the chest. This remedy is still used today as "vapor rubs."

On the other hand, some remedies were not effective. Physicians throughout the mid-nineteenth century practiced bloodletting with leeching and wet cupping on the throat, cutting a vein in addition to purgatives to rebalance the humors that likely weakened the patient. During the sixteenth to eighteenth centuries—adopted from North American First Nation people—healers recommended tobacco smoking and tobacco enemas to help cure a cough. Tobacco likely made coughing worse.

In the mid-nineteenth century, family physicians also suggested rye whiskey toddies and Jamaican rum to help nourish the patient and stop coughing at night. For CB, physicians recommended a hot water foot bath to "improve the circulation" along with Dover's powder that contained opium and ipecac, which thinned the mucus to make coughing easier but caused vomiting. A popular patent medicine for cough and bronchitis was Ayers Cherry Pectoral in the late nineteenth and early twentieth centuries, which contained an opiate. These opiates helped quell coughs. In the early twentieth century, inhaling the smoke from a "carbolic smoke ball"—a hollow rubber ball filled with carbolic acid and fitted with a tube for the nose—was also advertised as a remedy for coughs and bronchitis, which could cause tissue damage.

For hoarseness and laryngitis from shouting, fourteenth-century physicians recommended that patients take a bath and eat poached eggs and apple jam mixed with herbs, linseed oil, and bean flour. Sulfur added to eggs was also thought effective. They also warned that bloodletting should not be done for hoarseness "unless it is of hot and moist cause." If laryngitis was caused by moist conditions, healers suggested the patient ingest a mixture of calamint and resin from the tragacanth plant. This concoction acted as a purgative and was thought to bring the body humors back into balance. If hoarseness was caused from cold conditions and was severe, healers gave pills made from the asafetida plant, the Middle Eastern herb fenugreek, and orobus (a wild pea). Dried figs were recommended for all causes of laryngitis, which acted as a purgative and may or may not have helped.

In the nineteenth century, physicians suggested patients should inhale steam containing opiates and hops for laryngitis. For chronic laryngitis, they suggested a solution of silver nitrate be applied with a brush to the vocal cords every two days. Physicians also recommended counterirritants, such as croton oil, be rubbed on the throat every night until blisters formed to release the infection, which could have led to infections. Another treatment consisted of rubbing this area with warm aromatic oils. Some of these nostrums are still used today in folk medicine and can help the patient feel better due to the placebo effect.

Depression and Melancholia

Depression is one of two major "mood disorders." The others, bipolar I and II, are discussed in the "Mental Disorders ('Psychotic'): Bipolar Disorder and Schizophrenia" entry. These mood disorders sometimes are called "affective disorders." In the past, a depressive mood with abnormal beliefs was called melancholia. Depression has several forms. One of the most serious types is the major depressive disorder, also known as clinical depression. If the person cycles from normal to depression, it is termed Unipolar Depression. Major depressive disorder involves long periods of extreme sadness, sleep disturbances, hopelessness, apathy along with physical symptoms, decrease or increase in appetite, slowing of thoughts, reduction in physical movement, loss of energy, and fatigue. It generally lasts for more than two weeks. Other characteristics include feelings of worthlessness, suicidal ideation, and sometimes the inability to carry out daily responsibilities. A decrease in the neurotransmitter serotonin in the brain is thought to be the cause. In addition, stressful life events like death of a spouse or family member, divorce, or trauma can trigger depression.

Although everyone suffers from mild bouts of depressive moods that last a short time, clinical depression can last for months without treatment. About 7 percent of people in the United States per year will suffer from depression, and it is higher among females. The biggest ramification of depression is suicide, which is the second leading cause of death among young adults, but depression also interferes with work, school, and social relationships. Over the twentieth century depression has been treated with a range of methods, including rest, talk therapy, amphetamines in the 1930s–1940s, tranquilizers beginning in the 1950s, antidepressants, and some other therapies discussed in the "Unusual Treatments" section.

Another depressive disorder, seasonal affective disorder (SAD), is classified as a major depressive disorder with seasonal pattern. People with SAD exhibit depression symptoms and often gain weight due to carbohydrate craving and overeating during the fall and winter months. SAD usually improves with the arrival of spring and more sunlight. About 5 percent of people in the northern regions of North America and Europe experience SAD. It typically starts in young adulthood and is more common among women. Management of this disorder is intense light therapy and antidepressants.

Hormonal changes can also cause depression. About 40 percent of menstruating females experience premenstrual syndrome (PMS). Due to an increase in female hormones, they experience short-lived abdominal bloating, breast tenderness, irritability, and a depressed mood right before menstruation. However, approximately

3–10 percent experience premenstrual dysphoric disorder (PMDD). It is characterized by severe and disabling symptoms such as serious mood swings, anger, anxiety, and hopelessness that can disrupt their studies, work, interpersonal relationships, and daily living. This type of severe depression generally requires management with antidepressant or other medications.

Postpartum depression and perinatal mood disorder (PPD), sometimes called the "baby blues," can occur during, or up to a year, after pregnancy, but it's most common during the first four weeks after delivery. Symptoms are like other depressive disorders and may range from mild to severe. Crying is a very common symptom due to lifestyle change, stress, and the rapid decrease of pregnancy hormones. PPD occurs in 15 percent of women and from 1 to 26 percent of new fathers. The affliction can affect the mother's health and interfere with her ability to care for her family. PPD has been noted for centuries, and in a very few women, a postpartum psychosis develops where the woman is at risk of killing her infant.

The term "depression" was first used in the fourteenth century from the Latin *deprimere*, meaning "to press down." Its meaning of "dejection, state of sadness, or a sinking of the spirit" is from the early fifteenth century. Greek physician Hippocrates (c. 460–370 BCE) believed that depression, or melancholy, was caused by an imbalance of the four humors caused by too much black bile in the spleen. However, many ancient Greek and Roman physicians were divided in their thinking about what caused a depressed mood. Some thought it was due to spirit possession and others from an imbalance of one of the humors. After the fall of the Roman Empire in the fifth century, scientific thinking about the causes of depression and mental disorders regressed. A small minority of clinicians continued to believe that depression and mental disorders were from natural causes, but others, under the control of the growing power of the Christian Church, believed it resulted from demonic possession or sinful behavior and was a spiritual issue. Therefore, the patient was referred to the clergy for treatment who often performed exorcisms.

The term "melancholy" is derived from the Greek *melankholia*, from *melas*, *melan*, meaning "black," plus *kholē*, meaning "bile," and was first used in English in the fourteenth century. Melancholy was also considered to be associated with the creativity of the scholar and artist. In the seventeenth century, English physician Robert Burton (1577–1640) in his *Anatomy of Melancholy* (1621) suggested depression was a medical condition. As the classic authors before him, he claimed it was caused by an excess of black bile. During the eighteenth and early nineteenth centuries, melancholy, or depression, was believed to be an inherited and unchangeable weakness, which often meant people with the condition were put in asylums or rejected by their families.

However, in the late nineteenth century, Austrian neurologist Sigmund Freud (1856–1939) believed it was a person's unconscious anger over loss and recommended a "talking cure" or psychoanalysis. The behaviorist movement in psychology in the mid-twentieth century believed depression was a learned behavior. By the late twentieth century, mental health personnel embraced the biopsychosocial model that argued that biological, psychological, and social factors are all linked together to cause disease or health. Therefore, depression makes a person more likely to develop other physical conditions, and conversely, a person who has a physical malady, is more likely to develop depression. Risk factors include genetics, chemical imbalance in the brain, and severe stress or trauma.

Depression may have affected the course of American history. President Abraham Lincoln (1809–1865), for example, suffered from bouts of melancholia throughout his life. Some believed insight from these attacks allowed him to fight for the prohibition of slavery in the new northern states after the U.S. Supreme Court's Dred Scott decision (1857) championed by the South. Lincoln spoke out against this decision at the risk of his political career, which paradoxically allowed it to take off. His arguments eventually led to his presidency and the Civil War, or the War between the States (1861–1865). Depression has also been featured in literature. In American author Sylvia Plath's (1932–1963) *The Bell Jar* (1963), for instance, a bright, successful young woman becomes severely depressed and almost dies from a suicide attempt and is institutionalized. She is a woman who is attempting to find herself in a society of the 1950s, where women were often forced into the role of a married housewife or single career women.

UNUSUAL TREATMENTS

Treatments attempted for depression were often harsh. Trepanation—scraping or drilling a hole in the head—has been used for depression and other mental disorders since the dawn of history. Greek physician Hippocrates (c. 460–370 BCE) advocated trepanation to allow evil spirits, pain, or air to depart, thereby relieving various mental disorders. This treatment was used from antiquity into the eighteenth century. Some patients did survive, but it did not prevent or cure depression. Roman physician Celsus (25 BCE–50 CE) recommended starvation, shackles, and beating as remedies to eliminate bad spirits, which was sometimes still carried out into the nineteenth century. Clinicians believed that if depression was not adequately controlled it could develop into the more serious mania. Greco-Roman physician Galen (129–c. 216) advocated bleeding to rebalance the humors. This did little for depression and likely caused weakness from loss of blood.

During the Middle Ages, many people were locked up in "lunatic asylums" or executed by burning or drowning if they were thought to be possessed by demons. In the Victorian age of the early-to-late nineteenth century, hydrotherapy was tried. Clinicians plunged depressed patients into cold baths often for several days. Some clinicians advocated forcing patients to drink copious amounts of water to get rid of depression; neither of these remedies were very successful although fear of the treatment may have brought some patients out of their depression. Other physicians advised alcohol at most meals followed by opiates to help the patient sleep at night, which likely led to substance use disorder. Some tried cautery on the spine, which led to open blisters, or bleeding by placing leeches on the head to balance the humors. These could have caused infections and weakness. Clinicians also developed a special spinning stool in which the patient was spun to cause dizziness to "rearrange the brain" into its correct position. Enemas, purging, special diets, horseback riding, or traveling were also advocated. Riding or a change in scenery may have helped in some cases. In the late nineteenth and early twentieth centuries, Freud and other physicians prescribed cocaine to treat depression. It often helped to lift people out of their depressed mood but also led to cocaine dependency.

In the 1930s, lobotomy—destruction of brain tissue—was developed for the treatment of depression and was popularized by American physician Walter Freeman (1895–1972). Freeman would push a sterilized icepick-like instrument through the eye orbit into the prefrontal area of the brain, wiggle the pick around, and destroy the tissue. Lobotomy was extensively used in the 1940s and 1950s and fell into disrepute in the mid-1960s. Lobotomies sometimes brought the patient out of a serious depression but often led to detrimental personality changes such as apathy, lack of initiative, social disinhibition, poor judgment and decision-making, incontinence and sometimes coma or even death.

Another treatment developed in the 1930s was "shock therapy" or electroconvulsive therapy (ECT). It produced seizures and did bring most patients out of depression although permanent memory loss, confusion, and high blood pressure could occur. Electric fish were placed on the head in antiquity and electricity from batteries in the eighteenth and nineteenth centuries to cure mental disorders without much effect. ECT was not commonly used until the 1940s and is still considered a controversial treatment by some. Severe convulsions were common into the 1950s that sometimes resulted in broken bones. However, by the late 1960s patients were first given muscle relaxants and sedation to prevent seizures. Approximately 70 percent of ECT patients are women and more than a third are 65 and older. Its use declined in the 1960s due to stigma and the development of antidepressant medications.

By the early twenty-first century, some folk and alternative therapies became popular for treating depression. Sound therapy advocates, for example, believed that the correct frequencies could heal and balance the "energies" in the body. Although the research is mixed, some have suggested that high-frequency sounds might stimulate the brain to naturally increase serotonin and create a positive emotional state. Engaging in creative activities and artwork including knitting, painting, quilting, making music, or other activities has sometimes been effective in helping to reduce depression and anxiety and to increase self-esteem.

Aromatherapy became another popular remedy. For depression and insomnia, a 100 percent lavender oil solution inhaled or applied to the skin of young adult females, for example, was found to increase feeling of well-being and better sleep. Potential side effects, when the oil was applied to the skin, included allergic skin irritation and sun sensitivity. Some research suggests that ingesting cannabinols can increase mood and lessen anxiety. Crystal therapy proponents claim that the vibrations from the lilac-colored stone lepidolite can dissipate negative thoughts. No evidence suggests that crystals work to decrease depression other than the placebo effect. Detox foot pads were advertised to reduce depression, insomnia, and headaches by "removing toxins and heavy metals" from the body. These pads are put on the foot overnight and turn brown by morning due to body oils. However, there is no evidence that they decrease depression or draw out toxic substances from the body.

Diabetes

Several categories of diabetes exist. Type 2 diabetes—diabetes mellitus—is the most common form of this chronic condition. Over 90 percent of people with the disorder have this "adult onset" form of diabetes, although it is increasing among overweight children. The signs of diabetes include constant thirst, frequent urination, increased hunger, fatigue, rapid weight loss, and slow-healing sores. A key risk factor for the malady is "metabolic syndrome," which includes obesity, high cholesterol or triglyceride levels, and high blood pressure. Other risk factors include a lack of exercise and heavy alcohol consumption. In addition, genetic or even epigenetic (gene function being switched on or off) factors are probably involved.

Diabetes is currently the fifth most common cause of death in the world, and it is estimated that around 20 percent of people over 65 years of age have the condition. The illness affects many major organs, including the heart, blood vessels, kidneys, nerves, and eyes. In Type 2 diabetes, the beta cells on the pancreas do not produce enough insulin, or there is insulin resistance where the body does not properly respond to insulin. This causes abnormal carbohydrate metabolism resulting in elevated levels of glucose in both the blood and urine. High blood sugar can lead to coma and death. Sometimes enough sugar is secreted in urine so that it can be fermented into alcohol.

Around 5 percent of diabetics have Type 1 diabetes previously called "juvenile diabetes." In Type 1, the immune system mistakenly destroys the beta cells that produce insulin. With little insulin, death soon occurs without treatment. Another type of diabetes is gestational diabetes found in about 6 percent of pregnant women who later have a higher risk of developing Type 2. A rare condition is diabetes insipidus in which the pituitary gland does not adequately secrete the hormone vasopressin or the kidney does not have a normal response to the hormone. This leads to great thirst and much diluted urine.

The first use of the term "diabetes mellitus" in English was in the mid-sixteenth century. "Diabetes" is via Latin from Greek, from *diabainein*, meaning "go through," and "mellitus" from the Latin *mellitus*, meaning "sweet." It is an ancient disease. The *Papyrus Ebers* mentions plentiful urine, which may be in indication of diabetes mellitus. In the second century BCE, Hindu physicians noted that "black ants were attracted to honey urine" and used this as a diagnosis for the condition. Greek physician Arestaeus of Cappadocia (c. 100) was the first to describe the malady. He observed that patients had great thirst and that large masses of the flesh were liquefied into urine. He also noted ulceration and gangrene on the legs of diabetics.

Some physicians in the early modern age mentioned that individuals who were corpulent, ate much, and drank heavily were more likely to be diabetic. Early in the eighteenth century, Scottish military surgeon John Rollo (d. 1809) was the first to use diet and weight loss to manage Type 2 diabetes. Though the mid-nineteenth century, diabetes was thought to be a kidney issue. However, in 1910, a British physician suggested that diabetes developed when the pancreas failed to produce a particular chemical that he called insulin. Insulin was then isolated and researched by three Canadian and one Scottish researchers in 1921. The discovery of insulin revolutionized the treatment of diabetes and allowed children with Type 1 to live relatively normal lives. Insulin by self-injection along with a low-carbohydrate and low-fat "diabetic diet" became the standard for treatment. In the 1980s, synthetic human insulin became available and largely replaced animal insulin. About 40 percent of Type 2 diabetics need insulin while the rest rely on the "diabetic diet" alone. In the 1990s, an insulin pump was available to provide measured amounts of insulin and by the twenty-first century various oral drugs. Research on artificial pancreases, immunotherapy, and even genetic engineering was also being carried out.

Diabetes has been a plot line in the arts and fiction. The most famous is the play *Steel Magnolias* (1987) by American Robert Harling (1951–) and a film (1989) by the same name. It follows a young woman with Type 1 diabetes who gets married, has a baby despite her doctor's and mother's warning, has a kidney transplant, and dies a few years later. The film frightened many diabetic women who were afraid to become pregnant although diabetic women can have normal pregnancies without complications. The crime novel *One Step Behind* (1979) by Swedish author Henning Mankell (1948–2015) is part of his "Inspector Wallander" novels and later a British TV series (2008–2016). In one episode, Wallender's Type 2 diabetes causes issues as he hunts down a murderer. In Part III (1990) of the mafia movie series *The Godfather*, the film's murderous crime boss, who suffers from diabetic symptoms, dies of a heart attack likely because of his diabetes.

UNUSUAL TREATMENTS

Many folk treatments have been attempted for diabetes. Several are still advertised today as alternative medicine or home remedies to prevent diabetes or keep blood sugar levels low. The Persian physician Avicenna (980–1037), to reduce both sugar in the urine and copious urine outputs, administered a concoction of three crushed seeds: lupine, a diuretic, which would have been counterproductive; zedoary that had no effect; and fenugreek that has been found to reduce blood glucose, so the mixture may have sometimes been effective. He also treated diabetic ulcers with honey, which often stopped the infection as sugar in high concentrations (hypertonic solution) can kill bacteria.

Physicians from at least the seventeenth into the nineteenth century prescribed crushed red coral flowers, viper flesh, oil of roses, arsenic, and dates for diabetes, which had no effect and could have poisoned the patient. They also prescribed syrup of poppies (opium) perhaps to relieve painful neuropathy in the feet or gangrene. English herbalist and physician Thomas Sydenham (1624–1689), and other physicians, administered gum Arabic and caraway, which have been found to help control urination, at least in rats, and they prescribed nettle root, which has been shown to reduce blood glucose. Another helpful plant was sumac, which has been found to decrease both blood cholesterol and blood sugar levels. Some clinicians also prescribed eating a large nutmeg twice a day, which had little effect on diabetes but could cause hallucinations. They also recommended a lot of easily digested meats such as veal and mutton and a few carbohydrates. A high-protein diet may have helped prevent spikes in blood glucose in some cases.

By the late eighteenth century, physicians, such as William Buchan (1729–1805) from Scotland, recommended diabetic patients should lie upon a hard mattress as he believed a soft bed hurt the kidneys, which clinicians believed caused the malady. He also recommended flannel clothing, a "broad girdle worn . . . about the loins," and anything to promote perspiration. These had no effect upon the course of the disease. Another Scottish physician Alexander Macaulay (1783–1868) in the early nineteenth century, based upon humoral theory, in his *Dictionary of Medicine* (1831) suggested applying blisters to the lower part of back to help draw out fluid, but this could cause infections when they were opened. He also recommended riding horseback to jog the kidneys and to provide friction over the kidneys to halt the copious urination. This could have caused kidney damage.

On the other hand, American physician Henry Hartshorne (1823–1897) realized there was no cure for diabetes but did offer many suggestions for its management. He gave patients leaves of creosote or chaparral, which may have had some effect, as animal studies have found it improves insulin sensitivity but could cause kidney and liver damage. He noted that some physicians tried constant galvanic (electric) current with one pole applied to the back of the neck and the other over the liver, but it had no effect on the disease. In the early twentieth century, American physician John Harvey Kellogg (1852–1943) believed that heat and harsh light baths could treat diabetes, which was not effective.

Even after the discovery and use of insulin, some folk and alternative medicines to control diabetes were promoted into the 2020s. Many were based upon remedies used for centuries, and some do influence blood glucose levels or hypertension, a component of metabolic syndrome. Essential oils such as coriander and lemon balm have been found to reduce blood glucose levels when inhaled. Likewise, oils derived from cinnamon, cumin, fenugreek, and oregano have been found to lower blood pressure. Although the research is mixed, when these oils were combined, they also lowered blood glucose levels. The bitter melon herb has been shown to slightly reduce the fructosamine (glucose combined with protein) levels, and milk thistle has been shown to decrease blood glucose levels and cholesterol along with insulin resistance. Coriander has also been found to reduce blood glucose levels. Prickly pear cactus, which is part of the Mexican culture's diet, has been shown to control blood sugar in rats, and grape seed extract may block glucose in the intestine. However, the safety of any of these substances in combination with regular diabetic medication is not clear.

Other folk and alternative remedies have been promoted in the early twenty-first century. Some practitioners have used bloodletting with metabolic syndrome patients. Limited research indicates that this ancient technique reduces blood pressure and blood glucose and cholesterol levels. The use of colloidal silver as an antibiotic was used from the nineteenth century, but with the development of antibiotics, it was abandoned in the 1940s. However, it began to be promoted in the 1990s as a dietary supplement to treat diabetes. Another folk remedy is "oil pulling," in which oil is held in the mouth for up to 20 minutes to "pull toxins" out of the body and to treat diabetes. These are not effective.

By the twenty-first century, crystal therapy became popular. Proponents believe that vibrations from crystals can impact different organs and correct imbalances within the body. The primary crystal for diabetes is citrine, a greenish yellow stone, which supposedly promotes healthy function of the pancreas. The person carries it or uses it in meditation. No evidence exists concerning cures for anything from crystals unless by placebo effect.

Dislocations, Sprains, and Bruises

Joint dislocations (luxations), sprains, and bruises (contusions) can be painful and interfere with everyday activities of daily living. The signs and symptoms of shoulder and hip dislocations include severe pain, swelling and bruising near the joint, numbness and/or weakness in the extremity, difficulty in moving the limb, muscle spasms near the joint, and an out-of-place limb. Both shoulder and hip dislocations are generally the result of trauma including motor vehicle accidents, falls, battles, and contact sports such as North American football.

Dislocations

Shoulder luxations represent about 50 percent of all dislocation. Anterior luxations are the most common, in which the humerus comes out of the socket (glenoid) in the front. About 1.7 percent of people have a shoulder dislocation during their lifetime. Young men and elderly women are most prone to this issue. On the other hand, hip dislocations are typically posterior—the ball at the head of the femur comes out of the socket (acetabulum) in the back of the hip area. Active and athletic males between the ages of 16 and 40 years have the majority of hip dislocations. However, approximately 3 percent of individuals after hip replacement surgery—which is primarily done in individuals between ages 50 and 80—have dislocations. These injuries can affect movement activities and can sideline professional athletes.

For these luxation injuries, unless a reduction (putting the joint back into its normal position) is performed shortly after the injury, long-term dysfunction and constant pain in the joint can result. This is particularly true for hip dislocations as death of the femoral head from lack of blood supply can cause permanent disability. Hinge joints such as elbow and knees are less likely to luxate as opposed to shoulders and hips with round sockets.

The term "luxation" was first used in English in 1552, from the Latin *luxare*, "to dislocate," from *luxus*, meaning "dislocated." The term "dislocation" was first used in the early seventeenth century also from the same Latin root word. Dislocations have been noted for centuries in many writings. The *Papyrus Ebers* (c. 1500 BCE) discusses them as does Hippocrates (c. 460–370 BCE) in the collected works *Corpus Hippocraticum*. Physicians throughout the Middle Ages discuss and detail treatments for dislocations. In the eighteenth century, Scottish physician William Buchan (1729–1805), besides shoulders and hips, describes various other luxations including ribs, jaw, neck, and vertebra. Near the turn of the twentieth century, congenital hip dislocations became a concern. British physician Alfred Herbert Tubby (1862–1930), for example, in his surgical text details this condition.

Sprains

In addition to dislocations, sprains are also a common problem. A sprain is a sudden tearing of a ligament supporting a joint and generally occurs in the ankle, but it can happen in the wrist, knee, and other joints. It causes pain, swelling, bruising, and the inability to move the joint depending upon its severity. A sprained or twisted ankle is the most common injury in the North America and often occurs during sports or recreational activities. Although less than half of people seek medical attention for sprained ankles, it is estimated that around 2 per 1,000 people per year suffer from this malady. The term "sprain" was first used in English in the early seventeenth century of unknown origin. Sprains have been mentioned in the medical writings throughout written history going back to the time of the ancient Egyptians.

Bruises

A very common injury is a bruise, or contusion. It is an area of discolored skin on the body caused by a blow or impact to the skin that ruptures underlying blood vessels. It occurs when the small veins and capillaries under the skin break. After an injury, the area of impact will look reddish or purplish and may or may not swell. After a few days, it will turn bluish or black. A few days later, it will fade into a greenish or yellowing color and then turn to a light brown. The term "bruise" is from the Old English *brȳsan*, "crush or injure with a blow," from the Old French *bruisier*, meaning "break." The first-known use of the term "contusion" in English was in the fifteenth century, which is from late Middle English, from French, from the Latin *contus*, meaning "bruised, crushed." Bruises have been mentioned in some health and medical writings throughout history and were often treated by home remedies.

Although various treatments for these injuries have been around for centuries, modern management is often a combination of home and medical remedies. For joint dislocation, the medical professional accomplishes closed reduction, which is skilled manipulation of the bones and joint, to return the bones to their normal alignment. It is sometimes accomplished with traction and usually performed under sedation or anesthesia to minimize pain. In some cases, surgery is required. For sprains, health professionals generally recommend protection, rest, ice, compression, and elevation (P.R.I.C.E.) to minimize swelling. Ice or cold packs on a contusion will also minimize any swelling and improve the appearance of a bruise. However, some recent research suggests that ice may slow healing as it restricts blood flow.

Dislocations and other injuries have influenced the outcome of professional sports games. For example, Peyton Manning (1976–), the franchise quarterback of the highly successful football team, the Indianapolis Colts, injured the vertebrae of his neck resulting in a pinched nerve and surgery. He did not play for the 2011–2012 season, causing the Colts to finish last in their division. Starting forward, Ben Simmons (1996–) for the NBA (National Basketball Association) *79ers* partially dislocated his left kneecap and underwent surgery. His absence from the team in the 2020 season resulted in a slump in the team's performance.

UNUSUAL TREATMENTS

Treatment of dislocations was usually done by healers or physicians. However, poorer people—at least from the seventeenth into the nineteenth century—used itinerate or local "bonesetters" who also set fractures. But management of sprains and bruises was often based upon home remedies. As with other ancient diseases, remedies were often tried through trial and error or based upon a particular medical philosophy.

Dislocations

In ancient Egypt, the *Papyrus Ebers* (c. 1500 BCE) alluded to the treatment of dislocated joints. The healer caused the bones to go back to their normal position but did not explain how this reduction was accomplished. Then the joint was bound with stiff rolls of linen followed by a daily treatment with grease along with incantations to ward off evil spirits until it was healed.

Hippocrates employed several methods for reduction of shoulder dislocations. He placed the affected arm over a variety of objects, including a long stick, a ladder rung with a round object attached to it, or the back of a chair. He, or an assistant, pulled down on the arm until it was back in its normal position. Hippocrates also suggested that a strong physician, taller than the patient, should place his shoulder in the patient's armpit and lift the patient up while holding and pulling down on the affected arm. After a week, bandages of linen soaked in wax and oil were applied to the joint followed by splinting. These methods were generally successful in reduction of the dislocation but could cause nerve and other damage and were painful.

For a hip dislocation, Hippocrates suspended the person upside down by the ankle on the affected leg, and an assistant grasped the trunk near the joint with his hands and pulled the joint back into its proper place. Conversely, Byzantine physician Paulus Aegineta (c. 625–c. 690), based upon Hippocrates' method, for a hip dislocation suspended the patient from a rope under the arms and had a strong assistant grasp the affected leg and swing from it. Buchan in the eighteenth century, after he had reduced the dislocations, bled the patient and placed vinegar or cloths of camphor soaked in wine on the joint. Bleeding and these substances had little effect on healing. To reduce a dislocation of the jaw, Buchan describes a folk procedure practiced by peasants. A strong man put a handkerchief under the patient's chin and turned his back to the back of the patient. The man then stood up and pulled the patient off the ground by the cloth, thus setting the jaw back into its normal position. However, Buchan thought this was dangerous to do.

By the early nineteenth century, since muscle contractions made it difficult to reduce dislocations, Scottish physician Alexander Macaulay (1783–1868) suggested extending the limb with the use of a pulley to apply traction. He also recommended a warm bath and bleeding from the limb for difficult cases. After reduction, he placed leeches and cold water on the injury, which could have helped to prevent swelling. Some physicians suggested getting the patient intoxicated to help quell the pain.

Sprains

These injuries have also been treated since antiquity and cause problems in movement for several weeks. For example, the *Papyrus Ebers* for what likely was a sprain gave several recipes to treat the malady. These included warming a mixture of ox grease and wax, which was placed on the joint that was then bandaged. Another treatment combined the fat of a serpent, mouse, and cat and placed it on the joint. James Henry Breasted's (1865–1935) translation of the *Surgical Papyrus* (c. 1600 BCE) for a sprain in the neck wrote that the healer bound the neck with fresh meat the first day and then honey for several more days mixed with an unknown substance until the swelling decreased. These remedies likely had little effect but were harmless to the patient.

Paulus in his work suggested that for both sprains and bruises unwashed wool, or a sponge, be soaked in vinegar and oil and applied to the injury. After the inflammation and pains had subsided, he suggested that the sprained joint should be massaged. Several nineteenth-century medical texts detailed various remedies to treat sprains. American physician Henry Hartshorne (1823–1897) suggested electro-massage for joint sprains with a fine wire metallic brush connected to a battery. Macaulay recommended leeching and cooling laxatives and if the swelling did not decrease bleeding from a vein. Leeching may have reduced some swelling but could have transmitted infections. For bruises and sprains, some physicians mixed equal parts of chloroform, ether, and Vaseline and applied it to the sprain. This may have been dangerous due to the explosive nature of ether but may have relieved some pain. Others recommended a mixture of turpentine, camphor, ammonia water, olive oil, and mustard should be placed on the injury. These substances were also used for bruises and acted as a counterirritant and took away pain.

Bruises

Many folk remedies have been developed over the ages for bruises. The *Papyrus Ebers* mentioned using meat and honey on wounds. Honey does have antiseptic properties but was not likely effective for bruises. Paulus based upon Hippocrates' method recommended astringents from the inside of citrus fruit—such as lemons—should be applied over a bruise. He also endorsed a poultice of radish with crumbs of bread.

In the nineteenth century, several health and medical texts gave recommendations for the treatment of bruises. For example, Macaulay suggested that if there was no wound, the injury should be rubbed with camphorated oil or turpentine liniment. If there was a laceration along with the contusion, a poultice of bread and milk was to be applied to the injury. In addition, for much inflammation, he utilized leeches to reduce the swelling, which may have been effective.

A bruise remedy that emerged in the twentieth century advocated placing a piece of meat over the injury, especially a steak on a black eye. This myth was largely perpetrated by Hollywood and fiction. If cold meat was used, such as a refrigerated steak, it was the coldness that constricted blood vessels and helped reduce swelling. Since steak carries bacteria, it could have caused an infection.

Most health personnel today, recommend that a bag of frozen peas be placed on the injury to reduce swelling from a bruise or a black eye.

Home remedies by the twenty-first century included rubbing toothpaste on the bruise as it was supposed to reduce swelling. Rubbing vanilla extract and bandaging parsley or ginger over the bruise was also supposed to reduce swelling due to their anti-oxidation and anti-inflammatory properties. However, the evidence for the success of these remedies is mixed.

Drowning, Choking, and Asphyxiation

Drowning, choking, and asphyxiation can lead to unconsciousness and death from lack of oxygen (hypoxia). In drowning, the person's mouth and nose are immersed in a liquid, usually water, which is inhaled, thus cutting off oxygen. Worldwide, drowning is the third leading cause of unintentional death and causes grief for family members and friends. Drowning is highest in low- and middle-income nations. Risk factors are very young age, access to water, being male, or being under the influence of alcohol or drugs while recreating on or in the water. In flood disasters, it accounts for about 75 percent of all deaths. The term "drowning" was first used in the fourteenth century from the Middle English *drounen*, meaning "being drowned." Drowning deaths have been recorded since the seventeenth century. English civil servant John Graunt (1620–1674) reported that in London in 1632, out of the 9,535 deaths, 34 people died from "drowned." In the United States, the 1850 census reported that 84 per cent of the reported deaths were from drowning.

Choking is severe difficulty in breathing and lack of air and is a frightening experience for the person who is choking. It occurs when breathing is blocked by an obstruction in the trachea, such as a piece of food, or constriction such as an allergic reaction that swells the throat. If insufficient oxygen is delivered to the body, hypoxia, asphyxia, and death may occur. Risk factors for choking deaths include being under 2 years and over 75 years of age, and the primary cause is food. In the United States, less than 1 death per 1000 people is from choking. The first use of the term "choking" is from the fourteenth century from the Middle English *achoken*, from the Old English *ācēocian*, from *cēoce*, meaning "jaw" or "cheek."

Choking leads to asphyxiation, sometimes referred to as suffocation. It can result from mechanical, or nonmechanical, constriction of the airway or from a decrease in breathable air leading to hypoxia. In addition to choking, asphyxiation can result from drowning, strangulation, poisonous gas, or conditions such as lung cancer and infections such as COVID-19 leading to oxygen starvation. This in turn damages the organs such as the heart that stops beating, resulting in unconsciousness and death. Of people seen in emergency rooms for suffocation, the overall mortality is around 10 percent. Data for suffocation has also been recorded in the past. For example, the 1850 U.S. Census reports that out of all deaths, 3.3 of 1,000 people had died of asphyxiation. The word "asphyxia" is from New Latin, from the Greek *a*, meaning "without" plus *sphyxis*, meaning "to throb," or stoppage of the pulse, and it was first used in the late eighteenth century.

Drowning, choking, and asphyxiation have been part of the human condition since its beginning. First aid and medical procedures that we take for granted today for these issues were found in antiquity. The Egyptian *Papyrus Ebers* (c. 1500 BCE), for instance, recommends the surgical procedure of tracheotomy (cutting into the trachea below the vocal cords) for choking. But Hippocrates (c. 460–c. 370 BCE) cautioned against tracheotomy, fearing that it could damage the arteries in the neck. Instead, he recommended tracheal intubation, which would have been difficult without anesthesia. In ancient Greece, it was reported that Alexander the Great (356–323 BCE) performed a tracheotomy with the point of his dagger to save one of his soldiers from asphyxiation. Ancient Greco-Roman physician Galen of Pergamon (129–199) noted the procedure was used in life-threatening situations although the patient often died of infections caused by the procedure. Tracheotomy has continued to be used into the present in cases of life-threatening airway obstructions, coma, and certain surgeries.

For choking, instruments for extracting fish bones from the throat were common in ancient Egypt, Greece, and Roman societies. Some eating establishments in the twentieth century carried bent forceps to extract food from a choking person. In 1972, American physician Henry Heimlich (1920–2016), a thoracic surgeon, developed the abdomen thrust, sometimes called the Heimlich Maneuver for choking. From the mid-1980s until around 2005, abdominal thrusts were the only recommended treatment for choking. Now the person is encouraged to cough. If they are unable to talk, a bystander can administer five slaps on the back alternated with five abdominal thrusts. However, the maneuver is only recommended as a last resort as it was found to sometimes cause broken ribs. The abdominal thrust is not done with a drowning victim due to the risk of vomiting leading to aspiration and asphyxiation.

In antiquity, mouth-to-mouth resuscitation with expired air was used. It is mentioned in the Christian Old Testament and Hebrew Torah (c. 1200 and 165 BCE) scriptures, where the prophet Elisha gave the "breath of life" to a child who was revived, and Hebrew midwives may have performed the procedure on newborn infants. It was also likely performed on newborn animals. Throughout the Middle Ages, the technique was thought to be used secretly by midwives to resuscitate newborns. Known as the "biblical method," mouth-to-mouth resuscitation was fostered in the early eighteenth century for adult resuscitation after Scottish physician William Tossach (c. 1700–1771) in 1732 revived a suffocated coal miner. But the technique fell out of favor in the nineteenth century as being vulgar and unhygienic and was not reinstituted until the mid-1950s. Except in some circumstances, "rescue breathing" is not used today and has been replaced by cardiopulmonary chest compression when there is no pulse.

Moreover, into the seventeenth century, it was thought to be unlucky to rescue and revive a drowned person as it was believed to be God's will. Once this belief was abandoned, rescue societies were formed. In the mid-1770s, The Amsterdam Rescue Society and in England the Society for the Recovery of Persons Apparently Drowned were established, which did save lives.

Depictions of drowning and asphyxiation have been presented in art and fiction. For example, in William Shakespeare's (1564–1616) play *Hamlet*, Hamlet cruelly rejects his former lover, Ophelia, by telling her to go to a "nunnery" (convent/brothel). Being rejected by the man she loves, she commits suicide by drowning in a stream. In the painting, *Ophelia* (c. 1851), by John Everett Millais (1829–1896), Ophelia is shown dead lying on her back in the stream amidst flowers and greenery. The novel *Moby Dick* (1851) was written by American Herman Melville (1819–1891). A great white whale named Moby Dick bit off whaling boat's Captain Ahab's leg. Ahab stalks the whale and harpoons it, but he becomes entangled in the harpoon line and is pulled overboard by the whale and is drowned.

UNUSUAL TREATMENTS

Various treatments and folk remedies have been carried out for centuries for drowning, choking, and asphyxiation. A few were successful, but most were not. In terms of drowning, the ancient Egyptians used a method that became popular again in the 1700s in Europe. When a person was pulled out of the water, they were hung by their feet to drain water from the lungs and pressure was exerted on their chest to foster breathing. This may have been effective in some cases.

By the early Middle Ages, warming the body became a technique to revive someone from drowning. It had been noticed that a body becomes cold when lifeless. Therefore, to bring the person back to life, they were put near the fire, laid in warm ashes, bathed in hot water, or placed naked between two people in bed. This was unlikely to revive a person who had asphyxiated from drowning or choking. Another method was flagellation or whipping the person to stimulate breathing, which had little effect. By the fifteenth century, a fireplace bellows was used to blow hot air and smoke into the victim's mouth. Swiss physician Paracelsus (c. 1494–1541) noted this method was occasionally successful.

In the early eighteenth century, blowing air or tobacco smoke into the rectum was advocated. In one method, tobacco smoke was first blown into an animal bladder. Then a tube was placed in the rectum, and the smoke from the bladder was pushed into the victim. This was not successful as the gut was not connected to the lungs. Later in the century, both the Amsterdam and English Rescue Societies placed fireplace bellows at intervals along the waterline so bystanders could attempt to resuscitate victims pulled from the water. Tobacco was often included with the bellows kit for a tobacco enema. Another procedure was lighting a pipe and blowing smoke directly into the victim's mouth or nostrils, which sometimes may have worked as it was like the mouth-to-mouth technique.

By the eighteenth century, the victim was also rolled back and forth over a log; a large wine barrel was also used. This rolling action acted to compress the victim's chest cavity and forced the air and water out. When the log or barrel was rocked backward, a release of pressure allowed the chest to expand, resulting in air being drawn into the lung and sometimes reviving the victim. In other techniques, people rubbed the victim with ammonia, put snuff or smelling salts on the nose, and tickled the back of the throat with a feather. Sometimes victims were revived if they were not completely asphyxiated.

Ivory "nostril pipes" were used by the Royal Society. A tube was inserted into a nostril and into the lower part of the larynx to prevent air from entering the esophagus. The tube was attached to bellows and air blown into the lungs. This may have occasionally worked. A Russian folk remedy in the early nineteenth century for drowning was covering the victim in snow or ice to reduce the body's metabolism. It usually did not work.

In the nineteenth century, lifeguard stations were equipped with a horse. When a drowning victim was recovered from the water, the lifeguard would put the person onto the horse and have it trot. Due to the bouncing of the body on the horse, alternate compression and relaxation of the chest cavity helped to expel air and water and aided inspiration. A French folk remedy recommended tongue stretching. The rescuer held the victim's mouth open while forcefully pulling the tongue rhythmically. Rarely was it effective.

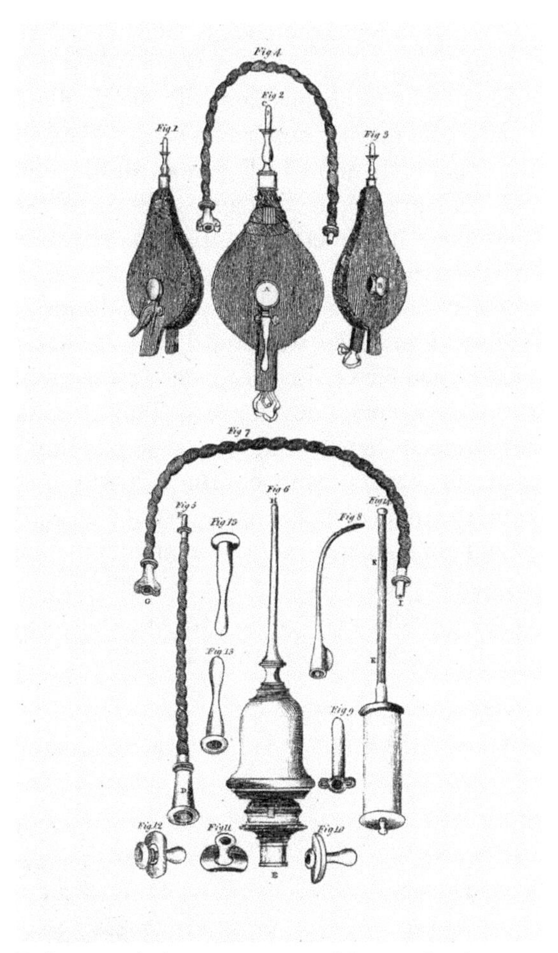

Bellows and clyster pipes used for artificial respiration for drowning victims in the eighteenth century. Smoke was also sometimes put into the bellows and blown into the anus. (Wellcome Collection. Attribution 4.0 International (CC BY 4.0))

Bellows had fallen out of favor by the mid-nineteenth century, which led the way to manual methods of resuscitation. In the 1850s, two English physicians developed different methods to resuscitate drowning victims that were commonly used into the early twentieth century. Marshall Hall's (1790–1857) "prone and postural respiration in drowning" method alternately repositioned the patient 16 times a minute from lying on the side to lying face down. While the victim was prone, the chest expired air, when the victim was rolled onto his/her side, pressure was taken off the chest and inspiration would occur. Henry Silvester (1829–1908), another London physician, created the "chest-pressure arm-lift method." In this method, the victim is laid face up and the arms are raised above the head to aid inhalation. Then the arms are lowered and pressed against the chest to aid exhalation. In 1903, still another London physician, Edward A. Schaefer (1850–1935) developed a method used into the mid-twentieth century known as the "prone-pressure method." The victim is placed prone with the arms above the head, and the face is turned to one side to let fluids drain out. Then the rescuer straddles the victim and pushes on the back with his/her arm extended to push out air. This is followed by the rescuer releasing the pressure, which causes the lungs to expand and pull in air. These methods were sometimes successful.

Similar to these methods is the Holger Neilson technique developed by Danish athlete Holger Louis Nielsen (1866–1955) in 1932. In this procedure, the victim is placed in the prone position with the head to the side resting on the palms of both hands. The rescuer applies upward pressure to the patient's elbows, which raises the upper body and which returns air to the lungs. The elbows are lowered, which puts pressure on the chest and which causes air to expire. This is done 12–15 times a minute. These methods were also used for victims of electrical or gas accidents and sometimes for choking victims.

Over the centuries, a few methods for choking were developed. In ancient Egypt, for instance, when a person was choking on a fish bone, an incantation was recited, and the victim was given a piece of cake or bread to force the bone into the stomach. Roman encyclopedist, Pliny the Elder (23–79), in his *Natural History*, recommends the choking person should immediately plunge his/her feet into cold water for a stuck fishbone. But if it was another type of bone, a bone was taken from the same dish and applied to the head. Bread was also put in the ears. These would not have been effective, and the person could have died. Byzantine Greek physician Paul of Aegina (c. 625–690) suggested sticking a finger or feather in the throat to induce vomiting. This could have caused the object to be pushed down into the trachea, suffocating the person.

Ear Issues and Deafness

Major ear problems include infections in the middle and outer ear canal, earaches, obstruction by earwax and objects, and tinnitus. In addition, various levels of loss of hearing and deafness are also found.

Earaches, Infections, and Tinnitus

Earaches, or ear pain, affects people of all ages and can feel like a dull, sharp, or burning pain. The ache can either be constant, or it may come and go in a short period of time. In adults, ear pain is often caused by infection in the external ear canal (otitis externa); injury; earwax buildup; change in air pressure; or referred pain from cavities, sinus infections, migraines, tonsillitis, or even strokes. Ear infections usually clear up after a few days or weeks without treatment. About 1 in 10 people will have ear pain at some point in their life. In children, the most common cause of an earache is a middle ear infection (acute otitis media), and around 83 percent of children will have at least one middle ear infection by age three. They are often accompanied by a fever, cough, and irritability, and they usually clear up on their own in a week or two. Sometimes pus runs out of the ear, which may come from a broken eardrum, but the membrane generally heals on its own. Although ear infections usually are not fatal, they can cause worry for parents, sleep reduction, and physical discomfort. Antibiotics were used in the 1950s, but by the late 1990s, they were less likely prescribed due to antibiotic resistance concerns and spontaneous healing without treatment.

The medical term for earache, "otitis," was first used in the eighteenth century from modern Latin, from the Greek *ous*, meaning "ear," and *-itis*, meaning "disease." Earaches and infections were first recorded in Egyptian papyri. Greek physician Hippocrates (c. 460–370 BCE) noticed that when the eardrum membrane ruptured it led to earache relief.

A common ear issue is tinnitus, or ringing in the ears, that sometimes makes it difficult to hear outside sounds. It affects about 15–20 percent of the population and is often a symptom of an underlying condition. These include an upper respiratory infection, ear wax buildup, ear injury, effects of medications such as aspirin and quinine, or disturbance of the auditory nerve. Tinnitus is commonly found in Ménière's disease, which is a disorder of the inner ear that causes severe dizziness (vertigo), hearing loss, and a feeling of fullness or congestion in the ear. The first-known use of "tinnitus" in English was in 1843 from the Latin *tinnire*, "to ring," and was originally coined by the ancient Roman scholar Pliny the Elder (23–79).

Hearing Loss and Deafness

Hearing loss and deafness ranges from mild to severe and may occur in one or both ears. Those who have hearing loss, or who are hard of hearing, have a reduced ability to hear sounds compared to a previous time in their life or compared to those with normal hearing. Those with mild to moderate hearing loss may be able to hear with sound amplification or lip reading. In profound deafness, a person is unable to hear sounds even with sound amplification.

Hearing impairments have both genetic and environmental risk factors. Two types of hearing loss and deafness are found, and both types can be present from birth or acquired throughout life. The first type, "conductive hearing loss," is a mechanical problem in which sounds are unable to be transmitted to the inner ear. Wax buildup in the outer ear canal or fluid or inflammation in the inner ear is often a factor in this type of loss. With wax removal and reduction of inflammation, hearing is usually restored. The second type, "sensorineural hearing loss," occurs when the cells or nerves in the inner ear cannot receive or transmit sounds. Prenatal or childhood diseases, trauma, and constant loud noise from industrial or recreationally activities can cause permanent damage to the hair cells in the inner ear, which leads to loss of hearing or deafness.

Most people who lose their hearing have "postlingual" deafness. In other words, they have acquired spoken language before their hearing declined. However, if they do not obtain sound amplification devises, they can become socially isolated and find it difficult to be part of conversations particularly in noisy places. Lack of hearing is also a risk factor for dementia. A major cause of hearing loss is age-related hearing decline (presbycusis). About 50 percent of those who are 75 and older have a disabling hearing decline. The biggest decrease occurs among those 60 and 69 years old. Adult men between 20 and 69 years of age are twice as likely to experience hearing loss compared to women due to occupations or recreational activities. Those who are born deaf often face difficult social challenges and as adults may feel socially isolated. Most people who are born deaf learn sign language, often form their own cultural group, and become part of a diverse deaf culture.

The first use of the term "deaf" was in the twelfth century from the Middle English *deef,* from the Old English *dēaf,* meaning "deaf." The earliest-known evidence for hearing loss was about 10,000 years ago in skeletons found in Iraq that exhibited bony growths in the ear canal that would have impacted their ability to hear. The first written record of hearing loss is found in the *Papyrus Ebers* (1500 BCE). The ancient Greeks considered those who were born deaf as both "deaf and dumb" as they were mute (could not speak). Since they could not speak, these deaf-mutes were thought to be unintelligent and not allowed inheritance and civil rights. This attitude prevailed throughout much of Western history. Under Christianity, St. Augustine (354–430), for example, contended that congenital deafness resulted from the sin of the child's parents or a punishment from God. In the

Middle Ages, deaf adults were sometimes thought to be possessed by demons. Although the rich often had tutors for their deaf children, by the early nineteenth century, education for poorer deaf children began. Today rehabilitation by sound amplification and cochlear implants in young children and adults is used for those with hearing loss. Experimental drugs and stem cells to regrow hair cells in the middle ear are also being investigated.

Ear and hearing issues have been represented in the arts. For example, an unknown illustrator of the fifteenth century drew *Jesus Christ Cures the Man Who Was Deaf and Mute* in an illuminated manuscript. An American romantic play (1977) and later film (1986) *Children of a Lesser God* show the romance between a new teacher at a deaf school and a young deaf woman who works as a janitor. They form a relationship that she breaks up. After many struggles and family issues, they reconcile. Another American film, *Mr. Holland's Opus* (1995), follows the life of a high school music teacher who aspires to write his own composition. However, he has a deaf child who brings tension into the family and has little time to compose. The school eliminates the music program, but on his last day at school due to a forced retirement, his wife and son secretly organize his former students to play his composition.

UNUSUAL TREATMENTS

Many procedures have been attempted to help ear infections, earaches, and tinnitus or to cure deafness. Some became folk or alternative treatments that are still used today. For example, dropping warm oil into an ailing ear has been common over the centuries as has been pouring a bit of wine or alcohol mixed with vinegar into the ear. These were marginally effective in soothing an ache, acting as an antiseptic, or clearing out earwax and were generally harmless.

Earaches, Infections, and Tinnitus

Remedies for earaches and ear infections have been found since ancient Egyptian times. For example, the *Papyrus Ebers* recommends a mixture of ground malachite, lanolin, oil, and honey—which has antibacterial properties—should be applied to the ear. For a discharge of foul-smelling material from the ear, the papyrus instructs the healer to apply ground-up frankincense and grain mixed with goose grease and cow's cream to the ear. Frankincense has some anti-inflammatory properties and may have helped. To dry an ear discharge, a mixture composed of an ass's ear, red lead, caraway, and olive oil was placed into the ear. Since lead has some antibiotic properties, it could have reduced an infection, but if used frequently, it could cause lead toxicity. If the ear became swollen, the head of a shrew-mouse, goat gallbladder, tortoise shell, and thyme were mixed and applied to the ear. It was likely ineffective but not harmful.

For an earache in Rome during the first century, Pliny the Eder placed ashes of burnt mice mixed with honey or boiled with oil of roses in the ear. For a ruptured eardrum, he boiled earthworms with goose grease and applied it to the ear. Another remedy was snakeskin added to oil of roses for all ear troubles. He also believed that dried lizards with salt placed on the ear cured injuries inflicted by trauma. These remedies likely had little effect but were largely harmless.

By the eighteenth century, based upon humor theory, physicians carried out bloodletting near the ear or placed leeches in the ear canal to cure earaches and infection. By the nineteenth century, physicians added a blistering agent, such as the wings of the cantharides beetle, and lard behind the ear twice a day to drain out bad humors in additional to bloodletting by leeches. These remedies were not effective and could have introduced infection.

Folk remedies for an earache today include blowing tobacco smoke into a child's ear, which does not clear up an earache or infection. Some claim that ginger juice or ginger heated in oil and applied around the outer ear canal can help soothe pain from earaches. Ginger does have anti-inflammatory properties and may have helped in some cases. Others assert that putting Vicks VapoRub on a cotton ball and placing it in the ear canal will reduce earache pain and prevent infection. No research, however, supports this claim. A recent folk remedy to sooth an earache is blowing a hair dryer set on warm on to the ear, which may, or may, not help.

The removal of earwax has been a concern since antiquity. The *Papyrus Ebers* recommends applying red ochre (iron oxide), cumin, an ass's ear, and oil to the ear. The oil may have dissolved the wax leading to better hearing. Pliny in his *Natural History* (77 CE) suggests dissolving natron—which is used in embalming and has a cleaning effect—in wine and placing it in an ear to clean out earwax. This may have been somewhat effective and generally harmless to the patient. In Roman times, ear spoons and ear picks were used to remove wax from ears. They continued to be used into the nineteenth century and were made of bronze or silver, but they could accidentally puncture the eardrum or cause an infection if the ear canal was scraped. Scottish physician William Buchan (1729–1805) suggested putting oil in the ear and giving the person snuff to cause sneezing and thus drive out wax or objects in the ear canal.

By the 1990s, "ear candling" to remove ear wax became popular. This supposedly ancient practice consists of putting a hollow cone-shaped device, made of cloth soaked in wax, into the ear canal and lighting it on fire. Suction from the fire was believed to remove the earwax. Waxy material formed inside the device, which was not wax from the ear but products of the cone itself. Proponents claimed that it not only cleaned out wax but also cured tinnitus, Ménière's disease, and many other illnesses. No scientific research supports any positive effect from the procedure. It can be harmful to the patient as wax from the candle may burn the ear canal or even eardrum.

Tinnitus has also been treated since antiquity. In ancient, Egypt it was thought to be caused by a "bewitched ear." Besides incantations to eliminate the spell, the Egyptians placed a mixture of frankincense mixed with oil in the ear. Roman encyclopedist Celsus (c. 25 BCE–c. 50 CE) suggested castoreum—from the scent glands of beavers—mixed with vinegar and laurel oil. Pliny in the first century believed that veal suet mixed with wild cumin put in the ear was a cure for tinnitus. The Romans also recommended boiling earthworms in goose grease and putting them in the ear to solve all kinds of ear issues. None of these were effective for tinnitus but were largely harmless to the patient.

Folk remedies have also been tried for tinnitus. An old British method instructs the patient to cut freshly baked bread into two pieces and then place them over both ears. This may have felt soothing but likely was not effective in relieving tinnitus. In the nineteenth century, passing an electric current (Galvanism) into the ear canal sometimes lessened tinnitus. Research in the early twenty-first century has found that a mild electric current did help to decrease tinnitus. Another folk method calls for placing the palms of the hands over both ears with the fingers resting on the back of the head. The index finger is then snapped against the middle finger. Proponents claim this drumming sound relaxes the suboccipital muscles, which reduces tension. It sometimes works and is harmless.

Hearing Loss and Deafness

Profound deafness is generally incurable, but many remedies have been tried over the centuries for the hard of hearing. Egyptian healers, for example, for the "ear-that-hears-badly," applied a mixture to the ear that included olive oil added to ant eggs, bat wings, or goat urine. Pliny the Elder applied dog's fat mixed with wormwood, old oil, or goose grease. If the cause was from wax buildup, these recipes may have helped to dissolve it. Celsus, in the first century, shaved the head and flushed out the ears with various juices or put a probe with turpentine-soaked wool into the ear canal and twisted it around, which could have caused damage.

By the thirteenth century, older individuals who developed hearing loss used hollowed-out cattle or ram horns to hear sounds more clearly. The device was placed in an ear and amplified sound waves. In the late eighteenth century, "ear trumpets," made of metal, wood, or shell, became fashionable to help the hard of hearing. These tubular, horn, or funnel-shaped devices were phased out when hearing aids were developed in the early twentieth century. In the nineteenth century, hypnotism was tried for hearing loss, and some claimed it was successful. In these cases, psychological issues may have been a factor in the original loss.

In the twenty-first century, "hearing loss pills" with antioxidants were marketed for "better hearing in 60 days," although the promoters claimed it might not work for everyone. The pills likely had little effect in restoring better hearing. Conversely, another remedy to reverse hearing loss from exposure to high decibel levels is eating a lot of cheese (about 5 pounds), which contains D-methionine, an essential amino acid for humans. Research does support this remedy, but the extra fat in so much cheese per day could cause other health issues.

Eye Conditions, Visual Impairment, and Blindness

Eye conditions and visual impairment range from inflammation of the eye, refraction abnormalities to diseases that cause serious vision impairment and blindness. The causes of many common eye conditions were only discovered in the mid-nineteenth century when the ophthalmoscope was invented in 1851, and other than vision correction by eyeglasses, few safe treatments were available until the mid-twentieth century.

Eye Maladies and Vision Impairment

A common eye condition is conjunctivitis—"bloodshot" eye or "pinkeye." Conjunctivitis is an inflammation of the conjunctiva, the clear membrane that covers the white part of the eye (known as the sclera) and that lines the inner surface of the eyelids. It is caused by viruses, bacteria, and allergens. Symptoms include redness and swelling of the conjunctiva, swollen eyelids, itchiness, and sometimes a mucous discharge that may stick the eyelids together overnight. Conjunctivitis affects up to 2 percent of the population per year with about 80 percent caused by a virus, and it is highly infectious. It is a frustrating condition that can affect work and social obligations. Virus and allergic reactions usually get better on their own; however, infections are easily transmitted especially among children. Allergic conjunctivitis occurs in up to 40 percent of the population. If an infection results from gonorrhea or chlamydia (see the "Sexually Transmitted Infections" entry), unless treated with antibiotics, serious vision loss often occurs. The first use of the term "conjunctivitis" in English was in the early nineteenth century. It is from the Late Latin *membrana conjunctiva* plus *-itis*, meaning "inflammation." The ailment was noted in the ancient Egyptian manuscript the *Papyrus Ebers* (c. 1500).

In another condition, "bleary eyes," the vision may be temporarily blurred or the eye reddened from exhaustion, eye strain, dry eyes, alcohol intoxication, or lack of sleep. This problem usually goes away on its own without treatment. Another vision issue is night blindness in which it is hard to see objects in dim light. It is caused by a vitamin A deficiency. Consuming colored fruits, vegetables, and cold water fish will reverse it. However, if it is caused by inherited Retinitis Pigmentosa, permanent low vision gradually develops.

Other eye conditions have also been noted since ancient times. Today about 43 percent of all vision issues are caused by uncorrected refraction errors such as myopia (nearsightedness), hyperopia (farsightedness), and astigmatism, which are usually inherited. These conditions result from the way light rays are focused inside the eye due to the length or shape of the eyeball or curvature of the cornea and cause fuzzy vision, which makes close reading difficult for hyperopia and far vision difficult for myopia. Another issue, presbyopia, is caused by a hardening of the eye lens, which loses focusing power by age 50. These eye problems may cause frustration but can usually be corrected with spectacles (eyeglasses), contact lenses, or refractive surgery.

Archaeological evidence suggests that pieces of crystal were used throughout the Middle East in antiquity as magnifying glasses. In classical Greece, although glass lenses were used to start fires, they likely were also used as magnifying lenses to authenticate seal impressions and in Rome for fine gold and artistic work. The first wearable pair of eyeglasses was introduced in the thirteenth century in Italy.

Low Vision and Blindness

Vision loss can range from mild (near normal vision) with corrective lenses to profound visual impairment (profound low vision) in which the vision is uncorrectable using conventional means. If there is no light perception, this is considered total visual impairment, or blindness. Vision that is 20/200 or worse, as measured on the Snellen Eye chart, is defined as "legally blind" in the United States. The major causes of vision impairment, such as cataracts, are higher in low- and middle-income countries, while in high-income countries, diabetic retinopathy, glaucoma, and age-related macular degeneration are more common. Most vision loss happens in older age groups and can lead to economic and social issues such as not being able to drive a motor vehicle and loss of independence. Since women live longer than men, they are more likely to be affected by these aging eye conditions.

For some of these conditions, early treatment can prevent blindness. For example, a cataract—clouding of the eyes lens, which causes blurred vision and glare—without treatment can lead to profound vision loss. Risk factors besides aging include genetics, sun and other radiation, trauma, and drugs such as corticosteroids. By age 65, over 90 percent of people will have some cataract formation. Surgery to replace the lens is carried out, resulting in near normal vision. The first use of the term "cataract" was in the fourteenth century from the Middle French *catharacte*, meaning "obstruction," from the Latin meaning "waterfall." Cataracts have been noticed since antiquity and were known to result in blindness. However, cataracts and glaucoma were not seen as separate conditions until the eighteenth century.

Glaucoma is caused by excess fluid in the eye, which increases intra-ocular pressure and which in turn damages the optic nerve. Two major categories of glaucoma are found. In "open-angle," the disease progresses over several years with a gradual irreversible loss of peripheral vision. In acute "closed-angle" glaucoma, without treatment, blindness can occur in a day or so. Symptoms include sudden severe eye pain along with nausea and vomiting, blurred vision, halos around lights, and a red eye. For both types, medicated drops and laser or regular surgery to relieve the pressure slow down the disease process. Risk factors include family history, severe myopia, diabetes, high blood pressure, eye surgery or injury, and corticosteroids use. In Western nations, from 2 to 3 percent of the population over 40 develops the condition. The term "glaucoma" did not become commonly used until after 1850 when the ophthalmoscope revealed optic nerve damage. Glaucoma is from the Greek *glausso*, meaning "to glow." Greek physician Hippocrates (c. 460–370 BCE) used the term to describe blindness in old age associated with a "clouded or blue-green" eye or the "glaze" of blindness.

Since the causes of some blindness were unknown until the 1850s, the term "gutta serena" became the code word for profound or sudden blindness due to an unknown cause in the posterior of the eye—now known to be retina, optic nerve, or brain damage. Vision loss with a normal looking eye was also called "amaurosis." This blindness usually resulted from glaucoma, macular degeneration, retinal detachment, or diabetic retinopathy. The first use of the term "blind," from the Middle Dutch *blint*, meaning "obscure," was in the early eighteenth century. In ancient Egypt and in the Middle Ages, blindness was thought to be a curse or a punishment from God due to sin.

In age-related macular degeneration (AMD), the macula that provides detailed central vision needed for daily tasks of close vision such as reading is damaged. Small fat deposits (drusen) form under the retina near the macula and cause the light sensitive cells to be displaced. This "dry" AMD makes up about 90 percent of all cases but often has little effect on vision. Ingesting antioxidant vitamins may slow the progress. However, in "wet" AMD, fluid leakage or hemorrhages from new abnormal blood vessels that grow in the macular cause loss of central vision. Laser surgery and certain drugs injected into the eye may slow the progress of this scourge. Around a third of people over the age 75 have some form of AMD. Risk factors include genetics, light eyes, bright sun exposure, and smoking. The term "AMD" was not coined until 1995.

Other conditions can also lead to profound low vision. In a detached retina, sudden onset of painless flashing lights, black floaters, or a dark veil covers a portion of a person's vision. The lifetime risk in normal individuals is about 1 in 300 per year, and it is more common among the aged and males. Risk factors are myopia and previous eye surgery for cataracts or glaucoma. It can cause permanent blindness without treatment, which is usually laser or other eye surgery that was not perfected until the late twentieth century. Another condition causing profound low vision is diabetic retinopathy (DR), a side effect of diabetes, which results in progressive damage to the blood vessels of the retina due to high blood sugar. DR affects both central and peripheral vision, causing blurriness and floaters in the eye. Some medications or laser treatments may halt the progression of the disease, but profound low vision and blindness is common.

A serious infection of the eye, trachoma, or "Egyptian ophthalmia" can lead to blindness but is rare in high-income nations. Repeated infections with the bacterium *Chlamydia trachomatis* cause the eyelids to turn inward, resulting in trichiasis (eyelashes rub the eyeball) that scratches the cornea and eventually leads to irreversible blindness. In endemic areas such as the Middle East, South Asia, and Africa, it is found in 60–90 percent of school-aged children. In earlier stages, it can be cured with antibiotics and prevented with good hygiene and sanitation. It has been known since ancient Egypt and was noted by Roman physician Galen (131–201 CE).

Blindness has been a theme in literature and the arts since antiquity. It was believed to be a punishment from the gods and used as a metaphor for not understanding. The Christian New Testament, for example, narrates parables of the blind "seeing" through faith. In the Greek tragedy, *Oedipus*, the King of Thebes blinds himself when he discovers that his wife is his mother and that he has killed his father. Blindness has been found in numerous illustrations. One example is by Flemish painter Pieter Bruegel the Elder (c. 1530–1569). In his *Parable of the Blind Leading the Blind* (1568), six blind men are walking in a row while each hold on to the man in front of him. The lead man falls on the ground, the next one stumbles, and the others begin to become unstable. Blindness has been featured in fiction. A blind protagonist is found in the science fiction television series *Star Trek: The Next Generation* (1987–1994). A leading character overcomes his vision loss by a device—the VISOR—which allows him to see things that his crewmates cannot see.

UNUSUAL TREATMENTS

Some folk treatments have been attempted to improve vision. None reversed total blindness. A common folk treatment throughout history has been placing honey in the eyes, which may have helped some infections as it has antibiotic properties.

Eye Maladies and Vision Impairment

Numerous recipes are found in the *Papyrus Ebers* (c. 1500) to prevent or treat eye ailments. Most of them also included magic spells. In one example, when "something evil" has happened to an eye, a human brain was cut into half. Honey, which has antibiotic properties, was added to one-half, and this was used to anoint the eye in the evening. The other half was dried and crushed into a powder and used to anoint the eye in the morning. The honey may have lessened the infection, but dried brain could have introduced pathogens.

For bleary eyes, the papyrus advises the healer to mix myrrh, onions, verdigris (poisonous copper acetate) antelope dung, entrails of a "Qadit-animal," and oil. This was painted on the eyes with a vulture's feather. This mixture, which contained gut bacteria had antibiotic properties from myrrh and verdigris, but likely had minimal effect and in some cases may have led to an eye infection. Healers also bathed the eye in urine, which became a folk remedy used for centuries. In the fourteenth century, for example, English physician John Arderne (1307–1392) added "warm acid urine of a man" to a bowl that had butter smeared on the bottom. The bluish sediment that formed was added to the fat of a capon and mixed in the sunlight. This mixture was smeared on the eyelids with a spatula and bandaged right before bed but was not effective and may have caused infections. Treating the eye with urine largely disappeared in the West. However, by the late 2010s it was used again as an alternative treatment for various eye problems but could have led to major eye infections and more damage to vision.

The *Papyrus Ebers* for bloodshot eyes (conjunctivitis) recommends placing on the eyes a poultice of frankincense, which has antibacterial and inflammatory properties, and crocus with anti-inflammatory property. If it did not work, a more complicated poultice composed of crushed onions, verdigris, and Arabian wood powder mixed with ink and water was tried. They may have sometimes been effective if redness was caused by a bacterial infection, but too much of verdigris could be poisonous. Based upon humoral theory, in the eighteenth century, Scottish physician William Buchan (1729–1805), for bloodshot eyes that did not resolve on their own, bled the patient, placed a poultice on the eye, and administered laxatives. Even into the nineteenth century, some physicians applied leeches near or even in the eye, bled from the temporal artery, or placed ground cantharis beetles ("Spanish fly") on the neck to cause blisters to draw out the "bad humors." These were not effective. Silver nitrate was also dropped in the eye for conjunctivitis, and if caused by a bacterial infection, this could have been successful.

For an eye infection with a discharge, the *Papyrus Ebers* recommends applying a poultice made from clay from a statue, leaves from the castor oil tree, and honey. To prevent eye infections and as a cosmetic to make eyes more beautiful, antimony and lead sulfite were ground together, sometimes with frankincense, and painted around the eyes and also on the eyelid. It is still used in the Middle East, Asia, and Africa for these purposes, where it is known as kohl. It could help prevent some infections, but lead and antimony are both toxic. To restore the luster of the eyes, the papyrus recommends applying a poultice of onions, resin (likely frankincense), and verdigris, crushed and mixed with the "milk of a woman who has borne a son." In addition to the antibiotic properties of honey and verdigris, both frankincense and castor oil have anti-inflammatory properties and castor oil is a lubricant, so this combination may have been helpful in some cases. Buchan in the eighteenth century for a discharge or "weeping eye" bathed the eye with brandy and water, or sulfuric acid diluted in water. The alcohol and acid could have acted as an antiseptic but could also have damaged the patient's eye in high concentrations.

In ancient Egypt, several treatments were used for night blindness. First healers treated the eyes with bat's blood as it was thought that the night vision of the bat would be transferred to the patient; it was not effective. However, one treatment was successful. Roasted ox liver was pressed and applied to the eye. Alternatively, the healer dropped liver oil into the eyes. This was carried out in many ancient cultures, and liver oil is still used today as a folk treatment. The oil in addition to eating cooked liver helped restore night blindness caused by vitamin A deficiency as animal liver stores the vitamin in high concentrations. However, raw oil could have caused an infection, and eating too much liver can cause vitamin A toxicity. Ancient Greek physicians suggested ingesting raw beef liver soaked in honey. Later Galen recommended goat liver. Eating liver and liver oil treatment in the eye became part of Canadian folk medicine and is still practiced today for night blindness among some First Nation's groups.

In the early twentieth century, several alternative remedies emerged to improve vision from refractive errors. For example, the "galvanic spectacles" from the first decade purported to improve vision by passing a mild electric current into the eyes. It did not work. A popular cure was advertised by American eye physician William Horatio Bates (1860–1931) who claimed that most sight problems were caused by eye strain. Therefore, to improve vision he recommended various eye exercises such as swinging the eyes back and forth or staring at the sun while doing eye exercise to strengthen eye muscles. Bates believed that full spectrum light waves from the sun were necessary to keep the eyes healthy. This folk method, now called "sunning," is still practiced today with the eyes closed. Based upon Bates' belief, British author Aldous Huxley (1894–1963) touted nose writing, in which the patient pretends his nose is a pencil and then writes in the air with the nose. Into the twenty-first century, "natural vision" methods were still advertised for better eyesight. For example, marketers claim that "pinhole glasses" (it has many small holes) along with eye exercise can permanently improve vision. However, no scientific evidence has shown that eye exercises, pinhole glasses, or staring at the sun can improve vision. In fact, staring at the sun may result in serious retinal damage and permanent central vision loss.

Low Vision and Blindness

The prevention of, or reversal, of blindness has been attempted since antiquity. For a trachoma infection causing trichiasis, for example, the eyelids and margins of the eyes were painted with a mixture of garlic, copper carbonate, ochre, and lead, all of which had antibiotic properties along with healing chants and rituals. Roman philosopher Celsus (c. 25 BCE–c. 50) in *De Medicina* recommended burning the eyelashes so they avoided scratching the cornea. Surgery was also attempted by the ancients by cutting the upper eyelid and attempting to cause the eye lashes to bend outward. They also plucked out the eyelash and cauterized the roots. These procedures besides being extremely painful likely resulted in further infection and blindness due to lack of sterile procedures. In another remedy, the healer removed the water from two eyes of a pig and added it to red lead and black eye makeup—both of which had antibiotic properties—to wild honey and injected it into an ear of the patient along with reciting magic incantations to the crocodile god, which was ineffective but likely harmless to the patient.

Various remedies were attempted to cure white, glazed, or light-colored pupils, which were likely caused by either glaucoma or cataracts. For example, the *Papyrus Ebers* recommends mixing dried excrement from a child with honey and fresh milk and applying it to the eyes. If this did not work, the healer made a paste that included "bile-of-the abdu-fish." If this did nothing, along with magic incantations, the brain of a tortoise was mixed with honey and applied to the eyes; these were not effective, and the patient remained blind.

However, successful cataract surgery called "couching" was carried out in antiquity, and even today, folk healers in some developing nations use the technique. Clinicians also used it into the late nineteenth century. In the procedure, the healer pushes a needle or lancet into the eye and pushes down against the damaged lens out of the line of vision. This procedure allowed some patients to regain some vision although many lost their eye to postoperative infections, and thick glasses were needed to be worn. In the nineteenth century, based upon humoral theory, physicians such as Alexander Macaulay (1783–1868) in his *A Dictionary of Medicine* (1831) recommended bloodletting, wet cupping, and cantharis blisters on the head or neck to extract bad humors, which had no effect on cataracts. He also rubbed the eyes with ether that likely caused further eye damage from chemical burns. For gutta serena blindness, he advised leeching, more cupping about the temples, or electricity, but he and other physicians realized there was no cure for profound low vision.

F

Fevers

Fever, also known as pyrexia, is an elevation of body temperature above 99.9 °F/37.7 °C (oral). It is generally accompanied by shivering, headache, sweating, flushing, aching muscles and joints, and rapid pulse. With very high temperatures (over 104 °F/40 °C), convulsions, hallucinations, delirium, or confusion are possible. A fever results from various factors such as infectious diseases, inflammation, wound infections, autoimmune diseases, reaction to various medications, cancers, and overheating. Research suggests that fever is a complex adaptive response to the immune system and in most cases is self-limited and resolves with no, or symptomatic, treatment.

The term "fever" is derived from the Latin *febris*, Old English *fēfor*, and Old French *fievre* and has been used since at least the sixteenth century. Fevers have likely been observed since prehistory in both humans and animals; the hand was used to determine warmth or coolness of the organism. The ancient Egyptians noticed different types of fevers as did later Greek and Roman physicians. Roman physician Celsus (25 BCE–50 CE) described the quartan and tertian, the four- and three-day cycles of malarial fever, respectively. He thought fever was a separate disease as did Greek physician Galen of Pergamon (129–210 CE) who influenced medical thought into the 1800s. Conversely, Hippocrates (460–370 BCE) believed that fever was a symptom of disease rather than a disease itself. Like other illnesses, they believed fever was related to an imbalance of the four body humors.

In the seventeenth century, English physician Thomas Sydenham (1624–1689), based upon classical descriptions, categorized fevers into three types: "continued," "intermittent," and "eruptive." From the mid-nineteenth century onward, these were renamed as "sustained/continuous fever" such as that found in lobar pneumonia, meningitis, urinary tract infections, and typhus; "intermittent/relapsing fever" in which temperature spikes at a certain time of day or every two to three days and then decreases to normal such as with malaria; and "remittent fever" in which the fever has daily fluctuations but does not fall to normal. Most infectious diseases fevers are remittent. Curiously in 1961, the concept of "fever of unknown origin" was coined for fevers with no obvious cause, which is like some ancient Egyptian and Greek beliefs that fever was its own pathological entity.

People for centuries have been concerned about fevers as they often led to death prior to the mid-twentieth century with various medical advancements. In 1632 in London, for example, the second leading cause of death—after infant deaths and TB—was "fever." One of the most unusual and deadly fevers in Europe was the

"English sweating disease" in the late fifteenth to mid-sixteenth centuries. The illness had a rapid onset of fever, much sweating, headache, nausea, pain, delirium, hallucinations, and profound stupor and was greatly feared by the population. It reached a critical stage within 24 hours and the patient either lived or died—about 50 percent died. For those who lived, it did not confer immunity against another attack. It was more common among the rich and more likely to kill strong young men and women but not old men and children. Physicians of the era thought it was caused by lots of mist, rain, and high humidity, and flooding along with gluttony and heavy drinking. Today some researchers suggest it may have been caused by an unknown hanta virus.

Although people have attempted to reduce fevers for centuries, many physicians now suggest letting a fever run its course—unless the patient is very uncomfortable—as it may help destroy the pathogen. Aspirin developed in the nineteenth century was commonly used to reduce fever. However, in the late twentieth century aspirin was found to be associated with Reye's syndrome, a serious swelling in the liver and brain of children and teenagers with flu-like symptoms. Over-the-counter medications such as ibuprofen or acetaminophen are now suggested instead if necessary. Antibiotics are only used for serious bacterial infections.

Fevers are sometimes found in fictional characters. For example, in English author Jane Austen's (1775–1817) *Sense and Sensibility* (1811), the protagonist, a vivacious and sometimes inappropriate Marianne, walks through a cold rain, suffering from a broken heart. She contracts a fever and almost dies as a result but realizes that she has not been sensible in her activities or been kind to others. Brain fever was a common illness among Victorians and was perhaps some form of meningitis. In French novelist Gustave Flaubert's (1821–1880) *Madame Bovary*—serialized in the *Revue de Paris* in 1856—Emma Bovary suffers from brain fever after reading a letter in which her lover breaks off their affair. She has several affairs but becomes bored with each relationship although she is married. After a convoluted plot, she finally commits suicide.

UNUSUAL TREATMENTS

To reduce fevers since circa 1500 BCE in ancient Egypt, healers would bleed the patient by cutting a vein to balance the body humors. Leeches were used with children or very weak adults. Classical Greek physicians also bled patients, and most also used clysters (enemas) and emetics to cause vomiting and cupping (making scratches on the skin and placing a hot cup over them to draw out blood). In some cases, these may have reduced the fever. These practices became standard fever treatments into the 1800s but were discontinued as modern biomedicine gained ascendency. However, cupping is still used in Chinese and some other alternative medical systems for mild fevers.

During a fever, many classical Greek and Roman physicians would withhold food as they thought it was injurious until the fever had abated. This may be an origin of the folk saying, "feed a cold, starve a fever." Because fever often caused weakness, British physician Sydenham in the seventeenth century suggested putting a young person of the same sex in bed with a sick old person so their vital energies would help reduce the fever and transfer strength.

From antiquity though the seventeenth century, physicians and folk healers gave patients human blood to reduce fever. It was drunk whole, dried and mixed into food, or cooked down into a jam. Physicians also administered various herbs to reduce fever, which had no effect. Western physicians prescribed feverfew and traditional Asian healers sweet wormwood (Artemisinin) for malaria-type fevers, which were sometimes effective. Physicians for centuries gave the poisonous herb belladonna and camphor, the sap from a plant, as a fever reducer. As part of folk medicine, cinnamon and garlic were used for centuries, ginseng was used by Europeans by the sixteenth century, and Native Americans used wild ginger to reduce fevers. These plant extractions are still used today in over-the-counter homeopathic medications and in alternative medicine systems, but there is inconclusive evidence as to their effect for fever reduction.

During the fourteenth to seventeenth centuries, doctors recommended patients drink pig urine as a cure for fever, which could have transmitted bacteria. Western physicians from at least 500 BCE into the seventeenth century gave patients clay to eat in the form of terra sigillata tablets, which were thought to absorb the "poisons" that caused fever. In addition, several poisonous metals have been used to reduce fevers over the centuries. For example, since 3000 BCE, antimony was used as an emetic. St. James fever powder in the eighteenth century included antimony, but this metal could cause convulsions and death. Arsenic was also used to treat fever into the nineteenth century, which also killed many patients.

To treat the English sweating sickness, physicians of the era found the less treatment the better. They recommended the patient be lightly covered, kept cool, no air or light be allowed to touch the body, and the patient be kept awake until the crisis had passed. To prevent the disease, the rich would make a "philosopher's egg"—made of an egg, saffron, mustard seed, and ground "unicorn horn." In Germany and Switzerland, patients were covered in feather beds, and most died from overheating. Therefore, Switzerland forbade the use of feather beds during an outbreak and recommended keeping patients awake by pulling their hair and putting vinegar in their eyes. The only medication allowed was cinnamon water.

Foot Problems

Many foot issues cause pain and a decrease in the quality of life. Some foot conditions such as blisters or immersion foot syndrome ("trench foot" or "jungle rot") can lead to gangrene, amputations, and death. Interestingly, many foot problems including athletes' foot and heel bone spurs were not considered important enough to be included in most medical texts until the early twentieth century, and their treatment was largely relegated to home remedies. Other ailments such as Achilles tendinitis have been noted for centuries along with treatments.

Achilles Tendinitis (Tendo Achilles)

A torn or severed Achilles tendon causes considerable pain in the heel that often radiates from the heel up the leg. It affects activities of daily living and can result in permanent lameness. A ruptured Achilles tendon most commonly occurs in runners, dancers, and individuals who have suddenly increased the intensity or duration of their activity. Risk factors for the affliction are advanced age, being male, steroid use, excess weight, and antibiotic use such as fluoroquinolones. In the general population in developed nations, the incidence of Achilles tendon rupture is approximately 5–10 per 100,000 people. Achilles tendinitis was described by the Greek physician Hippocrates (c. 460–370 BCE) and was named "tendo Achilles" in 1693 by a Dutch anatomist. The term comes from the Greek myth of Achilles, which says his mother dipped him as a baby by a heel into the river Styx to make him invulnerable. He was killed by a poisoned arrow shot into his heel that was not touched by water as this was his only weakness.

Plantar Fasciitis and Heel Spurs

Another painful foot ailment is plantar fasciitis, which is one of the most common reasons for heel pain and is caused by inflammation of the thick band of tissue (plantar fascia) that runs across the bottom of the foot and connects the heel bone to the ball of the foot. The pain is particularly severe with the first few steps in the morning. Plantar fasciitis can lead to a heel spur, which is a calcium deposit that forms a bony protrusion along the plantar fascia. These conditions are most common among active people between the ages of 40 and 70. Besides age, other risk factors are running or jogging excessively on hard surfaces, abnormal gait that puts too much stress on the heel bone, diabetes, standing for many hours, excessive weight, ill-fitted shoes, and flat feet or high arches. About 10 percent of people have plantar fasciitis or heel spurs, but only about 5 percent with heel spurs have foot pain. The term "fasciitis" caused by infection or injury was derived from New Latin in 1893. However, the term "plantar fasciitis" was not defined until 1940; it is from the Latin *plantaris*, from *planta*, meaning "sole," paired with *fasciitis*, "inflammation of the fascia."

Bunions and Corns

When high heels and pointy toe footwear became fashionable in the seventeenth century, another foot problem, the bunion, emerged. A bunion occurs on the first joint or base of the big toe. It forms a large bony bump that is sore, red, and inflamed and may cause painful walking. The top part of the big toe gets pulled toward the smaller toes. The main causes of bunions are wearing ill-fitting footwear, and sometimes it's arthritis or the shape of the foot. Since the 1800s, the condition has been more common among women. However, horseback riders and men who wear tight "cowboy" boots can also be affected. Corns (clavi) and calluses can also cause pain. These maladies are patches of hardened skin on areas of the feet (sometimes on the hands). The term "bunion" was first used in the early eighteenth century from the Old French *buignon*, from *buigne*, meaning "bump on the head." The use of the term "corn" was used by the early nineteenth century in some medical texts, including those of American physician Elijah B. Hammack (1826–1888), but its derivation is unknown.

Athlete's Foot, Blisters, and Immersion Foot Syndrome ("Trench Foot" or "Jungle Rot")

Another common foot issue is athlete's foot (*tinea pedis*), a fungal infection. The skin between the toes or other places on the feet becomes dry, red, flaky, and itchy. It can also spread to other parts of the body. The condition was first described by English dermatologist Arthur Whitfield (1868–1947) in 1908. From 15 to 25 percent of the population is affected with the malady at some point. In the early twentieth century, it was considered a highly infectious emerging disease due to the fungus found on the floors of locker rooms and pool changing areas. But, by the late twentieth century, tinea pedis was viewed as a mildly contagious disease with some people being more susceptible than others likely due to genetics. The name "athlete's foot" was coined by a pharmaceutical company in the late 1930s who manufactured treatments for the condition as it was common among athletes. The first-known use of "tinia," from the Latin for "worm," was in the fourteenth century. Tinia is also used to describe other fungus condition such as "ringworm." However, the term "tinea pedis" meaning fungus on the feet was not used until 1948.

Trench Foot, an immersion foot injury (IFI), has symptoms of cold and blotchy skin, numbness, tingling or itching sensation, pain, and swelling. When the foot warms up, it may be red, dry, and painful, and blisters may form. It is caused by prolonged exposure to dampness, cold water, and often unsanitary conditions. On the other hand, another IFI, Jungle Rot (chronic ulcerative skin lesion), tends to be found in tropical climates with warm water temperatures. The symptoms of the condition include ulcers on the feet or legs, which are often infected with several bacteria. In severe cases, untreated trench foot or jungle rot can involve the entire foot and lead to gangrene and amputations. Blisters from ill-fitting shoes can also cause pain and soreness that make it difficult to walk.

Medical management today for Achilles tendinitis, plantar fasciitis, heel spurs, bunions, and corns include exercise modification, heel lifts, arch supports, stretching exercises, nonsteroidal anti-inflammatory drugs, and avoiding pointy tight high-heel shoes or boots. Surgery for these conditions is recommended as the last resort. Athletic foot is managed with antifungal medication and immersion syndrome by drying the feet, wearing clean and dry socks, and in the case of infection the use of antibiotics.

These foot issues have been common among armies since wars first began and likely influenced the course of some campaigns. Blisters have always been associated with ground troops in wars. For example, the prevalence of blisters was 33 percent among soldiers during a 12-month block of Operation Iraqi Freedom (2001–2004). Blisters can become infected and like other foot injuries can lead to serious infections like gangrene. Trench foot was first documented by Napoleon's army in 1812. The term "trench foot" was coined during WWI (1914–1918) as it was associated with trench warfare. Jungle rot in WWII from jungle warfare in the Pacific and Southeast Asia (1941–1945) was also found in the Vietnam War (1955–1975).

UNUSUAL TREATMENTS

Treatments were developed by folk healers and families over the ages for many foot issues. Only a few conditions were managed by physicians.

Achilles Tendinitis (Tendo Achilles)

For injured Achilles tendons, Galen (129–c. 200) applied to the heel oils that produced a sensation of warmth such turpentine-rosin. He also recommended rubbing the heel with a concoction of quicklime, yellow iron sulfate, copper ore, arsenic, and sandarach (gum resin from cypress) mixed with oils or waxes. When tendons were cut, he directed healers to unite the ends together with sutures, but they often became infected. He also suggested bloodletting or purging to eliminate toxic humors, which was not effective.

Byzantine Greek physician Paulus Aegina (625–c. 690) heated stones such as pyrites and dissolved them in strong vinegar. The vapors from this concoction bathed the injury. Then oil boiled with marshmallow roots or wild cucumber was poured over the affected part once a day, and patients were advised to stay out of the Roman baths. As the wound healed, gum resin of the ammoniac herb was dissolved in vinegar and rubbed on the injury followed later by rubbing it with fat mixed with spicy resins such as sweet myrrh. These likely had little effect but were harmless to the patient.

In 1831, British physician Alexander Macaulay (1783–1868) developed an apparatus to keep the broken ends of the Achilles tendon together called a "footstock," which bended the knee and extended the foot or ankle joint. It was kept at this angle with straps and belts until it healed. Counterirritation by a blister of cantharis beetle (Spanish fly) was also employed to draw out toxic humors. By 1912, some physicians ordered absolute rest and applied a splint. When the most acute symptoms had subsided, they placed a plaster-of-Paris cast on the foot for several weeks. These likely increased the chance of DVTs. By the 1980s, "shock-wave therapy" with low-energy sound waves was aimed at the torn tendon. Some claimed this increased the blood flow to the injured area, which sped up the healing. The research is mixed regarding this remedy.

Plantar Fasciitis and Heel Spurs

Treatments for these painful foot ailments were like those for tendinitis. In addition, the painful foot and heel was rubbed with lavender and flaxseed oil. Some folk remedies suggested soaking the feet in baking soda and the herb arnica or Epsom salt and ingesting apple cider vinegar. Foot soaks may offer temporary relief, but the other remedies had little effect but were harmless to the patient. In another folk remedy, the patient walked barefoot in the cool morning dew or even snow in the winter. These remedies may have offered temporary relief from pain.

Bunions and Corns

Several methods for treating bunions had emerged by the seventeenth century. Some physicians recommended rubbing the foot with a stimulating ointment.

Macaulay in the early nineteenth century suggested placing a small blister with Spanish fly on bunions and when they had healed to firmly apply a piece of sheet lead over it to keep it straight. Other healers recommended placing a wedge-shaped object or cotton between the first and second toe to straighten the toes or separate the toes with adhesive or use a splint. Some of these bunion remedies are still advertised today, which are only minimally successful.

The Greek physician Paulus, for corns, recommended placing the lees of wine on it or burning it out with a hot iron rod. He also suggested that the quill of a vulture be placed into the corn and boiling water poured into it, which was painful. In the early twentieth century, nitric acid was put on corns to dissolve them. From the 1920s to the 1960s, to see if shoes were fitting well to prevent these foot issues, children and adults could view their feet in a fluoroscope, which exposed many individuals in Western nations to dangerous radiation.

Athlete's Foot, Blisters, and Immersion Foot Syndrome ("Trench Foot" or "Jungle Rot")

Foot infections such as athlete's foot, immersion foot syndrome, and blisters have often been treated with home remedies. People have rubbed garlic, camphor, and tea tree oil on the feet. They have placed wet tea leaves—now tea bags—on blisters, in which the tannic acid helps as an antiseptic and hardens the blister. For fungus infections in the nineteenth century, people rubbed arsenic on their feet. In the early twentieth century, American dermatologist Philip Lewin (1888–?) advised patients to sterilize their shoes with formaldehyde or expose them to sunlight. He also treated infected feet with x-rays, which likely caused an increased cancer risk to the patients. By the 1940s through the late century, patrons were required to walk through a disinfection solution at pools and locker rooms to contain the fungus that had little preventive effect and was abandoned.

During WWI, physicians tried various treatments to manage immersion foot syndrome. For this trench foot, they rubbed a mixture of lead and opium, mercuric chloride in alcohol, picric acid (used in battlefield explosives), and chloral hydrate (a sedative) on feet. Some of these treatments along with radiant heat and electrical stimulation often caused the condition to get worse. During the Vietnam conflict, soldiers used turmeric on ulcers and antiperspirants on their feet, which often helped.

By the late twentieth century, toothpaste was used as a topical treatment. Depending on the brand, the substance acted as a disinfectant, antiseptic, and fungicide, and some found it eased itching, reduced swelling, and helped dry up blisters. People also used a hairdryer on the feet to reduce foot moisture. Some alternative therapy systems suggest ultraviolet (UV) light for athletes' foot infections, which is effective in some patients.

Fractures and Bone Issues

Fractures and bone issues result in pain, loss of function, immobility, and difficulty in tasks of daily living or work. The most common bone maladies are fractures and metabolic bone conditions such as osteoporosis and osteopenia, which cause weakened bones that lead to fractures. These metabolic conditions, however, until recently were thought to be part of normal aging and were not treated.

Several types of fractures are found, and most occur without skin injury (closed fractures). In an open or compound fracture, the skin is wounded by severe trauma or by a sharp bone fragment that pierces the skin from within and exposes the bone. An incomplete, or greenstick, fracture occurs when the bone cracks and bends but does not completely break and is usually found in children. In a comminuted fracture, the bone is shattered into many fragments because of severe trauma such as an automobile accident. A stress fracture is an overuse injury that is common in individuals who participate in repetitive activities such as running, jumping, and marching and usually affects the shin or foot bones. When a fracture occurs, blood surrounds the break, and bone-forming cells (osteoblasts) deposit a new layer of bone called a callus that forms a collar around the fracture. It can take several months or longer for a mature bone to form and the thick callus to disappear.

Fractures are very common, and individuals in a developed nation can expect to sustain two fractures over the course of their lifetime. Wrist fractures are the most prevalent for those under age 75, while hip fractures are the most common over that age. By 2017, in English-speaking nations, over 6 million fractures occurred per year in the United States while in Canada 20 percent of the population had had a fracture. In the United Kingdom, 1 in 2 women and 1 in 5 men suffered a fracture after the age of 50. There were 20,000 hip fractures per year in Australia and over 84,000 fractures in New Zealand resulting from osteoporosis with 60 percent of these occurring in women.

A major cause of fractures in the elderly is osteoporosis—low bone mineral density or deterioration of bone. It is associated with a sedentary lifestyle, lack of vitamin D, and calcium and is most common among older women of northern European and Asian ancestry. Osteopenia is the reduction of bone, which often leads to osteoporosis. Osteoporosis affects 1 in 4 women and more than 1 in 8 men over the age of 50 years. In addition to fractures, these bone diseases may result in bone pain and loss of height due to compression fractures of the vertebrae.

The term "fracture" is from late Middle English, from the Latin *fractura*, from *frangere*, "to break." It was first used in English in the sixteenth century. "Osteoporosis" was not used until the mid-nineteenth century; it is from the Greek *osteo*, meaning "bone," and *poros*, meaning "passage." "Osteopenia" was not coined until the 1960s; the term is from the Greek *osteo*, meaning "bone," and *pena*, meaning "poverty," and it is used for bone density mineral lower than normal. Fractures have been identified since the first organism with a skeletal framework emerged. Healed fractures have been identified in Egyptian mummies and in historical archaeological sites throughout the world. Greek physician Hippocrates (460–370 BCE) records different types of fractures, which Roman physician Celsus (c. 25–c. 50) and then Galen (129–210) expanded. These descriptions are still used today.

The basic principles of managing fractures have changed little from antiquity. Hippocrates describes the three fundamental steps: reduction—manipulation of the fractured bone ends into their original anatomical alignment, immobilization—preventing the bones from moving by placing them in a ridged bandage so the bones grow back together, and rehabilitation—restoring the proper function of the body part.

From the Middle Ages, "bonesetters"—folk healing practitioners whose skills were generally passed down through apprenticeships or families—set fractures and reduced dislocations. Thus, few medical texts discussed fractures until the nineteenth century when physicians developed the field of orthopedics that was originally concerned with the care of "crippled children." Over the century, they pushed the "uneducated" bonesetters out of Western medicine. By the twenty-first century, bonesetters were primarily found in developing nations. By the late twentieth century, orthopedic physicians, besides plaster and plastic casts for simple fractures, used "internal fixation" of bone with metal plates, rods, nails, screws, wires, etc. to stabilize more complex fractures. This often allowed a patient to bear weight soon after surgery before bone union (healed fracture) as it was found that early ambulation after surgery helped to prevent bedsore, fatal pneumonia, and blood clots.

Throughout history, fractures have been associated with combat and battles. Wars have led to new techniques that changed the management of fractures and other injuries in the civilian population. The specialty of orthopedics was initially the surgery of war and became the domain of general surgeons. Often the standard of care at the beginning of a war was different from the care at the end of the war as surgeons learned about new treatments for each type of war injury as weapons advanced and injuries were different for each conflict. Treatment of war fractures has been featured in war diaries. For example, American author Louisa May Alcott (1832–1888) who worked as a nurse during the American Civil War describes physicians amputating open fractured limbs without attempting to save them. An open fracture was often thought to be a death sentence due to infection until the development of aseptic techniques and surgery in the late nineteenth and antibiotics in the mid-twentieth century.

UNUSUAL TREATMENTS

Various methods were developed in antiquity to set and immobilize fractures. For example, the Egyptian *Papyrus Ebers* (c. 1500 BCE) recommended bandaging the break daily with cloth, alum, and honey. Hippocrates, after reduction of the fracture, applied bandages of linen soaked in wax followed by splinting after a week. In the first century, Celsus attached thongs above and below the fracture, and two assistants pulled against each other to line up and set the fracture. Some physicians bound a broken humerus to the chest. For a simple foot or leg fracture, the healer fastened the ankle to a foot board to prevent muscle contractions, elevated it, and often placed it into a leg box. In the fourteenth century, an assistant was placed at the foot and head of the patient and he pulled on the leg while the healer pushed the fractured parts together. To relax muscles to align the bones, hot baths were used into the nineteenth century. Until the 1980s, patients with leg, ankle, or hip fractures were often kept on strict bed rest and those with vertebral fractures were sometimes put into a full body cast for weeks. However, this often resulted in debility and death.

Various substances were used to immobilize broken limbs. The ancient Egyptians used bark from trees to support the limb. Egg whites, in combination with lime or flour, were used for plaster casts in the Arabic world from at least the ninth century. In the fourteenth century, for an open fracture, linen was dipped in egg white, milk, mulled wine, hot water, and black ointment and wrapped around the limb. Felt was placed over this bandage to give more stability.

Healers in addition to setting and immobilizing the fracture have used various herbal concoctions over the centuries to treat them. The ancient Egyptians for simple fracture bandaged the limb with a mix of natron (salt used in embalming), grease, black flint dust, and honey daily. Another recipe included flakes from copper, ass's dung, grease of ibex, and other substances, as well as rubbing on the injury the clotted milk of a woman who has borne a male child. These had no effect on healing the fracture.

For an open fracture, Hippocrates suggested placing a pine pitch plaster over the wound. Celsus laid a wine and oil-soaked cloth over the fracture and poured hot water on the wound, which may have prevented infection. If a bone was protruding, he sawed it off or pushed it back with a lever and bled the patient to eliminate any bad humors based upon humoral theory. Greek physician Dioscorides (40–90) recommended the application of cerate (wax) to fractures and pouring of a decoction of elm roots or elm leaves over fractured bones to "knit them quickly." Many early physicians also recommended bleeding and purging when there was much pain to eliminate toxic humors. Healers in Anglo-Saxon England (fifth–eleventh centuries) laid warm bull's dung or dog brains over the fracture. Although dung and other substances likely led to fatal infections, wine and hot water may have acted as an antiseptic to prevent some infections and gangrene. But this ancient hot water treatment was lost in the Western civilization's dark ages and not practiced again until 1896. Due to this lack of knowledge, many limbs and lives were likely lost into the nineteenth century.

From antiquity into the twenty-first century, herbs have also been used to treat fractures through folk traditions or alternative systems. In Europe and North America, the horsetail plant that contains silicon is boiled and used as a tea for the early stage of fracture healing. The herb arnica is also reported to help in bone healing and pain. However, research is mixed as to the effectiveness of these remedies.

In the late twentieth century, surgery for leg lengthening and electric current to unite bone faster was tried. Until this time, individuals with limb length discrepancies—the result of a poorly healed fracture, disease, or a congenital defect—had few treatment options. A remedy was developed in Russia in which the orthopedic surgeon cut the leg bone to be lengthened. The bone was then stabilized using an external and/or internal fixation device or frame and gradually pulled apart until the desired length was achieved. New bone grew between this gap. Some individuals for vanity reasons, who wanted longer legs, had also undergone this surgery, which could be risky. Device such as electromagnetic stimulation, ultrasound, and extracorporeal shock waves—"bone growth stimulators"—were developed that claimed to hasten healing. However, the research is mixed as to their effectiveness.

By the twenty-first century, crystal healing for fractures was advertised. Proponents claim that vibrations from the crystals have healing powers and balance the bodies' energy. For broken bones, advocates recommend stones like blue kyanite, the blue lace agate, hematite, and several others. These can be worn as bracelets and necklaces or carried by the person in a pocket or kept on the bedside table at night. There is no evidence that crystals heal fractures.

Gastroenteritis, Dysentery, and Diarrhea

Gastroenteritis (GI) is also known as "food poisoning," "stomach flu," "Deli Belly," "the trots," "Montezuma's revenge" among other terms. Its main symptom is diarrhea and often vomiting, cramps, and fever. GI is usually caused by viruses, bacterial and parasitic infections, and toxins. The illness is contracted by ingesting contaminated food or water or person-to-person contact through the fecal-oral route generally as a result of poor hygiene and/or sanitation. Dysentery, diarrhea, and cholera (discussed in a separate entry) were viewed as distinct conditions into the late nineteenth century; however, other common diarrheal diseases were referred to as "cholera morbis." By 1900, most causes of GI had been identified. Today it typically does not result in serious complications, but it can result in dehydration and be fatal for infants, children, the elderly, and those who are malnourished or have compromised immune systems.

Dysentery is a type of gastroenteritis and is associated with *Shigella flexneri* and *S. dysenteriae* bacteria, which can cause epidemic bacillary dysentery. Symptoms are bloody diarrhea, and it is common in childcare centers, overcrowded refugee camps, and prisons. Amebic dysentery, caused by the *Entamoeba histolytica* amoeba, is primarily found in tropical areas. It also produces bloody diarrhea along with mucus. The amoeba can inhabit the intestines for years and, without treatment, sometimes leads to death. Protozoa such as *Giardia lamblia* causes watery, foul-smelling diarrhea and is one of the most common causes of waterborne disease in the United States. The protozoa are found in backcountry streams and lakes.

Viral gastroenteritis is the most common diarrheal affliction and is mainly caused by a rotavirus (norovirus) discovered in 1973 in Australian children, first described in 1929. Infections are frequently found in semi-closed environments such as schools, nursing homes, and cruise ships. The duration of the affliction is about two days. Some bacteria produce foodborne outbreaks. Campylobacter, for example, is one of the most common bacterial causes of GI worldwide, particularly in children under two years of age. *Escherichia coli* (*E. coli*, the common bacteria in human and animal intestines) is the leading cause of "travelers' diarrhea" and a major cause of diarrheal disease in the developing world. *Staphylococcus aureus* produces a toxin and is characterized by an abrupt and violent onset. Summer diarrhea was common in infants in hot weather in the past. It was often caused by salmonella bacterium, which is now a leading cause of hospitalization in young children, particularly in developing nations.

These infections were rampant in past centuries before adequate sanitation and good hygiene. In 1850, for example, the U.S. Census reports that 32,450 people died of diarrheal diseases whereas in the late 2010s only about 8,000 people died from these illnesses. On the other hand, the World Health Organization (WHO) notes that globally diarrhea is the second leading cause of death in children less than five years old, killing around 525,000 each year mostly in developing nations.

Diarrheal illnesses likely go back to the dawn of animal life. The term "diarrhea" was first used in the fourteenth century and is derived from Middle English via Latin; *dia* means "through," and *rheo* means "flow." "Dysentery" is from Old French, from the Greek *dusenteria*, meaning "afflicted in the bowls," also from the fourteenth century. "Cholera morbis" for other diarrheal conditions was first used in 1673, and "gastroenteritis" was first used around 1829 in the English language. GI and dysentery have been described since antiquity. For example, shigella was likely recorded both by Greek physician Hippocrates (c. 460–370 BCE), who noted that bloody dysenteries were epidemical during the late summer, and later by English physician Thomas Sydenham (1624–1689), who described both bloody and non-bloody diarrhea in late summer.

Over the centuries, GI and dysentery diseases have been blamed on many factors. Some ancient Greek physicians thought that after humid summer heat, cooler autumn weather constricted the skin and unbalanced the humors, which caused the intestines to discharge fluid. By the eighteenth century, other theories abounded. Scottish physician William Buchan (1729–1805) thought some foods such as rancid fats, cucumbers, raw fruits, or even "violent passions of the mind" led to diarrhea. In the late 1700s, German physician Johann Kampf thought all diseases came from impacted feces, including GI and dysenteries. In the nineteenth century, several American physicians, including Elijah Hammack (1826–1888) and Henry Hartshorne (1823–1897), claimed that besides rancid food, cold watery fruit, and sudden changes in the weather, bad drinking water caused GI and dysentery, but they did not yet accept the "germ theory of disease." In the late nineteenth century, several physicians theorized that poisonous "ptomaines" arose from constipated feces and produced food poisoning. This theory was popular among the public into the mid-twentieth century.

When the germ theory became an accepted fact in the early twentieth century, preventive measures such as washing fruits and vegetables, personal hygiene, and rehydration for diarrhea were introduced. In severe cases, antibiotics were given, but none worked for bacillary dysentery. Antiprotozoal drugs, however, are still effective against protozoa. The most common treatment today, which originated in 1901, is bismuth subsalicylate, a thick pink liquid that has anti-inflammatory and bactericidal action against diarrheal diseases.

Diarrheal diseases have shaped history. In the 1560s, natives in Brazil succumbed to a dysentery epidemic probably imported by Portuguese colonists. Since the natives were too sick to work, the colonists imported black Africans as slave workers, which helped spur the slave trade to the Americas. During the English Civil War (1642–1651), a combination of dysentery, typhus, and syphilis killed half of the English army, forcing its withdrawal. During WWI (1914–1918), dysentery was rampant in the trenches and killed hundreds of thousands on all sides of the conflict and may have been a factor in the war's outcome.

UNUSUAL TREATMENTS

Over the centuries, various remedies were attempted. Until the twentieth century, most healers used a purgative to get rid of unbalanced humors—even though this would cause more diarrhea—and an emetic to cause vomiting. In Egypt around 1500 BCE, for example, herbs and concoctions were developed to treat GI and dysentery. These included acacia, bayberry, henna, and myrrh. The *Papyrus Ebers* gives several remedies and incantations. One recipe called for fresh bread meal combined with lead-earth, white-of-egg water, which was taken for four days along with chanting, "O, Hetu, O, Hetu." Hippocrates recommended a quarter pint of beans and 12 shoots of madder mixed together and boiled with some fat. This may have made the condition worse.

In the mid-fifteenth century, English surgeon John Arderne (1307–1392) recommended castor oil and a type of prune, which first acted as a laxative and then as an astringent to treat diarrhea. For patients with dysentery and liver problems, not only did he purge them, but he also fed them hare (rabbit) cooked in rainwater and served with syrup made with plantains. This treatment may have helped hydrate the patient. Buchan purged and gave enemas for dysentery. If the person was not vomiting, he ordered an emetic to clean the stomach. Sydenham, also in the seventeenth century, treated dysentery with "chicken tea" and claimed to have success. It may have been effective as it rehydrated the patient. In the late eighteenth century, German physician Johann Kampf recommended colonic cleansing (enemas) still used today as an alternative medicine, but it was ineffective and may have dehydrated the patient.

During the mid-nineteenth century, some physicians at the onset of diarrhea gave a dose of castor oil, with 10 or 15 drops of laudanum (opium in alcohol). Laudanum would have quelled the diarrhea, but castor oil may have increased it, thus leading to dehydration. Hartshorne ordered a robust patient be bled with leeches on the abdomen. After this treatment, a warm poultice of flaxseed meal was placed. However, in obstinate cases of diarrhea, he recommended a large blister be made with a caustic substance on the abdomen. In chronic, or acute stubborn cases, he gave an enema with acetate of lead (a toxic compound), laudanum, and zinc sulfate (which is still used today orally to treat diarrhea in children), which he claimed resulted in "remarkable cures" and which may have helped quell the diarrhea and aided in rehydration. However, bleeding and blistering could have caused weakness and infection.

In addition, Hammack suggested the emetic ipecacuanha for vomiting and for thirst iced rice water, flaxseed tea, or an infusion of slippery elm bark that contains an aspirin-like substance. It may have helped soothe inflamed mucous membrane and rehydrated the patient. He also recommended treating diarrhea and dysentery with chalk mixed with laudanum, gum kino (from eucalyptus-type trees), and brandy. As a last resort for chronic diarrhea, he suggested that the patient should seek a "mild, dry, equable climate." In the mid-twentieth century, the inventor of a light therapy system, the Spectro-Chrome light box, claimed bathing in a yellow and then indigo light could cure diarrhea. The device was marketed into the 1950s but was not effective.

One treatment that has been successful is fecal microbiota transplantation (FMT). A fourth-century Chinese physician, Ge Hong, records that fecal material from a healthy person was administered as "yellow soup," a fecal slurry, to a patient for severe food poisoning and diarrhea, a practice that continued for centuries. Bedouin groups also consumed their camel's dung as a remedy for dysentery as did German soldiers in North Africa during WWII. Fecal transplantation in Western medicine was not fully recognized as useful until the early twenty-first century when it was found to successfully treat several bowel conditions.

Head and Hair Issues

Head and hair issues range from removing, restoring, and coloring hair to controlling parasites and skin diseases on the head and face. Concerns about these conditions have existed over the centuries.

Grooming, Removing, and Restoring Hair

Even before recorded history, being clean shaven or having a beard has come in and out of fashion. Archaeological evidence suggests that about 60,000 years ago men shaved off beards by using sharpened clam and obsidian shells. Growing or removing hair was so important to the Egyptians that the *Papyrus Ebers* (c. 1500 BCE) devotes a whole chapter on it. Hair on the body was considered uncivilized. To appear unshaven was a mark of low social status. In ancient Egypt, because of the hot climate shaving and hair removal was a regular part of daily grooming, but baldness was considered unfashionable. Wigs were primarily used by the upper classes while ordinary people wore their hair tied back or cut it short. Both male and female pharaohs wore fake beards made of precious metals symbolic of their stature.

On the other hand, in ancient Greek, beards were a sign of virility, manhood, and wisdom. Romans were clean shaven but left hair on their heads. However, the richer you were, the less body hair you had, and servants shaved or plucked out hair. Conversely, ancient Jews and Muslims were commanded not to shave off their beards. In Christianity, the acceptability of beards over the centuries has waxed and waned as has hair length for men, and often for women.

Girls were frequently allowed to have free flowing hair. After they married or were older, particularly in the Middle Ages, they were expected to cover their hair with a veil, hood, cap, or shawl to discourage unwanted attention. Even into the mid-twentieth century, women wore hats, although traditions about covering the head have waxed and waned over the centuries. Beauty in the Middle Ages demanded a high forehead among the upper classes. To maintain that look, women plucked or shaved their hairline as exemplified by Queen Elizabeth I (1533–1603) of England with an elongated forehead and covered hair. Hair coloring throughout the ages has also been popular. Henna was used to dye hair red and cover the gray in ancient Egypt and is still used today.

Hair loss and baldness often occurs in men and sometimes in women with age. Male pattern baldness (androgenic alopecia) is the most common type of hair loss and is generally genetic. Hair loss can cause psychological problems, including low self-esteem, low confidence, and negative influence on social interactions. By the age of 50, half of all people will experience some hair loss. Over 50 percent of women and about 30 percent of men have reported being upset by losing hair. The term "bald" is from the Middle English *balled*, meaning "white patch," from the fourteenth century. Today hair loss and baldness can be managed with some prescription drugs. In 1952, a dermatologist performed the first successful hair transplant, which helped bald men. On the horizon are techniques to clone hair cells to grow hair.

Head Lice

Head lice (pediculosis capitis) are small parasitic six-legged insects that feed on blood from the human scalp. They cause itching and sometimes sores primarily behind the ears and the hairline of the neck but can also infest beards. They have been around for about 5 million years. Although most common in children, they are also found in adults and are spread by close contact to an infected person. Each year, lice infestations still affect hundreds of millions of people worldwide. Today Western cultures usually respond in horror at lice infestations. From antiquity into the Middle Ages, lice were thought to result from spontaneous generation on the body or overindulgence in eating figs. Lice were common into the nineteenth century until better hygiene and sanitation developed. The first use of the term "louse" in English was before the twelfth century from the Old English *lūs*, and "lice" is from *lȳs*; both *lūs* and *lȳs* are Germanic in origin. Today lice are removed by several chemicals and nits are removed by a fine-tooth comb.

Eczema (Atopic Dermatitis) and Dandruff

Atopic dermatitis, commonly called eczema, causes red, dry, and itchy skin, particularly at night. It can lead to an oozing, crusting condition on the scalp or head, and in infants it is called "cradle cap." The malady is often accompanied by asthma or hay fever. It affects up to 5 percent of the general population, and slightly more men than women are affected. No cure for this scourge exists, and its cause is unknown although it is likely genetic. Today standard medical management includes creams to moisturize, halt inflammation and itching, and combat bacterial infections. The first-known use of the term "eczema" was in the mid-eighteenth century. It is from the Greek *ékzema*, meaning "to boil over." The term "atopic dermatitis" was not coined until the early twentieth century. The disease has been known since antiquity and Greek physician Hippocrates (c. 460–370 BCE) thought that skin conditions reflected internal humoral imbalances that allowed oozing and peeling to eliminate bad humors. This theory was still accepted into the nineteenth century.

Itchiness and flaking of scalp skin can also be dandruff (seborrheic dermatitis). Dandruff is usually the result of growth on the scalp of the *Malassezia* fungus. Other possible causes of dandruff include dry skin, genetics, or sensitivity to hair products. Dandruff affects up to 50 percent of people. The term "dandruff" is likely of Anglo-Saxon origin and is a combination of *tan*, meaning "skin disease," and *drof* or *ruff*, meaning "dirty." Dandruff today is commonly managed with shampoos or ointments containing selenium, aspirin, zinc, or steroids.

Head issues such as parasites and baldness have been found in the arts. For example, a detail of a painting by Flemish landscape painter, Jan Siberechts (1627–1703), shows a mother hunting for head lice in her child's head. Head lice were publicly stigmatized in the eighteenth century. Scottish poet, Robert Burns (1759–1796), for instance, wrote "To a Louse: On Seeing One on a Lady's Bonnet at Church." In the poem, Burns condemns the bug that is detested by saints and sinners alike for being on the curls of a fine young lady. Baldness is seen in fiction as a characteristic of both villains and heroes. For example, in the DC comic book and TV series *Smallville* (2001–2011), Lex Luther, the supervillain and Superman's nemesis, loses his hair in a lab accident blamed on Superboy who gave him the chemicals. On the other hand, some heroes are portrayed as completely bald. One example is Captain Jean-Luc Picard of the spaceship, *Enterprise*, in the *Star Trek* franchise, which included TV episodes and films from the 1990s into the 2020s. Another bald superhero is Luke Cage (Power Man) in *Luke Cage* first published by Marvel comics in 1972. It was turned into an internet TV series (2016–2018).

UNUSUAL TREATMENTS

Head and hair treatments for various issues have been known from antiquity. Today many are still based on alternative or folk medicine.

Grooming, Removing, and Restoring Hair

To remove hair for beauty, the *Papyrus Ebers* suggests the patient should frequently rub into the hair a mixture of crushed tortoise shells heated in "Fat-from-the Hoof-of-a-Hippopotamus." Although it was not effective, it was likely carried out because hairs are not found on tortoise shells or hippopotamus hooves. People also rubbed stubble with pumice stones and pulled out hairs with beeswax and tweezers made of seashells. The Romans and Greeks also used pumice stones, and if a dull razor caused nicks, they were treated with a mixture of spider webs soaked in oil and vinegar. This may have clotted the blood.

In the Middle Ages, to remove unwanted hair Italian noblewoman Caterina Sforza (1463–1509) in her book *Gli Experimenti* (c. 1500) recommends that women apply a mixture of cat dung and vinegar to remove hair. It likely did not work. Another recipe suggests applying a solution of arsenic and quicklime. A risk of this toxic mixture was skin removal along with the hair. If hair became tangled after washing, women used a conditioner of bacon fat and lizards. Then rose water, cloves and nutmeg, and other spices were combed into the hair to take out the bacon scent.

Restoring lost hair has been attempted for centuries. The ancient Egyptians combined the blood from the neck of a gabgu bird (unknown bird), mixed with balsam, and rubbed it on balding spots. The thyroid in the bird's neck may have stimulated some hair growth. Another recipe included lion, hippopotamus, crocodile, cat, and snake fat, and still another called for chopped lettuce to be placed on bald spots. To stimulate the scalp, porcupine quills were rubbed on the scalp. These remedies were ineffective but largely harmless.

Greek physician Hippocrates for male pattern baldness recommended a brew of cumin, horseradish, pigeon droppings, beetroot, and spices be rubbed on the head. It was likely not effective but harmless. Medieval monks, nuns, and alchemists concocted numerous recipes for restoring hair. One ineffective recipe included ashes of a hedgehog, burnt barley bread, horse fat, and boiled river eel while another included bruised peach pits boiled in vinegar. Peach pits release poisonous hydrogen cyanide gas when combined with an acid, so the treatment could have caused poisoning or death. Many oils have been tried and are still advertised in the twenty-first century as helping to restore hair, and some like lavender essential oil have been found to have some effect on hair restoration, at least in mice.

In the nineteenth century, advertisements for the "faradic" electric hairbrush, which usually contained a magnet, supposedly cured baldness by adding electricity to the scalp. Various devices using light and suction were developed in the early twentieth century. The "Thermocap" device, a cap with heat and blue light, was placed on the head and supposedly stimulated dormant hair cells to grow. In the 1930s, the Xervac machine used suction to spur hair growth. These did not work. Even in the twenty-first century, there were advertisements for a low-level laser therapy, sometimes called red light or cold laser therapy, which supposedly encourages hair cell growth, but research is inconclusive as to its effectiveness. Other folk and alternative treatments found in the twenty-first century include rubbing on the head onion juice, saw palmetto, sandalwood, lavender, rosemary, thyme oils, and various spices, but the research regarding their effectiveness is not clear—they are also relatively benign.

Various remedies for coloring hair have been common. In Mesopotamia (c. 3100–c. 535 BCE), now the Middle East, the priest healer, in addition to incantations, took gall (bile) of a black ox or the fat of a black raven and smeared it on the hair as it was thought that the color of the animal would transfer to the hair. The gall may have slightly darkened the gray hair. The Egyptian *Papyrus Ebers* recommends several concoctions to dye hair. One recipe calls for warming the womb of a cat in oil with the "egg-of-the-gabgu-bird." Another dye combines a black tapeworm with the "Hoof-of-an-Ass and the Vulva-of-a-Bitch." These had little effect but were harmless to the patient. A mixture of liver of an ass, warm oil, and opium made into little balls was rubbed on the eyebrows to take away gray hair, which may have eliminated pain but not the grayness. In the Middle Ages, women dyed their hair to cover gray and change its color. To darken hair, they used a mixture that contained burnt gall oak tree coals quenched in wine. To color the hair yellow, a concoction of ashes of the barberry tree and water was used. Since these recipes contained highly caustic lye from ashes and coals, they could have burned the scalp along with changing hair color. Lead acetate from antiquity into the twenty-first century has been used to dye gray hairs. Today it is included in over-the-counter gels, creams, and shampoos. Since they are used on the skin, the lead is generally not absorbed, and the products are relatively safe.

Head Lice

The first recorded treatment for head lice is found in the Egyptian *Papyrus Ebers*. It recommends filling the mouth with warm date meal and water and then spitting it on the skin to drive away fleas and lice, which was ineffective but harmless. Archaeological evidence suggests that mercury was used as a treatment to kill head lice. However, a major way of eliminating head and beard lice was shaving the heads clean and wearing wigs. They also used a mixture of vinegar, spices, and terebinth (turpentine) oil along with a fine-tooth comb, which was effective, although the terebinth could catch on fire.

Many toxic substances have been used over the centuries to eliminate hair lice that were often effective but dangerous to humans. For example, Roman encyclopedist Pliny the Elder (23–79) suggested rubbing the poisonous powdered seeds of the stavesacre (lupine family) plant on the head. In 1300, Italian Marco Polo (1254–1324) brought pyrethrum powder to Europe made from chrysanthemums, which is highly toxic to many insects, is a possible human carcinogen, and is still used today. From the fifteenth century into the mid-twentieth century, arsenic and, in the nineteenth century, mercuric oxide powders rubbed on the head successfully killed lice. But they were toxic to humans and could lead to poisoning and death. During the mid-twentieth century, DDT, a powerful insecticide, was developed, which killed lice and other pests, but the insects developed resistance to DDT, and by the twenty-first century, insects had developed resistance to pyrethrum.

Numerous folk and alternative remedies for eliminating lice are still used in the twenty-first century. These include kerosene, turpentine, petroleum jelly, garlic, vinegar, salt water, mayonnaise, gasoline, and hair dryers. Vinegar and salt water are not effective, but kerosene, turpentine, and gasoline can not only kill lice but also cause respiratory issues or ignite, resulting in serious burns. Petroleum jelly mixed with sulfur has been found to be effective in smothering lice. Some research suggests that heat from a hair dryer may kill both the lice and their eggs.

Eczema (Atopic Dermatitis) and Dandruff

A few treatments for eczema and dandruff have been used over the centuries. In ancient Egypt for eczema, healers painted the head with a mixture of durra-meal (sorghum) and fruit of the Dom palm warmed in soft fat and covered the head with a cloth. The second day, the head was anointed with fish oil, the third day hippopotamus oil, and the fourth day abra oil. If this was not effective, the head was smeared with rotten cereals. These may have softened scaling skin, but bacteria in the spoiled grains could have caused infections.

By the twelfth century, treatment for head problems were based upon humoral theory and treated systemically. Remedies included wet cupping (heated cups placed over the skin that had been scratched), bloodletting from a vein, and emetics to cause vomiting along with ingesting sulfur, arsenic, or mercury, which likely poisoned the patient. These treatments in addition to leeching, diuretics, and laxatives to eliminate "toxic" substances from the body were used well into the nineteenth century. They had little effect on the dermatitis and likely caused physical weakness.

Folk remedies are popular for eczema and dandruff, but the research is inconclusive as to their effectiveness. For example, by the twenty-first century, applying apple cider vinegar to the head was popular as proponents claim that the acid can kill the fungus infection. Other remedies include soaking the head with a diluted bleach solution or placing ice cubes on the scalp several times a day. To manage dandruff, proponents claim that applying egg yolks (or egg whites), cold green tea, which contains antifungal properties, or mixing willow leaves or bark (contains aspirin-like properties) in wine (antibiotic properties) and applying these to the scalp reduce inflammation, itching, and dandruff. These may be useful in some cases and are generally harmless.

Headaches and Head Injuries

Headaches can range from a mild dull ache to excruciating pain that is incapacitating or life-threatening. Head injuries can range from mild concussions to fractures or serious traumatic brain injuries.

Headaches and Head Pain

Head pain is classified as one of three types based upon the source of the pain. These classifications include primary headache, secondary headache, and neuropathic and facial pain. Primary headaches are the most common and include tension, migraine, and cluster headaches. They are not a symptom of an underlying disease, but they all can affect the quality of life. The most frequent primary headache is a tension headache in which pain is usually steady, mild, and dull and often at the temples and above the eyebrows. However, most people can function normally with a tension headache. The cause is not known, but it may be from stress. Up to 90 percent of all people experience a tension headache at some point in their life, and they are more common in women than men. They can be treated with over-the-counter pain medication.

A migraine, or "sick headache," is generally on one side of the head but can spread to both sides. Before the pain begins, an aura such as tunnel vision, flashes of light, or other visual changes occur in about 30 percent of all sufferers. The pain is intense, throbbing, and pulsating and may last for hours or even days and is so severe that it interferes with daily activities. Extreme sensitivity to light, along with nausea and vomiting, is also common. Moreover, it is the third-highest cause of disability worldwide among people under age 50 and is three times more common among females. The cause is suspected to be a genetically induced hypersensitivity involving neurons located within the central nervous system that are affected by a change in the external environment such as hormonal changes. Into the late nineteenth century, they were thought to be caused by toxic digestive problems and into the late twentieth century by vascular spasms. Today along with rest, some medications can help migraines.

In cluster headache, pain is generally excruciating and located around or behind one eye but may radiate to other areas of the face, head, and neck. The pain generally occurs in a cyclical pattern at a certain time and often awakens people in the middle of the night. It is one of the most painful types of headaches, is more common in men, and occurs in about 0.5 percent of the population. These headaches can last for weeks or months. The cause is not known but may be related to release of histamine and serotonin in the brain or abnormalities in the body's biological clock. Risk factors are smoking, heavy alcohol use, and a family history. They are difficult to manage.

Other primary headaches include those that result from exercise, sexual activity, eating cold food such as ice cream, and "rebound headache" from overuse of headache medication or alcohol or drugs. Lastly, neuropathic is pain caused by a lesion or disease of the trigeminal nerve resulting in head and stabbing, episodic facial pain.

The term "headache" was first used in English before the twelfth century from the Middle English *hevedeche*, from the Old English *hēafodeċe*, meaning "pain in the head, headache." Migraines, in particular, have been noted since antiquity. Greek physician Hippocrates (c. 460–370 BCE) first described visual symptoms of migraines and an association between a headache and various activities such as exercise and sexual intercourse.

Secondary headaches are those that are due to an underlying medical problem in the head or neck. They may include infected sinuses, strokes, brain tumors, and traumatic head injuries. Strokes discussed under the "Cardiovascular (Heart) Diseases" entry often have severe sudden onset of pain. Treatment depends upon the cause of the malady such as anticoagulants for strokes caused by clots and surgery to alleviate bleeds and swelling.

Concussions and Traumatic Brain and Head Injuries

Traumatic brain injuries (TBI) can range from mild to severe. They are usually caused by a blow to the head from contact sports, falls, transportation accidents, fights, and combat. A concussion is a mild traumatic brain injury (MBI) that can temporarily affect brain cells. The person may or may not lose consciousness, but can have issues with concentration, memory, balance, and coordination afterward. Repetitive brain injuries suffered in sports may cause dementia. By the 1980s, many sports had developed better head protection gear to decrease these injuries. More serious traumatic brain injury caused by fractures that indent the skull or when an object pierces the brain can result in physical damage to the brain and may cause long-term disability or death. In the United States, about 2 percent of the population are living with disability because of a traumatic brain injury. Surgery or methods to release pressure on the brain and medications are usual management for these distressing conditions.

Another type of head injury is a broken, fractured, or dislocated jaw, which can result in the lower jaw sticking out, bleeding from the mouth, breathing difficulties, and problems with eating. In a dislocated jaw, the jaw is pulled away from one, or both, of the temporomandibular joints (TMJs). Hippocrates and other ancient physicians provided a procedure that is almost identical to manual reduction of the jaw used today. Clinicians placed their thumbs against the lower back teeth inside the mouth with the remaining fingers under the jaw. The jaw was then wiggled back into place. For fractures, bandaging it to support the jaw or wiring it shut could also be used.

Head trauma and TBI has always been found in warfare and may have led to battles lost or won. In the late twentieth and early twenty-first centuries, they were a major cause of disability in veterans deployed to the Middle East. TBI has been found to interfere with duty performance in over a third of all soldiers many of whom experienced migraines. Only one-third of soldiers sent home due to headaches returned to duty. Headaches have been depicted in film and theater productions. For example, in the convoluted science fiction/horror film *Vanilla Sky* (2001), the protagonist may be lucid dreaming in a coma from either attempted suicide or an auto accident. In several scenes, he is described as having a migraine. Trepanation or drilling holes or scraping a hole in the head to relieve head pain has been found in illustrations. For example, printer Peter Treveris (active in London 1525–1532) in his *Handywarke of Surgeri* (1525) has an engraving of a surgeon performing trepanation on a terrified patient.

UNUSUAL TREATMENTS

Many folk cures using food or spices were employed over the centuries to cure headaches, and some are still used today; however, in most cases no reason is given as to why they might be effective. Remedies include placing on the head raw potatoes, cabbage, and onions in addition to cloves, ginger, and lavender oil and drinking chamomile tea. For migraine and cluster head pain, a vinegar and mustard plaster is placed on the head. Studies are mixed as to their effectiveness. They are harmless to the patient, and relief may be due to the placebo effect.

Headaches and Head Pain

One of the oldest treatments for head pain, trepanation (trephining), goes back to prehistory. It was performed to release demons from the head, which were thought to cause headaches. Trepanation is a dangerous cure although archaeological evidence suggests that up to a half of those treated survived. Others died from infections and bleeding. In the late twentieth century, alternative medicine proponents claimed that trepanation increased the blood flow in the brain and gave spiritual insight, but these claims are unsubstantiated by research. Some individuals have drilled holes in their skulls or traveled to developing countries to obtain this risky procedure.

Other methods were tried to relieve headaches. For example, the *Papyrus Ebers* gives several ointment recipes that were smeared on the head to relieve head pain. One was composed of hippopotamus skin and poppy seeds. These opium seeds may have helped to reduce the pain. Various poultices and lotions were also tried. One consisted of crocodile dung beaten with ostrich eggs placed on the head. These may have worked by placebo effect and likely were not harmful.

Electricity to the scalp is another ancient treatment for headaches. Greco/Roman physician Galen (129–c. 210), for example, recommended placing an electric torpedo fish on the head. He claimed that the numbing force of the electric

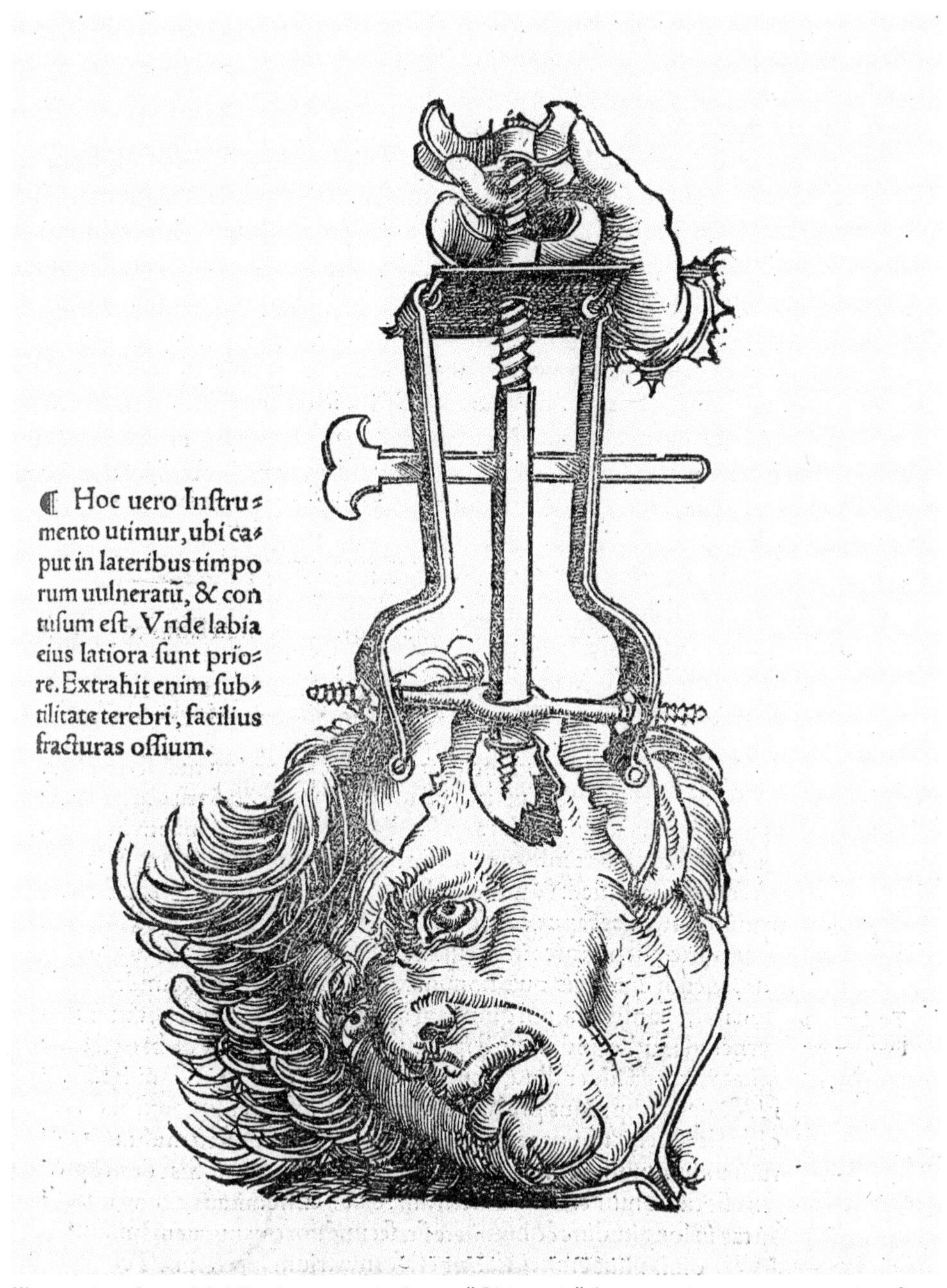

Illustration from Middle Ages surgical text "Chirurgia" for carrying out trepanation for headache and head injuries, depression, or mental illness. (Wellcome Collection. Attribution 4.0 International (CC BY 4.0))

discharge took away pain. Although it was shocking, it was usually harmless to the patient and may have acted as a counterirritant as do transcutaneous electrical nerve stimulation (TENS) units today for pain relief. "Electricity" to cure headaches was marketed in the late nineteenth century. One popular method was "Scott's Electric Hairbrush" filled with a magnet, but it had no effect on reducing headaches other than by the placebo effect. In the twenty-first century, a headband with an electrode pressed against the middle of the forehead to deliver an electric impulse to nerves above the eye was approved by the FDA. About a half of all people found it reduces migraine attacks.

Cautery has also been used since antiquity. For example, Roman philosopher Celsus (c. 25 BCE–c. 50) in *De Medicina* recommended—based upon humoral theory—cauterizing the head with a hot iron to cause a blister and drain "toxic humors." This was carried out in the Middle Ages and into the nineteenth century. This would have been very painful and could have resulted in infection and death but was not effective in reducing headaches.

In the eighteenth century, physicians, also based upon humoral theory, bled the patient by using leeches or cupping (heated cups placed over skin that had been scratched) or blistering agents such as cantharis beetle shells to cause oozing ulcers on the patient's head and neck to balance the humors. Some tension headaches likely were reduced by these treatments due to placebo effects or fear of the treatment, but they could have also caused infections. Snuff to cause sneezing was also used but was ineffective and relatively harmless to the patient.

For a migraine, English physician and herbalist Nicholas Culpepper (1616–1654) mixed crushed cherry stones, a handful of elder flowers, and three pints of muscatel wine and kept them in a closed pot in a bed of horse manure for 10 days after which it was distilled. The patient drank it. Oil from the stones were also boiled in vinegar. The clinician placed this mixture on the patient's temples and into ears or used it for an enema. However, when these pits are heated with an acid it can result in deadly hydrogen cyanide and could have led to poisoning and death.

For migraine and likely cluster headaches, Canadian physician William Osler's (1849–1919) textbook, *The Principles and Practice of Medicine* (1942), into the mid-twentieth century recommends strong coffee and a course of *Cannabis indica*, a near relative of *Cannabis sativa* (marijuana). Indica supposedly results in a numbing experience and is used to relax or relieve pain. These remedies are still used as an alternative medicine and may have been helpful.

Other alternative and folk treatments for headaches today include light, color, and crystals. In the 1960s, a technique called "colorpuncture" arose. Proponents claim that certain colors bring the mind and body back into balance. In this technique, various colored lights are placed on acupressure points. Another alternative treatment is crystal therapy, which had a resurgence in the late twentieth century. Proponents claim that certain crystals vibrate at a frequency that can interact with the human body to heal. For headaches and migraines, amethyst and rose quartz are often used. The patient places them against the skin on the head, body, or hands. Lapis lazuli is used to reduce vomiting and vertigo associated with migraine. However, research does not substantiate their effectiveness, and relief from symptoms is likely due to placebo effect.

Concussions and Traumatic Brain and Head Injuries

In the case of a serious concussion or traumatic brain injury, trepanation was used to relieve pressure and remove blood clots, which often saved the patient.

Other methods were also tried. Healers in ancient Egypt would not treat a deep head wound in which the brain was deeply cut as the person would likely die. If it appeared that the patient might live, grease was applied to the wound. For a broken jaw, if the patient was feverish and had a red face, they were not treated as they usually died. However, if the face was pale and they had no fever, a wooden device was placed into the mouth to keep it open for food and water and the patient's arms were draped over bricks to keep them in a sitting position. In Rome, Celsus in the first century put vinegar on a head fracture and Paulus Aegineta (c. 625–c. 690) poured oil and wine into the head fracture and bandaged it. These may have prevented infection if the fracture was open from deep cuts or trepanation. Trepanation fell out of fashion by the eighteenth century; however, debridement (picking out bone fragments in the wound) has been common for centuries, and some patients survived, but others often died of infection or bleeding.

Health Maintenance

The definition of health and how to maintain it has changed over the centuries. In the distant past, it was considered the absence of illness. However, the ancient Egyptian *Papyrus Ebers* (c. 1500 BCE) noted that interference with the normal health of a person, which was not due to any observable cause, was likely caused by some evil spirit that had obtained possession of the body. Therefore, priests recited incantations to drive the demons from the person. Ancient Greek physician, Hippocrates (460–370 BCE), considered the father of Western medicine, and many of his followers, and also later physicians such as Galen (129–210) of Pergamon, viewed health as a balance of four humors or bodily fluids (blood, phlegm, yellow bile, and black bile) and balance between the internal and external environment. Sickness resulted when these balances were upset. To keep a body in equilibrium, exercise, a proper diet, and bathing were necessary. This was expressed by the Romans as *mens sana in corpore sano* (a healthy mind in a healthy body). The theme of exercise and diet today is a major part of health promotion to maintain health and prevent illness. The term "health" was first used in the fifteenth century and is from the Old English *hǣlth*, of Germanic origin, meaning "whole."

In fifth-century BCE Greece, to maintain health and cure illness, people visited the sanctuary of Asclepius. Some people visited to maintain their health while others to cure various diseases. In the Roman Empire, some healers still believed in spiritual reasons for health and disease, and this was carried into Christianity. After the fall of the Western Empire in Europe in the late fifth century, the Church in the Middle Ages believed that possessions by evil spirits or punishment from God due to sin were responsible for ill health. Still, humeral theory remained as an acceptable theory for the cause of disease into the early nineteenth century.

From the nineteenth into the twentieth century with technological developments and discovery of microorganisms, a "biomedical" model of health emerged. Health was now defined as the freedom from disease, pain, or defect, making the "normal" human condition "healthy." It focused only on biological factors, but this model led to immunizations, hygiene, sanitation, and other public health methods to prevent diseases. This definition of health shifted in the immediate post–WWII era in 1948 when the World Health Organization (WHO) defined it as "a state of complete physical, mental, and social well-being and not merely the absence of disease or infirmity." This definition is still accepted by some today.

However, in the post–WWII years, to address the lack of fitness among American youth, President Dwight D. Eisenhower (1890–1969) in 1956 established the President's Council on Youth Fitness, and President John F. Kennedy (1917–1963) expanded the program. Exercise and later diet to maintain good health was emphasized. By the early 1960s, the definition of health became broader. It was called "wellness," a term coined by American physician Halbert L. Dunn (1896–1975) in his book *High-Level Wellness* (1961). Wellness was defined as the optimal state of health for an individual and included continuous growth and balance of physical, psychological, social, spiritual, and economical aspects of a person's life to realize the fullest potential of functioning. To maintain good health and prevent disease, living a "healthy lifestyle" that included exercise, avoidance of tobacco use, minimal alcohol consumption, and eating a low meat and many vegetables "Mediterranean diet" was encouraged. In 1986, the WHO made further clarifications on health and its maintenance through "health promotion," which was "the process of enabling people to increase control over, and to improve, their health." By the late twentieth century, it was realized that lack of exercise, in particular, was detrimental to the overall functions of the body. It contributed to many chronic disorders, such as Type 2 diabetes, colon cancer, obesity, coronary heart disease, and ischemic stroke, and that exercise was the key to maintaining good health no matter the person's age or disabilities. Many community and nongovernmental organizations sponsored a variety of exercise programs for all age groups to encourage this activity.

By 2020, the American Public Health Association's definition of health included "adapting to evolving health needs over the life course and optimally managing disease as a means to physical, mental, and social well-being." Over the centuries, philosophies of health have gone through changes, but the twenty-first century's view of health is rooted in ancient advice.

Health, and maintaining it, has appeared in the arts as shown in numerous images and writings, particularly about baths or "taking the waters," which for centuries was synonymous for maintaining good health. Greeks and Romans bathed almost daily, in private facilities or in large community baths. This is illustrated in the ornate Roman bath that still exists in Bath, England, which has been restored and is now a museum. Bathing fell out of favor in the Middle Ages but was revived again in the eighteenth century. Baths and spas have been noted in a few fictional works that have looked at health and how to maintain it. A historical fiction comedy novel, by American T. Coraghessan Boyle (1948–), *The Road to Wellville* (1993)—and made into a film (1994) of the same name—tells the story of clean living advocate John Harvey Kellogg (1852–1943). Kellogg ran one of the famous spas/baths in North America, the Battle Creek Sanitarium, called "the San." Wealthy patients went there not only for various illnesses but also to maintain their health. Treatment included soaking in pools, colonic irrigation, walking barefoot, dry breakfast cereals, electrical stimulus, exercise, fasting, and no meat diets.

UNUSUAL TREATMENTS

Various remedies have been tried over the centuries to maintain health. The ancient Egyptians worried about the possibility of ill health brought on by malevolent spirits, especially in the anus, resulting in frequent purging. Greek historian Herodotus (c. 484–c. 425 BCE) observed that each month for three days in succession the Egyptians ingested purgatives and emetics to clean out the system. This folk cure was found in the early modern period when enemas were considered essential to maintain good health. In the late nineteenth century, several researchers theorized that poisonous "ptomaines" arose from constipated feces and could cause diseases due to "autointoxication" or self-poisoning, leading people to take almost daily enemas or laxatives. This practice continued into the 1930s but faded in the post–WWII years as the treatment was not supported by research. Nonetheless, in the early 2000s, internet marketing for "colon cleansing" to "detox" the body increased, and it became a popular folk remedy to maintain health. However, the treatment has led to life-threatening conditions including bowel perforations, lethal colon infections, and kidney and heart failure.

Folklore suggests that people in the Czech Republic have been bathing in oak tubs full of warm beer for centuries for good health. Moreover, by the 2010s beer spas or beer baths became a new fad, particularly in Europe. The person sits in a bath of warm beer, which also has a spigot for drinking a cooler version to drink. Mud baths have also been popular in recent years for the same reasons. Another bath is a "forest bath" or a walk in the woods, which is popular in Japan. Some research has suggested that people who walk in the woods with the sounds of nature had lower levels of stress hormones compared to a comparable walk in an urban area, which likely led to better health. Another generally harmless remedy is an air bath. It has been advocated for general well-being from the eighteenth century into the current times. The air bath is often accomplished in the morning upon rising. The person sits or walks around a cool room without clothing for a half hour or so. At the turn of the twentieth century, Kellogg recommended a hot-air bath in a closed chamber that was like a sauna, which was popular in the late twentieth century. Although these remedies may have made the person temporarily feel good, there is little evidence that they improved health and could cause cardiac problems in some people.

An ancient remedy to maintain health and cure illness was bloodletting that may have originated in ancient Egypt. In the second century, Greco-Roman physician Galen expanded on Hippocrates' earlier theory that good health required a perfect balance of the humors and carried out this procedure to maintain balance. When Christian monk/physicians were forbidden to practice bloodletting in the twelfth century, barbers often carried out the technique before the medical profession adopted it. Bloodletting remained a primary cure and method to maintain good health into the nineteenth century. It was also believed that bloodletting, especially at the change of seasons, helped to keep the body healthy. Bloodletting was performed by cutting a vein in the arm or neck, using leeches, or by wet cupping—placing a heated glass cup over an incision that caused a vacuum—to draw out the blood. Leeching became a popular remedy in the nineteenth century but was discredited as a treatment in the early twentieth century. The procedure often caused weakness and sometimes infections and death from loss of blood. President George Washington (1732–1799) likely died from overzealous bleeding on the part of his physicians. By the 2010s, leeches and bloodletting were again part of medical practice. Sterile leeches were raised and used to prevent swelling in delicate plastic surgery procedures, and bleeding was accomplished as a treatment for hemochromatosis—chronic iron overload disease.

In the eighteenth century, German physician Franz Anton Mesmer (1734–1815) suggested that the body contained an invisible natural force called "animal magnetism," similar to Eastern religions' qi, or energy, flow. Mesmer claimed that the magnetic action of his hands on the diseased part of the body could treat the patient by interacting with this force, and he touted this as a cure for many diseases and as a method to maintain health. Sometimes people felt better after "mesmerism," later known as hypnosis due to placebo effect. Chinese folk medicine uses acupuncture (the insertion of needles at points along "channels" in the body), which became popular by the late twentieth century to help the qi flow, and claims it keeps people healthy and free from pain; it may be effective for some people.

Various herbs and tonics have been advertised that claim to keep people healthy. Patent medicines in the late nineteenth century, such as "Fellows Compound Syrup of Hypophosphites," claimed to restore vitality. It contained the rat poison strychnine (from the tree *nux vomica*); over-the-counter medicines containing strychnine were available until the 1970s. Fowler's solution containing the poison arsenic to restore youth was also popular. These substances were dangerous and sometimes fatal. Tonics that contained cocaine were popular in the late nineteenth and early twentieth centuries. Pope Leo XIII (1810–1903) promoted *Vin Mariani*, a mixture of coca leaves and wine. The manufacturer claimed the tonic would restore health, strength, and vitality. Coca-Cola in the United States soon followed, which was a mixture of cocaine, kola nut, and seltzer water that was also advertised to maintain health and to feel good. The cocaine-laced tonics resulted in drug dependency, and the sugar in the drinks was not good for tooth enamel. These and other cocaine-based tonics were banned by the 1930s in most Western nations. The health tonic Geritol, an alcohol-based iron and vitamin B mixture, was advertised from 1950 to the 1970s for "tired blood" and to increase vitality primarily for older individuals. However, in the late 1970s, supplemental iron products were found to be dangerous, especially those with hemochromatosis, as it could result in liver, kidney heart, and other organ damage.

Drinking one's own urine has been a folk remedy for centuries to maintain health. In the late twentieth century, some "natural health" advocates recommended the practice. But research suggests that drinking urine can introduce bacteria, toxins, and other harmful substances and put stress on the kidneys. To boost stamina and keep people healthy, another folk remedy is the consumption of human fetal tissue. Fetal tissue pills have been marketed since at least the early twenty-first century. In 2012, it was discovered that these pills were smuggled from China to South Korea and maybe North America. They do not improve health and can contain super bacteria.

Mega doses of vitamins became popular to maintain health in the 1970s. For example, American chemist Linus Pauling (1901–1994) claimed that mega doses of vitamin C could prevent illness and help keep a person healthy. Others touted mega doses of vitamin D and E; however, research has shown that high doses of vitamins did not necessarily keep people healthy, and that vitamin D and E could lead to toxicity, liver damage, and other problems.

By the early twenty-first century, folk healers advertised "liver cleanses" and "detoxes" to "detoxify the liver" and "rid the body of impurities." These "detoxification" substances include dietary supplements, special diets, chelation infusions, and enemas and can sometimes result in electrolyte imbalance. Detox foot pads have also been advertised to maintain health and remove toxins and heavy metals from the body. These pads are put on the foot overnight and turn brown by morning due to body oils but do not extract toxins from the body. "Hexagonal water" or clusters of water that form hexagonal shapes are supposed to balance the body and keep a person healthy. However, there is no scientific evidence that these remedies are useful although foot pads and hexagonal water are usually harmless.

Other alternative remedies were advertised in the late twentieth and early twenty-first centuries. Crystal therapy became a popular folk cure. Proponents believe that "vibrations" from crystals correct "imbalances" in the body. They claim that clear white quartz crystal is a "master healer" and helps to stimulate the immune system and balances the entire body to maintain health. They argue that rubies help restore vitality and energy. To use crystals, the person carries them as jewelry or in a pocket, uses them during meditation, or places them near their bed to absorb the vibrations. Health bracelets made of copper or magnets or those that contain holograms or are ionized have also been advertised to improve health. Proponents claim they enhance the natural flow of energy around the body. No scientific evidence exists about the effectiveness of any of these remedies other than placebo effect.

Other folk treatments to maintain health include color. In the 1960s, a technique called "colorpuncture" arose. Proponents claim that certain colors bring the mind and body back into balance. In this technique, various colored lights are placed on acupressure points. However, research does not substantiate their effectiveness, and relief of symptoms may likely be due to placebo effect. Likewise, for color light therapy. Advocates claim that different colored light balances energy lacking in the person. In the late twentieth century, a few alternative medicine proponents claimed that trepanation (drilling holes in the skull) increases the blood flow in the brain and gives spiritual insight, but this is unsubstantiated by research. A few people have drilled holes in their skulls or traveled to developing countries to obtain this risky procedure. On the other hand, a few twenty-first-century alternative and folk cures have shown modest scientific support including eating daily 100 mg dark chocolate per day, listening to classical music, and meditation for general well-being.

Hernia

The symptoms of a hernia, also called rupture, include a bulge in the inner groin (inguinal), through an incision (incisional), the belly button (umbilical), upper stomach (hiatal), and a rarer femoral hernia more common in older women. In many cases, ruptures are pain free and do not cause problems. However, they can also result in sharp pain, burning sensation, constipation, nausea, vomiting, and even fever. Hernias are caused by the protrusion of an organ through a weak spot in the surrounding muscles. The bulge often occurs as a result of coughing and long-term heavy lifting or straining. Other risk factors include obesity, smoking, chronic obstructive pulmonary disease (COPD), pregnancy, and previous open abdominal surgery. Minimal invasive laparoscopic surgery, for conditions such as an appendectomy, rarely results in a hernia.

Hernias come in two forms: reducible and irreducible. In a reducible inguinal, incisional, or umbilical hernia, the intestine can be pushed back into the abdomen, whereas in an irreducible hernia, the intestine cannot be pushed back into the abdomen though manual pressure. In irreducible incarcerated hernias, the inside of the intestine can become obstructed or the gut can become twisted in what is termed a strangulated hernia. A strangulated hernia can be life-threatening as the blood supply is cut off to the intestine and surrounding tissue.

Inguinal hernias are the most common type and are sometimes referred to as groin hernias. They arise in the lower abdomen and the upper thigh area primarily in males since males have a larger inguinal canal that is present from the testicles descending from their abdomens prior to birth. When the person lies down, the bulge generally can be pushed back into the abdominal cavity. Globally between 10 and 15 percent of males and 2 percent of females will develop an inguinal hernia at some point. They are most common in teenagers and people over 50 years of age. About 25–30 percent of both males and females will develop an incisional hernia after open abdominal surgery with a wound infection. A hiatal hernia is a condition in which the upper part of the stomach bulges through an opening in the diaphragm and can cause heartburn, pain, and burning under the breastbone (epigastric) area. About 50 percent of people over 60 may develop this type of hernia.

The first-known use of the term "hernia" in English was in the fourteenth century. It is from Middle English and Latin from the Greek word *hernios*, meaning "bud" or "sprout." Hernias have been described since Egyptian antiquity. For example, the mummy of Ramses the fifth (c. 1151 BCE) had a huge hernia sac in the groin and the mummy of the pharaoh Merneptah (c. 1224 BCE) had an incision over the inguinal region with one testicle removed, which was a common practice in hernia surgery. The Egyptian *Papyrus Ebers* contains a description of a hernia as a swelling that comes out during coughing. Greco-Roman physician and encyclopedists, Aulus Cornelius Celsus (c. 25 BCE–c. 50 CE), describes eight types of "hernias." Some of his descriptions are still used today.

Causes of hernia have been mentioned by healers since antiquity. In ancient Egypt, healers believed that a collection of one of the humors, phlegm, plugged or caused a twist in the bowel. Byzantine Greek physician Paulus Aegineta (625–c. 690) claimed hernias were caused by previous violence, such as a blow, a leap, or loud crying in children. Scottish physician William Buchan (1729–1805) believed ruptures were caused by exertions of strength, leaping, lifting, and carrying heavy weight, in addition to being overweight and eating "a fatty and moist diet."

Hernias from antiquity have been treated with binding the affected area with bandages or later a truss—a supportive undergarment for men—which helps to keep the gut inside the abdomen and is still used today in mild cases. Surgery attempted from antiquity through the mid-nineteenth century was not always successful due to infection, which led to many deaths. With anesthesia and sterile conditions, rapture repair became safer and is the main treatment today. Open surgery in which an incision is made over the location of the hernia by the late twentieth century became rarer as minimal invasive or laparoscopic surgery with several small incisions became common. Mesh to support the abdomen wall was also used by the late twentieth century in hernia repair. Although surgery is generally successful today, the death rate for emergency hernia repair is about 4.5 percent primarily among elderly individuals. Today uncorrected inguinal and other abdominal wall hernias are disqualifying conditions for military services.

The topic of hernias has been found in war novels. American writer and surgeon Richard Hooker's (Hiester Richard Hornberger Jr., 1924–1997) novel *MASH* (1968) was later made into a TV comedy war series (1972–1983). Throughout the TV series, there are a lot of jokes about hernias. In addition, during one episode, the temporary commanding officer, Frank, develops a hernia and needs surgery. Hernias have also been mentioned in science fiction such as *Valis* (1981), a novel by American writer Philip K. Dick (1928–1982). During a religious mystic experience, the protagonist has a vision that his son is sick, but a routine exam shows no illness. After more tests, on the instance of the father, the physicians find the child has an inguinal hernia, which would have killed him without immediate surgery.

UNUSUAL TREATMENTS

Many of the surgical techniques developed in antiquity were still used into the eighteenth century. However, in addition to inguinal hernia surgery, the *Papyrus Ebers* of ancient Egypt gave various recipes to treat hernias. Healers administered purgatives to clear the bowels. One recipe contained goose grease, seeds of the colocynth (bitter cucumber), honey, and dates that were ground together and ingested, which had no effect. They mixed lye or soap obtained from a "washerman," red natron (salt to preserve mummies), honey, juice of the acacia tree, and oil and placed it on the protrusion for four days. This did not work but could have caused skin irritation.

Hippocrates (c. 460–370 BCE) recommended enema therapy to produce numerous defecations in order to empty the bowel as a method to heal a hernia, which may have helped to make it easier to push the protrusion back into the abdomen. Injection of air into the rectum was also one of the clyster therapies, which had no effect. Paulus in the seventh century, for older patients who had pain, vomiting, and retention of feces, recommended bleeding from the arm, a tepid bath, warm poultice over the bulge, and a spare diet but gave no purgatives. This likely did not help the patient, and bleeding could have caused weakness.

Various herbal remedies were also recommended for a rupture, which were likely ineffective. For example, in the fourteenth century noted English surgeon John Arderne (1307–c. 1392) crushed together the plants comfrey, bugle, daisies, and the polypody fern in a mortar with oak bark and rubia major (an anti-inflammatory herb). He then added this mixture to red wine and let it stand for four days before he gave it to the patient to drink. The lees from this concoction were also placed on the rupture with a bandage for nine days. English physician Nicholas Culpepper (1616–1654) prescribed the plant colewort (herb bonnet) to drink for ruptures and pain so as to expel "crude and raw humors" from the belly, which was not effective in reducing the rupture.

In the sixteenth century, some physicians advocated liquid tobacco enemas—considered a cure-all at the time—for treating hernias. However, others such as Buchan in the eighteenth century first bled the patient and then gave an enema of mallows and chamomile flowers, butter, and fat along with warm water poultice made of the same materials. If this did not work, he suggested a "clysters of the smoke of tobacco must be tried," which had no effect on the rupture. After the gut had been pushed back into the abdomen, Buchan recommended a "steel-bandage" to keep it in, which was very uncomfortable to wear. He also suggested that patients should avoid foods that produced "wind" and strong liquors. This may have temporarily relieved some pain

In the nineteenth century, American physician Henry Hartshorne (1823–1897), for a suspected strangulated hernia, administered a tobacco or an ice water enema and placed ice on the abdomen. A needle was also inserted into the intestine to relieve gas, which were ineffective in relieving the strangulation. He also suggested that to save the patient an operation should be tried. This would have saved the patient, but before antiseptics, many patients likely died of infection. Hartshorne also proposed that an electric current be passed through the abdomen to restore peristalsis, which also was not effective in relieving the strangulation.

Some alternative treatments were attempted for hernias. For example, Indian American physician Dinshah P. Ghadiali (1873–1966) developed the "Spectro-Chrome Therapy" (method of healing using different color light waves), which was a popular alternative treatment in the mid-twentieth century. He claimed bathing in a yellow and then lemon light would cure reducible hernias. None of these methods were successful.

Hysteria and Mass Hysteria

A person with hysteria exhibits emotional behavior that is excessive or out of control. In addition, victims of a mass hysteria exhibit shared symptoms.

Hysteria

Symptoms of hysteria include shortness of breath, anxiety, insomnia, nervousness, fainting, amnesia, paralysis, convulsive fits, uncontrollable crying, and bizarre uncontrollable twitching or movement. Over the centuries, hysteria has been viewed as a medical disorder, demonic possession, and psychological problem and in Western culture primarily as a female affliction caused by a "wandering uterus." Any woman's behavior that deviated from the norm was blamed on this abnormal womb. However, in the 1800s men were also diagnosed as having the condition. The medical diagnosis of hysteria in Western culture was replaced by other terms such as conversion or functional or dissociative disorders briefly discussed in the "Mental Disorders ('Neurotic'): Anxiety and Functional Disorders" entry. Hysteria was a diagnosable medical condition until 1980 but was removed from the *Diagnostic and Statistical Manual of Mental Disorders* (*DSM*) in that year. Various aspects of the condition can cause personal, family, and social issues for the person exhibiting the behaviors.

The term "hysteria" was not coined until the early nineteenth century and is derived from the Greek *hysterikos*, from *hystera*, meaning "womb." One of the earliest references to symptoms of hysteria is found in the Egyptian *Papyrus Ebers* (c. 1500 BCE). Greek physician Hippocrates (c. 460–370 BCE) believed that the wandering womb that caused hysteria and other female problems was caused by an imbalance of humors due to the lack of sex. Greco-Roman physician Galen (129–216 CE), whose work was largely drawn from Hippocrates, also acknowledged the wandering uterus theory, which was accepted by some clinicians into the eighteenth century.

From the fifth through the thirteenth century, the increasing influence of Western European Christianity changed the perception of hysteria. It was now seen as the result of sin, satanic possession, and curses. However, in the seventeenth and eighteenth centuries, it again began to be thought of as a medical condition. English physician, Thomas Sydenham (1624–1689), for example, thought hysteria could have somatic or psychological causes and that the brain and not the uterus was the primary cause of the condition. The diagnosis of female hysteria reached its cultural zenith in the Victorian era, when 75 percent of all women were diagnosed with the condition for a wide array of symptoms. During the nineteenth and twentieth centuries, studies began to demonstrate that hysteria was not exclusively a female disease and was likely of psychological origins. Austrian neurologist/psychiatrist Sigmund Freud (1856–1939), for example, believed that memories too unpleasant were repressed and converted into physical symptoms. Psychotherapy and, later in the 1950s, tranquilizer drugs were used to treat symptoms of hysteria, and the condition has steadily declined in the West.

Mass Hysteria

When hysteria affects groups of people, it is called mass hysteria, mass psychogenic illness, or epidemic hysteria. The phenomena usually occur under stressful conditions when people believe a curse, bite, pathogen, toxic odor, chemical, or another environmental factor has caused an illness when no cause can be found. Symptoms among members of a group are triggered by an index or first case. The behavior is then exhibited by others upon seeing the index person displaying the behavior or by rumor about the behavior. Mass hysteria incidences have been recorded throughout history.

German physician and medical writer Justus Heckert (1795–1850), for example, describes several mass hysteria events including St. Vitus's (also called St. John's) dance and tarantism in his *The Epidemics of the Middle Ages* (1835). In Saint Vitus's dance, groups of men, women, and children twitched, jumped wildly, and danced until exhausted. It was attributed to a curse from St. Vitus, St. John, or possession by a demon because of sinfulness. People with this dancing mania often ended their dance processions at places dedicated to St. Vitus. This social phenomenon occurred primarily in mainland Europe between the fourteenth and seventeenth centuries. Also, sporadically from the thirteenth to the seventeenth century in southern Italy, the bite of a "tarantula" or even a scorpion was popularly believed to be highly venomous and to cause a dancing mania or tarantellism. People, generally women from the lower classes, would fall into a stupor if they believed they had been bitten and were revived by music.

Other types of mass hysterical epidemics have been common that take on the form of mass paranoia or "conspiracy theories." They transpire when a group of people believe that a threat is occurring, whether real or imaginary, and is often based upon rumors. For example, in Salem, Massachusetts in 1692, young girls exhibited hysterical "convulsions," which they blamed on witches' spells. Community members succumbed to the belief that local witches were placing spells on the girls. This in turn led to innocent people being put on trial and executed as witches. In the 1980s–1990s, a mass hysteria theory suggested satanic cults, often at day care centers, were ritually abusing or sacrificing children in the United States. The Federal Bureau of Investigation (FBI) found no evidence of this.

Mass hysteria is most common among groups of young females. In recent years, girls and young women in schools, colleges, and even a hospital have had headaches, felt light-headed, fainted, been nauseous, or even had convulsions from "toxic fumes," "poisonous food," or "mysterious viruses." Female garment workers became ill from being bitten by a "June bug." No evidence of toxins or bites in these incidences was found. In modern mass hysteria events, the probable cause is investigated, the victims are medically examined, and if nothing is found, they are reassured that they are well, and it is diagnosed as an "epidemic hysteria event," which sometimes brings hostile reactions from victims, parents, and institutions.

Examples of hysteria and mass hysteria have been found in the arts and literature. For example, Flemish painter Pieter Brueghel the Younger (1564–1638) in *Dance at Molenbeek* (date unknown) shows both men and women peasants exhibiting dancing mania along with a musician playing for them. The film *The Falling* (2014) takes place in a strict English girls' school. When one of the two friends dies, a mass hysteria fainting epidemic breaks out causing problems at the school and also uncovers mentally unhealthy family dynamics. The British romantic comedy *Hysteria* (2011) depicts the Victorian Age Dr. Granville whose hand cramps from massaging his female patients' clitoris to produce a "paroxysmal hysteria" to treat their hysteria. So, with help from a friend, he invents the vibrator to bring his patients more quickly to orgasm.

UNUSUAL TREATMENTS

Folk treatments for hysteria in women into the twentieth century included throwing cold water on their face, slapping their face, or putting them in an ice-cold bath. Smelling salts were also used based upon the ancient medical theory of wandering wombs to push it back where it belongs.

Hysteria

In ancient Egypt, to cure hysteria, the *Papyrus Ebers* recommended that if the uterus had moved upward, it could be brought back to its proper place by placing malodorous and acrid substances near the woman's mouth and nostrils, while sweet-smelling ones were placed near her vagina to attract the uterus. Conversely, if the uterus had lowered, the papyrus recommended placing foul-smelling substances near her vagina and sweet ones near her mouth and nostrils to raise it up.

Hippocrates maintained that the uterus would stop wandering around if a woman had satisfactory sex. Therefore, he urged his hysterical patients to engage in sexual activity. On the other hand, Galen gave patients purgatives and extracts of hellebore, mint, laudanum, belladonna, and other herbs to repress sexual urges. He also recommended that they get married. However, hellebore and belladonna are poisons and could kill the patient. Laudanum (an opiate) could take pain and anxiety away.

In the eighteenth century, German physician Franz Anton Mesmer (1734–1815) suggested that the body contained an invisible natural force called "animal magnetism." He claimed that the magnetic action of his hands on the diseased part of the body could treat the patient by interacting with this force and touted this as a cure for hysteria in both individuals and groups. Sometimes hysterical symptoms improved with "mesmerism," later known as hypnosis, again due to the placebo effect.

In the 1870s and 1880s, surgery was utilized to cure hysteria. Many hysterical women were forced to have healthy clitoris, uteruses, and ovaries removed. However, none of these surgeries cured the hysteria. Due to lack of sterile environments, many women sickened or died, and some women were also put into "insane asylums." In the late nineteenth century, American physician Silas Weir Mitchell (1829–1914) treated hysteria by confining women to strict bed rest. He required them to lie in the same position without any activity for several weeks or even months. Nurses massaged their muscles to prevent atrophy, and patients were forced fed a high-fat diet to fatten them up. Mitchel believed fat women had fewer mental issues. He also used electricity on the muscles. This regime was probably not successful in most cases and often led to serious complications such as blood clots in the limbs and severe calcium loss.

In the eighteenth and nineteenth centuries, to treat hysteria unrelieved by sexual activity, clinicians advised horseback riding, which provided enough clitoral stimulation to trigger "hysterical paroxysm." Midwives may have applied vegetable oil to women's genitals and then massaged the clitoris until the woman experienced orgasm. This was not considered sexual as the male medical community and others questioned the extent of women's sexual desires. Controversy exists as to whether male physicians also used this technique to cure hysteria. This treatment method, however, was marketed by quack operators such as "Dr. Swift" who advertised "health through magic of fine gentle massage." This advertisement shows a picture of a male physician with his hands under the skirt of a patient. British physician Joseph Mortimer Granville (1833–1900) invented the vibrator in 1869 for muscle cramps. Other physicians may have used it to create orgasm among their patients, but there is little evidence for this, and Granville denied using it for this purpose. However, vibrators did start being mass marketed in the 1950s for sexual stimulation.

In the late twentieth century, alternative practitioners suggested that amethyst was useful in the treatment of anxiety, stress, and hysteria as it calmed and relaxed the nerves if placed on the body or held in the hands. It may have been effective due to placebo effect.

Mass Hysteria

Regarding mass hysteria, Melampus, a legendary Greek healer, introduced the cult of Dionysus, the god of wine and vegetation, into ancient Greek culture. This was transformed into Bacchus in Rome. During festivals, drinking worshipers exhibited frenzied dance in a mass hysterical state. It was believed that women could be released from their anxieties if they had sex with strong young men during the rites.

Individuals participating in St. Vitus's or St. John's dance, in the Middle Ages, could be cured after dancing to exhaustion and worshiping at a church of these patron saints. For tarantellism, after falling into a stupor, various musicians would surround the supposedly bitten woman. When the woman heard their favorite tune, they jumped up, often ripped off their clothing, and wildly danced to sweat out the poison until they fell to the ground exhausted, and they were "cured" of their condition. These wild dances were later transcribed as Italian folk music tarantella dances. Some victims also found that rocking in a swing or cradle or having someone hit their feet would cure the condition.

I

Infant Issues, Illnesses, and Death

Acute illnesses have been a leading cause of infant deaths for centuries. In addition to "birth defects," discussed in a separate entry, the major causes of infant deaths until the late nineteenth century were respiratory and gastrointestinal issues, poor nutrition, convulsions, and sudden death. This loss of life during a baby's first year was so great in premodern times that the child was not named in some cultures until it had survived for at least a week or even a year. Until the mid-nineteenth century, approximately a fourth of all newborns died within the first year of life. By 2020, this was around 3.0 percent globally. In developed nations, since 1915 the infant mortality rate has decreased more than 90 percent due to better sanitation, hygiene, pasteurized milk, immunizations, prevention of preterm and birth defects, and better prenatal care and medical management. Now the major causes of death under a year are birth defects, preterm and low birth weight, pregnancy complications, and sudden infant death syndrome (SIDS).

Respiratory Illnesses

Throughout the late nineteenth century, whooping cough (discussed in the "Childhood Diseases" entry), bronchitis, and pneumonia discussed in the "Pneumonia and Lung Conditions" entry, along with croup, often killed infants and young children. Croup results from swelling inside the trachea, which causes difficulty in breathing, a "barking" cough, stridor, and loss of appetite. In the past, parents feared this condition as it could be fatal. Croup usually results from a virus infection, and today about 3 percent of children get the illness. In the past, croup and other respiratory infections were thought to develop from cold draughts, damp winds, or cold weather. The first use of the term "croup" is from mid-eighteenth century and is of Germanic origin, from *croup*, meaning "to croak." In London, out of a total of 9,535 deaths during 1632, the demographer John Graunt (1620–1674) reported 98 children had died of croup. Today it is usually a mild disease, and management of severe croup includes steroid or epinephrine to reduce airway inflammation and strictures, and it is rarely fatal.

Gastrointestinal Issues

Another potentially lethal condition that concerned parents in the past was diarrhea, sometimes called *cholera infantum*. Although death from diarrheal diseases today is rare in developed nations, in developing regions, they are the second leading cause of infant mortality particularly during weaning (6–11 months) due to lack of hygiene and contaminated food. Constant diarrhea and vomiting result in dehydration and loss of electrolytes, leading to death. The illness is caused by various microorganisms discussed in the "Gastroenteritis, Dysentery, and Diarrhea" entry. Deadly "summer diarrhea" during the hottest four months of the year (June–September) has been mentioned for centuries. In the early twentieth century, clean water supplies did not reduce the mortality from this condition, leading health officials to affirm that transmission was usually through contaminated foods or person-to-person contact. The term "diarrhea" was first used in the fourteenth century and is derived from Middle English via Latin: *dia*, meaning "through," and *rheo*, meaning "flow."

Teething and Convulsions

In ancient Greece, physicians such as Hippocrates (c. 460–370 BCE) observed that emerging teeth could kill since they caused inflammation and ulcers in the gums and fevers and led to deadly convulsions. By the nineteenth century, physicians such as American Elijah B. Hammack (1826–1888) believed infant convulsions or "fits" were due to overeating, fright, irritation, and an undeveloped brain or nervous system. Today most convulsions in infants occur when an infection results in a high fever, and from 2 to 5 percent of children will experience a febrile convulsion. The first use of the term "convulsion" in English was in the mid-sixteenth century from the Latin *convulsio*, meaning "cramp" or "spasm." To reduce fevers, clinicians now recommend cool sponging baths, light clothing, plenty of fluids, and in some cases antibiotics to treat bacterial infections.

SIDS

A most disturbing and wrenching death is the sudden death of a healthy infant. This occurs when a baby is sleeping in bed with a parent, or in its own crib or cot, and is found dead for no apparent reason. The cause of this syndrome is unknown but may be associated with a defect in the infant's brain that controls breathing and sleep arousal. For centuries, infanticide has also been suspected. In the past, folk beliefs attributed these sudden deaths to demons, breath-sucking cats, demigods, witches, etc. From the late nineteenth into the mid-twentieth century, an enlarged thymus gland, an undeveloped brain or nervous system, or cow's milk formula, which was becoming increasingly popular with mothers rather than nursing, were also thought to cause SIDS. By the mid-twentieth century, infants in most Western cultures slept in a separate bed. If an infant suddenly died, researchers suggested it was smothered from too much clothing or blankets or it was lying on its stomach. If an infant died while sleeping with its mother in the same bed, it was believed that the parent accidentally rolled over and smothered the child.

On the other hand, since the beginning of humankind, mothers have been sleeping with their babies. In most of the world today—except North America, northern Europe, Australia, and New Zealand—mothers and babies routinely co-sleep together until the infant is weaned. Moreover, in areas where co-sleeping is the norm, such as

China, Japan, and Hong Kong, SIDS rates are the lowest in the world. Globally in the late 2010s, about 35 deaths from SIDS per 100,000 live births occurred, and it was more common among the poor. The term "SIDS" was first used in 1969 to replace the older "overlaying," "crib," or "cot" death. By the mid-1990s, to reduce these deaths, many Western health professionals instituted a "Back to Sleep," now called "Safe to Sleep," campaign, which included placing an infant on its back, on a separate surface with minimal clothing, and without a blanket or pillow. Since the 1990s, the incidence of SIDS has declined, but smoking has also declined and breastfeeding—as opposed to formula feeding—increased, which are probable factors in the decreased incidence of SIDS. Moreover, by the early twenty-first century many Western breastfeeding mothers were co-sleeping in the same bed with their infants but kept this hidden from their health-care provider for fear of being reprimanded.

Infections

Infants often acquire infections from their mothers. For example, during the birth process chlamydia and gonorrhea may infect the newborns' eyes and cause blindness without treatment (see the "Sexually Transmitted Infections" entry). A genital herpes infection transmitted during birth is potentially life-threatening as it can lead to brain damage, blindness, and other organ damage. Syphilis can be transmitted in utero to the fetus and can result in congenital syphilis that may lead to face disfigurement, along with tooth and brain injury. Infant syphilis was first seen in the 1490s, and the mothers' or wet nurses' milk was blamed for the condition. Until the nineteenth century, until weaning, most diseases were attributed to a nurse's defective milk, which was thought to upset the infant's humoral balance. The human immunodeficiency virus (HIV) that causes acquired immune deficiency disease (AIDS) emerged in the early 1980s. It can be transmitted to infants in utero, through childbirth or breastfeeding, and usually results in death without treatment. Thrush, a yeast infection of the mouth, is common in infants but generally not life-threatening. Environmental conditions can also cause fatal conditions. For instance, if the cord is not cut in an aseptic manner, newborns can die from tetanus or other infections. Since the mid-twentieth century, physicians have managed many of these infections with antitoxin and antimicrobial medications.

In the arts, illness and death of an infant has been featured. In one painting, *The Legend of St. Stephen*, by Martino di Bartolomeo (active 1384–c. 1435), a demon takes away a baby and leaves a sick changeling. French artist Jean-Francois Millet's (1814–1875) painting *The Sick Child* (1858) shows a mother cradling her ill baby. In the British television series *Call the Midwife* (2012–early 2020s), a mysterious death of a newborn infant is depicted in one episode. The midwife is blamed for the death, but the postmortem shows the infant died of lung collapse (atelectasis) and not infanticide or negligence.

UNUSUAL TREATMENTS

Various methods to keep babies healthy and alive have been practiced for centuries. For instance, in antiquity physicians believed that a newborn body was like wax, and to keep the body straight, to prevent distortions of the limbs, and to foster mobility, the child should be swaddled, which had the opposite effect. The Romans also believed that unless the newborn cried vigorously and was in perfect condition that it should be exposed to the elements. The *pater familias*, usually the father, would place the weak infant in a public place, where anyone could take the child to adopt or train as a slave. If the infant was too weak and unlikely to survive, it was exposed in an isolated area to die. However, the wealthy were less likely to expose their infants.

Respiratory Illnesses

In treatment for croup, the Egyptian *Papyrus Ebers* (c. 1500 BCE) recommended heating poisonous black henbane weed, which has antispasmodic properties, on a hot brick for the child to inhale. This may have helped but could be fatal. Inhaling steam has been used for century for respiratory issues, but evidence is mixed as to its effectiveness. In 1811, French physician François-Joseph Double (1776–1842) wrote a book on croup and recommended the infant inhale a mixture of ether and opium through an inhaler, which may have helped reduce the spasms. He also suggested giving the child a mustard footbath to draw out poisonous body humors, which had no effect on respiratory issues.

By the mid-nineteenth century, American physicians suggested several remedies to draw out bad humors and rebalance the body as a treatment for croup. The first step was an emetic such as ipecac and alum to induce vomiting. Then the caretaker applied a warm poultice of roasted onions and cornmeal on the child's breast and neck to help in breathing. Physicians also applied blisters with cantharis beetle on the chest to draw out poisons along with footbaths. In more serious cases, they resorted to bloodletting by leeches and gave the child calomel (contained toxic mercury) as a mild laxative. These remedies were not effective, and the infant often became sicker or died of weakness or mercury poisoning.

Gastrointestinal Issues

Diarrhea has been treated since antiquity. The *Papyrus Eber*, for example, to balance the humors, recommends feeding the infant henna, sandalwood, bayberry, and acacia gum. To expel bloody diarrhea, which was likely dysentery, healers administered strained porridge, sweet gum from the manna tree, oil, and honey. In some cases, these recipes might have been effective as they added electrolytes and sugar and prevented dehydration. In the eighteenth century, Scottish physician William Buchan (1729–1805) gave infants a dose of rhubarb, which can cause diarrhea, along with substances to induce a "gentle vomit" before he administered astringent medicines to bind the diarrhea such as alum. This was likely ineffectual and increased deaths from dehydration.

By the mid-nineteenth century, physicians such as American Henry Hartshorne (1823–1897) gave infants moderate doses of calomel for summer diarrhea, which would have instead caused more diarrhea and dehydration. He also placed a spice plaster over the abdomen to "draw out the poisons" and rebalance the humors.

Opium was also given to stop the diarrhea, which would have helped. Physicians prescribed blue mass (a mercury compound) or hydrargyrum cum creta (mercury with chalk), both of which could have led to mercury poisoning, but magnesia with charcoal or chalk might have stopped the diarrhea. Folk medicine remedies recommend onion juice and aromatic syrup of rhubarb (which is a purgative) for diarrhea, which could have worsened the condition. They also administered ginger or cinnamon for nausea, which are sometimes effective and generally harmless to the patient.

Teething and Convulsions

Since it was believed that teething caused convulsions, various remedies for teething pain were tried. In Egypt, healers placed belladonna, camphor, and opium poppy on the gums, which would relieve the pain but could have been toxic to the infant. They also gave the infant a fried mouse to chew on to help with the pain. Nursing was also believed to prevent convulsions in the child as the disorder often began with teething and weaning. By the eighteenth century, some physicians believed convulsions were the result of tight swaddling and ordered the wet nurse to take off the bandages. They claimed this cured "convulsion fits." For teething, they also gave the infant a piece of hard metal or coral to chew on. Due to theories that the humors became out of balance during teething, some clinicians cut the gums to help the teeth erupt, release toxic materials, and prevent convulsions, but this could have caused infections. A folk method in the twenty-first century was to hang an amber teething necklace around an infant's neck. It was believed that the body heated the amber, causing it to release painkilling oils that are absorbed into the infant's circulatory system. No evidence for this has been found.

In the nineteenth century, since "teething convulsions" were thought to be a cause of infant deaths, physicians gave sick babies purgatives such as castor oil or emetics to induce vomiting to balance the humors. If the infant was constipated, physicians, in addition to purgatives, recommended emetics such as ipecac to cause vomiting. These were not effective as convulsions were likely caused by fever. By the 1850s, to help the infant sleep they gave opiates, wine, and even chloroform, which would have worked but could have depressed breathing. By the twentieth century, mothers gave teething powders that contained mercury to infants, which caused "pink disease" (acrodynia) or mercury poisoning in which the infant's hands and feet turned pink. This could cause brain damage and was removed from the market in the 1950s.

SIDS

Many methods to prevent SIDS, or punish the parents for their infant's sudden death, have occurred over the centuries. In Mesopotamia, for example, people recited incantations to an evil spirit, Lamashtu, or Larbatu, who was believed to steal away infants and caused sudden deaths. She was given a dog to care for and told to leave the child to prevent SIDS. Greek historian Diodurus Siculus, in the first century BCE, records that in Egypt mothers were sentenced to hug their dead infant continually for three days so she would experience this horror and prevent it from happening again. By the tenth century, if the Catholic Church believed that a parent or wet nurse intentionally suffocated an infant, they excommunicated or even executed her. By the nineteenth century, physicians, rather than punishing the parents for SIDS, attempted to prevent it by performing surgery to reduce an enlarged thymus in healthy babies as this was thought to be a cause for the syndrome, but it often resulted in death. After the discovery of x-rays in 1895, physicians used this invention for both diagnosis and treatment to reduce the size of the thymus into the mid-twentieth century. However, many deaths often occurred later in life from cancer due to the radiation.

Infections

In ancient Greece, to assure a healthy infant and keep away evil and diseases, the first food given to a newborn was a mixture of honey and boiled water, and after this first meal, the newborn received maternal or a wet nurse's milk whose antibodies likely helped prevent infections. Healers also believed that breast milk helped treat infections until weaning. In antiquity, tightly swaddling the newborn was also considered important for good health, but it did not prevent infections. If the infant acquired an infection, some ancient Greek healers after treatment would then take the child to a "sleep temple." At this temple, they believed that the mythical daughters of the demigod physician, Asclepius, would arrive with two holy snakes that would cure the illness. For various infections, particularly ulcerations of the mouth from thrush, healers applied honey to the lesion, which has antibiotic properties and may have been effective. By the nineteenth century, to prevent infections, physicians sometimes drew out "bad blood" by leeches or cupping, which had no effect. From the late nineteenth century into the late-twentieth century, clinicians treated thrush and other skin infections with gentian violet, a dye with antifungal properties. However, it has been found to be carcinogenic, but it is still used by some as a home remedy.

Inflammation and Pain

Pain often arises out of inflammation and is the body's signal that something is wrong. Inflammation is the tissue's immunologic response to injury, or an illness, to help the body fight off infections and speed up the healing process. It is characterized by mobilization of white blood cells, antibodies, and swelling. Often the inflamed area feels warm to touch. Besides inflammation, physical pain may be caused by trauma and acute and chronic illness. Pain is perceived as throbbing, searing, burning, stabbing, or aching depending upon the issue. It can lead to insomnia, social withdrawal, and mental depression.

The individual feels pain, such as when stubbing a toe, because pain receptors are triggered by an abnormal condition associated with actual or potential tissue damage. Pain can be temporary such as while touching a hot stove, a tension headache, strained muscle, or surgery, which generally dissipates over time. It can be chronic such as in cancer, phantom limb pain from amputations, arthritic back and joint issues, traumatic injuries, fibromyalgia, and gastroenteric issues such as colitis. In addition, chronic pain often has a psychological aspect. About 20 percent of the adult population has chronic pain, which is pain most days or every day for at least six months. Chronic pain is most common in people over the age of 45, and 70 percent of them are women. Both acute and chronic pain can interfere with a person's work, family, relationships, physical activity, and social life.

The term "inflammation" is from late Middle English, from the Latin *inflammatio*, from the verb *inflammare*, meaning "inflame." The term "pain" is from late-thirteenth-century Middle English *peine*, meaning "punishment" or "torture," from the Old French *peine*, from the Latin *poena*, meaning "penalty." Pain is also found in other vertebrates such as mammals, birds, and some reptiles. The ancient Egyptians thought that pain—other than injuries—was caused by evil spirits that entered the body through the nose or ears. They believed that certain materials or substances moved in a system inside the body that could become stuck to different sites and cause inflammation and pain and that these substances could also cause mucus or pus. Greco-Roman physician Galen (129–c. 210), however, based upon Hippocrates (c. 460–370 BCE), taught that pain could be used for diagnosis and identification of underlying disease and that it was a warning signal of important changes in the body, especially the imbalance of the four body humors. In Christianity, pain by the fifth century was often considered a punishment from God due to sins or a "divine trial" that the person needed to endure.

A major treatment for pain since early times has been opium from the poppy *Papaver somniferum*. Although opium was mixed with wine for pain and surgery in antiquity, Swiss physician Paracelsus (1493–1541) mixed it with distilled alcohol to develop laudanum, a tincture of opium. The drug was occasionally used into the early twenty-first century for diarrhea, but rarely anymore for pain. From antiquity, however, healers realized that opium was addictive. English chemist Joseph Priestley (1733–1804) isolated nitrous oxide gas around 1775, but it was not used for pain relief until 1800 when British chemist Humphrey Davy (1778–1829) discovered that inhaling it stopped pain, particularly for surgery. Willow bark, which contains salicylic acids, has also been used for centuries for pain relief. Its active ingredient was synthesized in the late 1890s in the form of acetylsalicylic acid (aspirin) at the German pharmaceutical company, Bayer. From this time into the twenty-first century, to manage pain many other synthetic analgesics have been developed. However, all the opiate-like medications sometimes led to drug abuse discussed in the "Substance Use Disorders" entry. Today, for pain management, a wide assortment of techniques are used from drugs to meditation.

The subject of pain has been found in the arts. For example, in the epic poem the *Odyssey*, ancient Greek author, or group of authors, Homer (c. late eighth or early seventh BCE), describes Helen of Troy putting a drug in wine and serving it to the son of Ulysses and his companions. This concoction took away pain, sorrow, and memory. The fairy tale "The Little Mermaid" by Danish author Hans Christian Andersen (1805–1875) is about physical and psychological pain. The mermaid gives her beautiful voice to a witch in return for her tail being changed into legs so she can become human and gain a soul. However, she constantly feels sharp pain when she walks or dances. In the WWI painting *Gassed* (1919), by expatriate

American artist John Singer Sargent (1856–1925), a group of blinded soldiers who have been gassed with mustard gas are led off the battlefield while others are writhing in pain on the ground from gas and other wounds. It shows the terrible physical pain and suffering of war.

UNUSUAL TREATMENTS

From prehistoric times, people have attempted to eliminate pain. Some archaeological evidence shows bones being set to eliminate the pain from fractures, trepanation (drilling a hole in the head) for head and other pain, and scarification to allow bad fluids, spirits, and demons to escape. Healers also used various herbs, barks, plants, and oils for pain relief. In ancient Egypt, healers and priests were one and the same. They used magic spells, incantations, and amulet to drive out evil spirits that were thought to be causing inflammation and pain. When pus came to the surface of the body, they would put willow leaves and thyme on the wound and bandage it. These plants would help reduce inflammation, pain, and infection due to their chemical properties and were relatively harmless to the patient. The essential oil of thyme is still used in the twenty-first century as a folk treatment for infections and in mouthwashes. Egyptian healers administered emetics to cause vomiting and herbs to cause sneezing and urination to encourage disease causing demons to leave the body. Besides opium, in ancient Egypt, frankincense made from the resin of the Boswellia tree was used for inflammation. This spice, containing boswellic acid, has been found to reduce inflammation and pain in joints and the gut. Negative effects are nausea and acid reflux.

In ancient Greece, Hippocrates recommended a host of recipes for pains in various parts of the body, which are discussed in other entries concerning a particular disease or condition. In general, his common remedies for pain included purging with emetics, laxatives, and enemas, and bloodletting through cutting a vein to eliminate bad humors. Cauterizing open wounds was also common as a counterirritant to relieve inflammation and ultimately pain. As part of the treatment, physicians would send their patients to "sleep temples." In these temples, they believed the daughters of Asclepius—historical healer made into the god of medicine—would arrive with two holy snakes that would cure pain and illness.

Hippocrates and later Galen, in the first century, also used opiates, henbane, mandrake, and nightshade plants, which besides pain relief caused hallucinations and unconsciousness and could lead to death. Physicians in the Middle Ages used many of these same substance as anesthetics. Boar's gall and hemlock were sometimes added to these plants. Hemlock, a deadly poison, not only eliminated pain but also likely caused death in many cases.

Galen and other Roman physicians prescribed some foods that helped to relieve inflammation and mild pain including cherries and mint leaves. Broadleaf plantain was placed on a painful area and used as a local anesthetic. These were usually harmless to the patient, but cherries could cause diarrhea. Galen also formulated a "theriac" or "cure-all" that acted against pain. It was made up of different substances such as mandrake, opium, viper's flesh, cinnamon, saffron, rhubarb, pepper, and ginger. Many of these have antiseptic and analgesic properties. These compounds were mixed with wine and honey to a pulp and then a small amount ingested. This universal remedy with different substances was commonly used until the late eighteenth century. Sometimes they were toxic to the patient and could cause liver and other damage and death.

In the seventeenth century, some folk treatments for inflammation and pain came from the human body. The "spirit of skull," "Goddard's drops," or "kings drops" were the distillate of a mixture of dried snakes, pieces of skull from a recently hanged man, ammonia, and sometimes nitric acid boiled in a glass container. It did not counteract pain but was largely harmless. "Man's grease," or the fat from a boiled executed convict, was also used as a skin salve against inflammation and pain. It was also worthless other than placebo effect.

Counterirritation to remove pain was practiced for centuries and in some complementary medicine practices in the twenty-first century. Blistering was carried out by placing cantharis from the cantharis beetle (or "Spanish fly")—sometimes mixed with mustard—near the inflamed or painful area to cause a blister. It was used for a variety of illness. The blisters were sometimes sliced open to allow fluid to drain out "toxic humors" based upon humoral theory. The process did little to reduce pain and often resulted in infection. By the early eighteenth century, other treatment of counterirritation for pain included hard rubbing of the painful area and flagellation.

Electricity has been used from antiquity for pain. The electric catfish was used in Egypt for the treatment of inflammation of head and joint pain. The fish was placed on the painful area. The Roman physician Scribonius Largus (c. 1–c. 50) recommended using live torpedo fish for the treatment of pain. This electric fish treatment was used for pain till the early nineteenth century, when new techniques to produce and accumulate electricity were developed. By the late twentieth century, this evolved in the transcutaneous electrical nerve stimulation (TENS) unit. As a counterirritant, electricity helped ease pain. However, too much electricity could cause burns.

Many folk and alternative methods to reduce inflammation and pain have been used for centuries. Some in the twentieth century would have been considered unusual treatments but became more accepted by the early twenty-first century. For example, in the eighteenth century, German physician Franz Anton Mesmer (1734–1815) suggested that the body contained an invisible natural force called "animal magnetism," similar to Eastern religions' qi, or energy, flow. Mesmer claimed that the magnetic action of his hands on the diseased part of the body could treat the patient by interacting with this force and touted this as a cure for pain and many diseases. Sometimes people felt better after "mesmerism," later known as hypnosis due to placebo effect. It along with "mindfulness meditation" is used today and is sometimes effective in reducing pain.

Some practices from Asian medicine have become more acceptable in the twenty-first century in the West. These ancient alternative systems have been found on the Indian subcontinent, China, and Southeast Asia. For example, acupuncture is found in traditional Chinese medicine (TCM). In this technique, a needle is inserted at points along "meridian channels" in the body where qi flows for pain relief. Some practitioners use acupressure, or pressure on the meridians, to accomplish the same ends. Proponents claim it helps to eliminate pain. Similarly, dry cupping in which heated cups are placed on painful areas, or on the meridians, to cause a vacuum have been claimed to reduce pain. Another alternative technique is moxibustion that involves burning moxa—a cone of ground mugwort leaves—on or near the body's acupuncture points that proponents claim relieves pain. This procedure can sometimes result in burns. Little research has been done on these procedures as to their effectiveness in pain management; the pain relief that is reported may be due to placebo effect.

Several herbs, spices, and essential oils have been advertised to reduce pain and inflammation. The spice turmeric with the active ingredient, curcumin, has anti-inflammatory properties but can cause stomach issues. Musk oil obtained from the glands of musk deer—or made synthetically—is used in expensive perfumes and also has anti-inflammatory properties. It has been used on the skin to combat inflammation and pain but can cause allergic reactions. A "cure-all" for many disorders and pain that was popularized in the 2010s was cannabidiol (CBD) oil, a derivative of the hemp plant. However, little research has been done on these substances to determine their effectiveness. Overall, they usually do not cause harm to the patient.

Several folk remedies have been tried for pain. For example, one remedy suggests wrapping a painful part of the body with red flannel or eating black beans. This might work by placebo effect. Some claim that applying raw potato peel on the painful area will reduce pain or inflammation. The coolness of the potato might help to reduce pain. A cold treatment that advocates claim to decrease chronic pain is sitting in a cryogenic chamber. The patient remains for a few minutes in a chamber, which is extremely frigid. Proponents claim that the extreme cold shocks the system into releasing endorphins, which in turn removes the pain. However, this treatment can cause frostbite or heart issues, and research is mixed regarding its effectiveness.

By the twenty-first century, crystal therapy was advertised to treat various conditions. Proponents believe that "vibrations" from crystals correct "imbalances" in the body. The primary crystal for pain is the brilliant but rare purple stone sugilite. Advocates claim it promotes positive emotions and peace of mind, which helps to reduce chronic pain by dispelling negative energies that makes it worse. The person carries it as jewelry or on their person, uses it in meditation, or places it on the painful part while resting. No scientific evidence exists about pain relief from crystal therapy other than by placebo effect. Another therapy promoted in the early twenty-first century is "earthing" therapy to ease pain by being in direct physical contact with the ground such as walking barefoot in the grass, walking in the woods, or being in nature to gain "excess electrons" from the earth that supposedly help eliminate inflammation and pain. These activities may reduce stress, which in turn reduces pain.

Influenza

Symptoms of influenza, or its short version "the flu," are coughing and sneezing—which usually spreads the disease—sore throat, high fever, head and muscle aches, disorientation, profound prostration, and sometimes vomiting and diarrhea. These symptoms weaken the victim and can lead to fatal secondary infections, such as pneumonia. Moreover, it had been noted for centuries that the disease has a high mortality rate among the elderly and very young although all ages and social classes are affected. Worldwide influenza is still a major cause of morbidity and mortality, and combined with pneumonia, it is one of the 10 leading causes of death.

The flu is thought to have emerged among humans during the domestication of animals around 9,000 years ago. Like many infectious diseases, it was carried to new areas by traders, migrants, explorers, and the military. Diagnosing influenza from antiquity is difficult as its first signs are like many other respiratory diseases. Most scholars today agree that the first well-described outbreak of influenza in Europe was the epidemic of 1510. Since that time, around two or three global influenza pandemics have been recorded every century.

The disease is caused by several subtypes of the *orthomyxoviridae* virus, including influenza A (H1N1), (H3N2), and the less serious influenza B and C. Influenza A circulates among humans, birds, and swine. The human form of the virus was not isolated until the mid-1930s. From 3 to 5 million cases of severe influenza and up to 500,000 deaths are reported each year from seasonal influenza that is generally found during the winter in temperate climates and sporadically throughout the year in tropical regions. The virus rapidly mutates from year to year, and sometimes a new virulent form of the Influenza A virus emerges, leading to a lethal global pandemic.

The disease is known by several names in English. It was first called *influenza* in the mid-fourteenth century from the Italian for "influence," as it was thought to be caused by misaligned celestial objects. The illness was first clinically described by British physician Thomas Willis (1621–1675) in the mid-1650s and was called catarrh, from the French *catarrhe* (to flow down), and in the late eighteenth century, it was called *grippe*, from the French *gripper* (to seize). In addition to celestial influences, throughout the late 1800s, influenza was thought to be caused by miasmas (bad air) or atmospheric and seasonal conditions. By the 1900s, with the acceptance of the germ theory of disease, it was attributed to some sort of microorganism.

The worst outbreak in modern history of influenza was the 1918–1919 "Spanish flu" pandemic during WWI, which was especially lethal to young adults. The epidemic came in three waves, and although still debated, it likely originated from a Midwestern pig that acquired the infection from a bird or possibly from Chinese immigrant workers in the United States. A military recruit from a farm was infected, who in turn infected other recruits at a training camp with a mild flu-like illness in March 1918. Beginning in May 1918, thousands of soldiers who had been in contact with troops sick with this flu were deployed to the European western front, and the disease soon spread worldwide. The virus mutated, and a highly lethal second wave emerged in the late fall of 1918. The third wave in the winter of 1919 was milder, and the disease had subsided by the fall of 1919. An estimated one-third of the world's population was infected with the virus. At least 50 million died worldwide and 675,000 in the United States.

No cure has been developed for influenza, but in the mid-twentieth century, an inactivated virus vaccine is developed each year against the strains most likely to cause an infection during the following winter, but it is only partially effective. In 2010, U.S. public health officials recommended that individuals six months of age or older be vaccinated annually. In addition, some antiviral drugs have been developed for prevention and treatment of influenza with limited success.

This Spanish flu pandemic influenced the course of history, particularly in Western Europe. Like most pandemics, it led to social unrest and economic crises that included a shortage of medical personnel and closures of worksites, schools, and public gatherings, and most communities required citizens to wear masks when in public places. The outbreak killed more people than WWI—both civilians and military. During the war, all combatants kept the lethality of the flu secret, and it was rarely discussed in newspapers. Spain was neutral, so it was free to report the seriousness of the outbreak, thus leading to the disease being called the Spanish flu. Although the Allies (the United Kingdom, France, Italy, and the United States) had serious causalities from the flu, some have speculated that it may have weakened the German military to the extent that they were unable to effectively fight. Near the end of the war during negotiations at the Treaty of Versailles, American President Woodrow Wilson (1856–1924) suffered from serious post-flu confusion and weakness and was unable to press his point for a "just peace" as the Europeans wanted revenge. Likewise, Wilson was unable to convince Congress to ratify the treaty. This harsh treaty resulted in a severe economic depression in the former German Empire and set the stage for the rise of Adolf Hitler (1889–1945) and WWII.

UNUSUAL TREATMENTS

From antiquity through the mid-1800s, if various herbs did not help the patient, physicians often bled from a vein, cupped (hot cups placed on skin to draw up blood) and purged to rebalance the body "humors." They also recommended wearing a necklace of garlic to prevent the disease. Some people during the Spanish flu resorted to this ineffective method. If taken internally, however, garlic could help reduce phlegm. In sixteenth-century England, wine; nut oils; and herbs such as rue, marjoram, and rosemary were mixed with lard and rubbed on to the chest to relieve congestion, which may or may not have been effective.

In the nineteenth century, one medical book recommended a steam bath or mustard foot bath after which the patient was put to bed with a hot brick or jug of hot water at the feet although the reasons for most treatments were not given. Physicians suggested various concoctions such as Dover's powder (a mixture of ipecac to cause vomiting and opium); opium would have helped to quell the cough. Physicians also prescribed a few drops of Fowler's solution (contains the poison arsenic) mixed with sarsaparilla, peppermint, or camphor for labored breathing. For debilitated patients, a hot brandy toddy or chicken soup were recommended, which may have relieved congestion and hydrated the patient. Doctors even prescribed quinine, which is only effective with malarial-type fevers, and mixtures of various tree barks or iron shavings dissolved in alcohol to little effect, but they were largely harmless to the patient. They even administered vaccines such as tetanus antitoxin, which were ineffective.

Inasmuch as physicians' treatments did not cure or prevent the disease, numerous folk remedies emerged. For example, to prevent catching the Spanish flu, Chicago families would shut all the windows and doors and boil ripe red peppers to stave off the disease. Wearing red was also thought to prevent the disease as people believed the flu did not like the color. If someone in a household caught the flu, to prevent it from spreading, families would place sliced onions around the house to "absorb the virus." A hospital in Louisiana recommended people sleep under a quilt made of wormwood, sewn between layers of flannel, and dipped in hot vinegar. A "carbolic smoke ball"—a hollow rubber ball filled with carbolic acid and fitted with a tube for the nose—was advertised as a remedy for the flu. Drops of kerosene on a sugar cube were also tried, but these remedies were not effective.

Insomnia and Sleep Disorders

Insomnia and sleep disorders are common health problems that can lead to mental, physical, and social issues. Insomnia is the most common form of sleep dysfunction and affects around 30 percent of adults. In insomnia, a person may have problems in falling asleep, staying asleep, or both. Insomnia can be either acute or chronic. Acute insomnia lasts only a few days and often occurs during times of stress or anxiety. Almost everyone suffers from acute insomnia at some point, and about 75 percent of people with acute insomnia return to normal sleep patterns. Chronic insomnia is a long-term condition. It is defined as lasting at least three months and occurring at least three times a week. Two main types of chronic insomnia are found. Primary chronic insomnia generally has an unknown cause. Secondary chronic insomnia has underlying medical and other conditions. These include depression that may cause a chemical imbalance in the brain that affects sleep patterns, advanced age (almost 50 percent of people over 60 years old experience symptoms of insomnia), mood disorders, being female, drugs, alcohol, obesity, chronic pain, and other conditions. The first use in English of the term "insomnia" is from the early seventeenth century from the Latin *insomnie*, from *insomnia*, meaning "want of sleep." The ancient Greeks believed that sleep was restorative for disease and everyday living and helped balance the humors. Hypnos was their god of sleep and dreams. Healers often sent patients with a variety of problems, including insomnia and sleep disorders, to sleep temples for dream interpretation. To manage insomnia, people throughout history have used opiates and alcohol, but these can lead to addictive disorders.

Sleep-related breathing disorder is another sleep disorder classification. In this category, breathing difficulties are common during sleep. The most frequent is snoring; about 40 percent of women and 57 percent of men snore. Snoring occurs when air moves around the uvula (loose tissue near the back of the throat) and causes it to vibrate. It can be highly annoying to a bed partner and cause them sleep deprivation. A critical breathing disorder is sleep apnea. The most common is obstructive sleep apnea. It is characterized by loud snoring and brief pauses in breathing. Sleep apnea can affect the balance of oxygen and carbon dioxide in the blood and lead to low-quality sleep and health consequences. Obstructive sleep apnea is estimated to affect around 5 percent of adults, and it is more widespread in men. By the late twentieth century, a Continuous Positive Airway Pressure (CPAP) machine that keeps air pressure constant was used to manage the dysfunction. Dental devices were also used for obstructive sleep apnea.

Another common sleep disorder is restless legs syndrome (RLS), which is also discussed in the "Leg Problems" entry. From 7 to 10 percent of the American population has RLS. It is most common among older individuals and women. Narcolepsy is a sleep disorder characterized by excessive sleepiness, daytime sleepiness, sleep paralysis, hallucinations, and in some cases partial or total loss of muscle control called a "sleep attack." These episodes are often triggered by strong emotions such as laughter or crying. During these attacks, a person may fall asleep in dangerous situations such as driving a car, causing an accident, or may fall and hurt themselves. It is found in about 0.5 per 1,000 people. In this chronic disorder, the brain loses the ability to the control "sleep-wake" cycles, and symptoms usually first appear in childhood or adolescence. Medications can help control the attacks. The term "narcolepsy" was first used in the late nineteenth century. It is from French, from the Greek *narkē*, meaning "numbness," and *lepsis*, meaning "attack." In the past, it was sometimes confused with epilepsy discussed in another entry.

Circadian rhythm sleep-wake disorders include "non-24 sleep-wake disorder" that is found with individuals who are totally blind. They experience insomnia symptoms at night and excessive daytime sleepiness because their sleep schedule is not in sync with a normal 24-hour circadian rhythm. Shift workers and people with jet lag also experience this.

Parasomnia is a sleep disorder category that includes abnormal sleep-related behaviors that occur during sleep, transitions from wakefulness to sleep, or within transitions between sleep stages. These sleep disorders can also cause psychological, social, and physical issues. During a normal sleep cycle, an individual progresses through stages of "non-rapid eye movement sleep (NREM)" into "rapid eye movement sleep (REM)." During REM sleep, dreams are most likely to occur. The eyes move back and forth, and major body muscles become paralyzed to prevent people from moving or acting out their dreams. However, in some REM sleep disorders, such as REM sleep behavior disorder (RSBD), the person may act out their dreams, talk or yell in their sleep, or injure themselves or their bed partner. It is more common among people over 50 and among men. In opposition to this condition is sleep paralysis (SP). During a SP episode, the person is unable to move any part of the body and is often terrified. Hallucinations can also occur, and this disorder is thought to be the basis of "alien abductions." Nightmares are also part of this REM sleep dysfunction category. Nightmares are frightening dreams that usually awaken the sleeper from REM sleep.

NREM parasomnia disorders include sleepwalking (somnambulism), being confused on arousal, and night terrors. Individuals who sleepwalk can seriously injure themselves and are often not aware of their surroundings. Around 7 percent of people over their lifetime will experience this condition. During night terrors, people can scream in their sleep. They are nonresponsive to outside stimuli and have no recollection why they screamed upon waking. About 2 percent of people suffer from this disorder.

The term "parasomnia" is from the early seventeenth century and is from the Latin *para*, meaning "adjacent to," and *insomnis*, meaning "sleepless." Various medications such as antidepressants can help quell these activities. The third category of parasomnia disorders includes bedwetting, sleep-related eating disorders, hallucinations, and numerous other abnormal sleep activities.

Common remedies for insomnia and some sleep disorders today include folk remedies such as warm milk that contains calcium, a muscle relaxer, and the amino acid tryptophan. This amino acid fosters the production of melatonin and serotonin. Melatonin helps regulate the sleep-wake cycle, and serotonin is thought to help regulate appetite, sleep, mood, and pain. Various herbal teas have been common for centuries for sleep such as chamomile, lemon balm, and valerian root. Relaxation, breathing exercises, repetitive sounds, or images such as "counting sheep" are also common methods for countering wakefulness. Clinicians can also prescribe hypnotic medications, but there is a risk of drug dependence.

Sleep dysfunctions have been shown in the arts and literature. For example, playwright William Shakespeare (1564–1616) in several plays depicts insomnia, sleepwalking, nightmares, and sleep apnea with snoring. He uses sleep-related themes for character development and to advance the plot. In *Henry IV* (c. 1598), for instance, the obese and heavy-drinking Falstaff is described as sleeping behind a curtain and "snorting like a horse." He likely suffered from obstructive sleep apnea. In *Henry V* (c. 1599), the king contrasts his insomnia to the sound sleep of a servant, whom he perceives as having few problems. The irritable Othello, in *Othello* (c. 1603), who has insomnia, after being made to believe that his wife has been unfaithful, kills her in a fit of jealousy. Several sleep disorders are found in *Macbeth* (1606). Macbeth experiences insomnia and has terrifying nightmares after he has committed multiple murders, and Lady Macbeth both sleepwalks and sleep talks. SP has been illustrated in paintings. For example, Swiss painter Henry Fuseli (1741–1825) in *The Nightmare* (1781) shows a woman in a state of SP. Her torso and head are draped over the side of the bed. A devil-like creature sits on her abdomen while to the left in the shadows is a horse-like beast. Films have also depicted insomnia and sleep disorders.

UNUSUAL TREATMENTS

Insomnia and sleep disturbances have likely plagued humans since they first roamed the African savannas. To remedy this common scourge, various folk and other treatments have been attempted to manage the problem. Many of the solutions were based upon trial and error with little scientific testing and have evolved into folk medicine—some are still practiced today.

A method in ancient Egypt, during the hot months of the year, to induce sleeping was to soak bed coverings in cool water right before going to bed. The evaporation of the water would keep the person cool. A sleeping potion was made from the psychoactive milky fluid that seeps out of wild lettuce stems and is known as lettuce opium. It contains *lactucin*, a mild hypnotic. Until the early twentieth century, it was listed in both the U.S. and British pharmaceutical standards for use in cough drops and in syrups as a mild sedative for sleeping.

As mentioned, sleep temples were found in Egypt and ancient Greece. Physicians in Greece would send their patients to these sanctuaries, as they believed the daughters of Asclepius—historical healer made into the god of medicine—would arrive with two holy snakes that would cure sleep disorders, pain, and other illnesses.

In the Christian *Old Testament* and the Hebrew *Torah* (c. 1200 and 165 BCE) scriptures, the mandrake root is mentioned as a sleep aid and was often used along with opium. However, mandrake is toxic and can produce hallucinations and nightmares. Hippocrates (c. 460–c. 370 BCE) and later Greco-Roman physician Galen (129–c. 216) also used opiates, henbane, mandrake, and nightshade plants for sleep. But these toxic plants could cause unconsciousness and death. "Drowsy syrups" to help with sleep in the Middle Ages were also common. For example, one recipe from the Benedictine monastery in Monte Cassino, Italy, in the early ninth century, used a mixture of opium, henbane, mulberry juice, lettuce, hemlock, mangrove, and ivy. This concoction was used as a sleep potion and also for anesthesia. A potion called "dwale" was described by English poet Geoffrey Chaucer (c. 1343–1400) for sleep. It was primarily a folk remedy made by women taking care of their families. It was also used by barber surgeons in higher doses as an anesthesia for surgery. It contained hemlock, opium, henbane, bryony (a mandrake-like root), boar's bile, wild lettuce, and vinegar. These were boiled together, stored in a bottle, added to wine, and administered to the patient. The toxic plants hemlock, mandrake, and henbane along with wild lettuce, opiates, and alcohol did cause sleep and unconsciousness. But the mixture could also cause hallucinations, and too much of the concoction could lead to death.

During the Renaissance, Italian mathematician and physician Gerolamo Cardano (1501–1576) noted in his memoirs that he suffered from early-morning waking insomnia four times a year. When this occurred, he smeared poplar ointment, bear's grease, or water lily's oil on 17 places on his body, including his feet to encourage sleep. It was unlikely to induce sleep other than by placebo effect.

In the nineteenth century, magnetism and magnets were thought to cure everything. British author Charles Dickens (1812–1870) believed that pointing his bed to the north was a cure for his insomnia. In opposition to this magnetic direction, twenty-first-century alternative practitioners suggest that sleeping with the head to the north causes the earth's magnetic pull to exert pressure on the brain and that people should sleep with their heads facing east. No scientific evidence exists supporting these hypotheses as the earth's magnetism effect on the body is extremely low. Nevertheless, Victorians embedded magnetite, a naturally magnetized mineral, in their pillows. In the early 2020s, magnets in pillows and other medical equipment were still advertised for curing everything from pain relief to a sleeping aid.

In late nineteenth century, a cure for sleep disorders was soaping the head with the ordinary yellow soap of the time. When the hair was lathered all over, it was wrapped in a napkin, and the person went to bed. It was washed out the next morning, and the procedure was done for two weeks. Any positive effect was likely due to placebo effect. In the nineteenth century, an easily available sleeping remedy was laudanum—a tincture of alcohol and morphine. However, opiates could cause serious addiction issues discussed in the "Substance Use Disorders" entry.

By the early twenty-first century, some alternative and folk food remedies were recommended for insomnia and sleep disorders. Proponents claim that eating onions before bedtime would lead to a better night's sleep due to the amino acid L-tryptophan in onions that causes sleepiness. However, onions can increase heartburn and acid reflux. Lettuce, especially romaine lettuce, contains lactucarium that has sedative properties and may help in getting to sleep. Tart cherries for baking pies contain melatonin, a hormone that causes sleepiness and regulates the sleep-wake cycle. Tree nuts such as walnuts and almonds also contain tryptophan, which converts to melatonin. Chickpea in hummus is also a source of tryptophan and may help with sleep. These foods are recommended to be eaten an hour or two before bedtime. They are generally harmless and may in some cases lead to better sleep.

Some folk treatments for snoring that might be related to sleep apnea or a dry throat have been recommended. One is to sew a rock in the back of the night shirt to prevent the person from sleeping on their back. Nettle leaf tea has antihistamine properties. Drinking it before bedtime may help reduce congestion by inhibiting the release of histamines. Essential oils have also been used including thyme for snoring. Eucalyptus oil placed in the nose may help liquify mucus to keep airway clear; it also has anti-inflammatory properties, which may reduce swelling in the nostrils and nasal passageways.

Aromatherapy became another popular folk remedy by the twenty-first century. For insomnia, lavender oil when inhaled or applied on the skin of young adult females, for example, was found to increase feeling of well-being and better sleep. Potential side effects, when the oil was applied to the skin, included allergic skin irritation and sun sensitivity. Another alternative sleep remedy is crystal therapy. Proponents claim that the vibrations from the howlite crystal, a white stone with fine gray marbling, have soothing energy that helps alleviate stress and aids sleeping. Promoters recommend placing one of these stones under a pillow at night. Advocates also allege that the opalescent moonstone's vibrations reduce emotional tension and help the person sleep. However, no evidence suggests that other than placebo effect.

L

Leg Problems

Leg maladies cause pain, spasms, decrease in the quality of life, and lack of mobility, and some like a deep vein thrombosis (DVT) can lead to death. (Fractures and dislocations are discussed in other entries.) Some leg issues including leg muscle cramps were not considered important enough to be included in many medical texts until the early twentieth century, and their treatment was largely relegated to folk and home remedies. Other ailments such as thrombophlebitis (clots in leg veins) have been noted for centuries.

Nocturnal Leg Cramps and Restless Legs Syndrome

Nocturnal leg cramps (NLC)—also called a "muscle spasm" or a "charley horse"—is an extremely painful uncontrolled contraction of muscles in the legs usually at night. These cramps mostly happen in the calf muscles but can also occur in the front and back of the thighs or feet or toes. Symptoms may last from several seconds up to several minutes, and muscle soreness often remains after the spasm goes away. Although the cause is not known, some theories suggest that over-exercise, or conversely, lack of exercise, standing for long periods of time, pregnancy, some medications, and chronic diseases may be factors in this malady. NLC affect up to 60 percent of adults and are more common among older adults and women. They can disrupt sleep, break up a person's sleep cycle, and lead to insomnia. The first use of the term "cramp" is from the fourteenth century from the Middle English *crampe*, from Anglo-French, of Germanic origin. "Charley horse" for cramps in English was first used in the late nineteenth century of unknown origin.

Restless legs syndrome (RLS) causes an intense and uncontrollable urge to move the legs while a person is attempting to sleep. Symptoms can include prickling, crawling, tingling, tugging, and sometimes painful feelings in the legs. Moving the legs often provides short-term relief. However, the syndrome can interfere with daily activities due to sleepiness. RLS may be genetic and is associated with some chronic diseases, iron deficiency, heavy alcohol use, and medications such as antihistamines and anti-nausea drugs. During pregnancy, some women experience RLS especially in the last trimester; however, symptoms usually go away a few weeks after delivery. From 7 to 10 percent of the American population has RLS. It is most common among older individuals and women. RLS is generally a lifelong condition for which there is no cure. The first-known medical description of RLS was in the late seventeenth century. In the early nineteenth century, it was called "Fidgets in the legs." However, it was not until 1945 that Swedish neurologist Karl-Axel Ekbom (1907–1977) coined the term "restless legs."

Varicose Veins and Thrombophlebitis (Blood Clot with Inflammation)

Other common leg conditions include varicose veins (VVs) and thrombophlebitis. VVs are enlarged veins near the surface of legs. They may form a zigzag pattern up the leg and sometimes feel hard. With this affliction, the legs often feel tired, achy, and heavy. VVs are caused by vein valves that have become weakened or damaged, allowing blood to collect in the leg. They can affect daily activities and quality of life. Risk factors for the malady include pregnancy, aging, obesity, hormonal imbalances, occupations involving much standing or sitting, lack of exercise, and traumatic injury. It is estimated that around 55 percent of women compared to 45 percent of men suffer from the malady in developed nations. They are most common in people over 40.

A thrombus (clot) with inflammation, which forms in a varicose vein near the surface of the leg, is termed a "superficial thrombophlebitis." It has symptoms of pain, warmth, redness, and swelling over the affected area and may lead to ulcers that are difficult to heal. Sometimes a clot forms in veins deep in the muscles. A DVT is a potentially deadly form of thrombophlebitis and cause much pain and swelling. It often occurs during pregnancy and after childbirth. Its signs and symptoms include pain, swelling, leg stiffness, warmth, redness, fever, a leg cramp often starting in the calf, and a bluish or whitish skin discoloration of the whole leg. DVTs are precipitated by trauma, surgery, prolonged inactivity, and hormonal changes. In the past, it was thought that during the postpartum period a DVT was due to the presence of unconsumed breast milk within the legs and called "milk" or "white leg." If a clot breaks off and travels to the lung, this pulmonary embolism (PE) can be fatal. The chances of developing DVT are about 1 in 1000 per year, and in the United States, up to 100,000 die from PE per year.

These conditions are ancient maladies; however, unlike varicose veins that were mentioned as early as 1500 BCE in the *Papyrus Ebers*, the first well-documented case of a DVT was reported during the late thirteenth century. The use of the term "thrombophlebitis" in English dates to around 1895–1900 from New Latin from the terms "clot" and "vein inflammation." The abbreviation DVT was not commonly used until the late twentieth century.

Medical management today for leg cramps includes massaging the muscle along with heat or cold. For varicose veins, compression stockings and mild exercises are recommended. There are no known cures for restless legs syndrome. Since DVTs are dangerous, medical personnel generally prescribed anticoagulants ("blood thinners").

Illustration of DVTs have been found in the arts. For example, Parisian artist Guillaume de Saint-Pathus's (1250–1315) *Vie et miracles de saint Louis* (1330s) contains 92 miniatures. One set shows a man with a swollen leg with likely ulcers from VVs and DVTs. The leg goes back to normal size by a miracle.

UNUSUAL TREATMENTS

Treatments were developed by folk healers and families over the ages for many leg issues. Only a few issues, such as DVTs, were managed by physicians.

Nocturnal Leg Cramps and Restless Legs Syndrome

In the early eighteenth century, for nocturnal leg cramps (NLC), one medical text suggested putting the foot on a cold stone, on the floor, or applying a cold cloth to the foot or leg; on the other hand, some folk remedies recommended heat. A few physicians also attempted galvanism or passing an electric current from a battery through the leg. From the eighteenth century, quinine (primarily effective for malarial-type fevers) was used to prevent or treat NLC, which often prevented cramps. However, because of adverse drug reactions, the American Food and Drug Administration in 1994 prohibited quinine use for NLC. But to prevent cramps, some alternative systems have advocated drinking a glass of tonic water before bedtime that contains a small amount of quinine. Home remedies in addition often recommend calcium, magnesium, or potassium supplements that research suggests are no better than a placebo effect.

Although there is no known cure for restless legs syndrome, some alternative systems suggest that anti-inflammatory spices such as turmeric and spirulina algae may be helpful. Over-the-counter concoctions are advertised, which besides spirulina include vitamin C, folic acid, iron, magnesium, and valerian, but the research is inconclusive regarding their effectiveness.

Varicose Veins and Thrombophlebitis (Blood Clot with Inflammation)

Greek physician Hippocrates (460–370 BCE), as found in his *Corpus Hippocraticum*, recommended compression bandages for varicose veins, swollen legs, and leg ulcers. By the thirteenth century, physicians employed bloodletting with leeches on the swollen leg and purging to eliminate "bad humors" along with placing flannels soaked in hot vinegar on the leg. When the acute swelling had decreased, they also massaged the calf with camphorated oil (think chest vapor rubs), which could be very dangerous as massaging could dislodge the clot. By the seventeenth century, healers encouraged women who had just given birth to breastfeed to prevent DVTs, which had no effect. Furthermore, people used laced stockings, elastic bands, and tight bandages with resin to harden them for enlarged veins in the leg. However, it was not until the late nineteenth century that these compression devices were used to treat DVTs.

In addition, from the late nineteenth century until the mid-twentieth century, the cornerstone of DVT treatment was strict bed rest due to fear of the clot moving into the lungs. Medical personnel often placed the patient's lower limbs in iron splints to prevent any movement; however, these measures likely caused more clots. Physicians, such as the American Elijah B. Hammack (1826–1888), suggested treatment should begin with a cathartic such as magnesium citrate and other purgatives to keep the bowels open as it was thought to reduce inflammation from DVTs. After the inflammatory symptoms had begun to subside, warm bran poultice or chloroform was applied to the limb, which likely had little effect. Between 1940 and 1944, a rat poison made from a powerful anticoagulant—coumarin—was synthesized. By the 1960s, dicoumarol (warfarin) that had been derived from this rat poison was used as an anticoagulant in humans to prevent DVTs, PE, and strokes.

Leprosy

Leprosy, also called Hanson's disease, can take years to develop. Some patients have mild symptoms while others suffer from a debilitating disease and loss of limbs. About 30 percent of patients have nerve damage. The World Health Organization (WHO) classifies the disease into two broad types: "Paucibacillary Leprosy" has no bacteria in the skin lesions, causes few damaged nerves, and is only mildly contagious. "Multibacillary Leprosy" has many bacteria in the skin lesions, causes much nerve damage, and is more contagious. If lesions are on the hands or feet, sensations to temperature extremes may be lost, which can lead to destruction of fingers and toes. The disease over time can cause deep ulcers and affect nose cartilage and bones. However, in some people spontaneous remission occurs. Leprosy is caused by the *Mycobacterium leprae* bacterium, which was first identified by Norwegian physician G. H. Armauer Hansen (1841–1912) in 1873 for which the disease is named. Most current researchers hypothesize that the disease is transmitted from person to person through respiratory droplets. Some also contend that contaminated soil or insects or handling infected armadillos or English Red squirrels might transmit the microorganism.

Worldwide, for unknown reasons, less than 10 percent of people are susceptible to leprosy, and children are more likely to be infected than adults. The WHO reports about 200,000 new cases each year primarily in South and Southeast Asia and Latin America. Worldwide, 2–3 million people are estimated to be permanently disabled from Hanson's disease.

Leprosy has been described since antiquity and was called *elephantiasis graecorum* by the Romans. In many cases, the disease may have been confused with other skin conditions discussed in another entry. However, through DNA analysis, leprosy is believed to have originated in East Africa or the Near East and traveled with humans along migration and trade routes. Skeletal remains in northern India from around 2,000 BCE have found evidence of the disease. It was described in an Egyptian Papyrus written around 1550 BCE, in Hebrew scripture by 600 BCE, in a classical Greek medical text around 460 BCE, and in a Chinese medical text around 400 BCE. DNA analysis has found mycobacterium in human remains, from 900 through the 1300s in Sweden, where the disease was widely found. The word "leprosy" comes from a classical Greek word for a disease "that makes the skin scaly" and was first used in the English language in a thirteenth-century manual for nuns.

Various cultures ascribed different causes to the disease. The ancient Abrahamic religions (Christians-Jews-Muslims) believed leprosy was a punishment from

God due to sin and that only God could cure the disease. Leprosy sufferers, called "lepers," were considered unclean. In the Middle Ages, physicians thought the disease was caused by imbalance of body humors that resulted from a bad diet or immoral lifestyle in addition to God's punishment. Through the 1800s, leprosy was also considered both an inherited—as it ran in families—and a communicable disease.

Throughout history, people with leprosy were feared, stigmatized, or shunned. In ancient Israel, for example, they were required to yell out "unclean, unclean" and warn people of their passing on the street. In the European Middle Ages, sufferers were required to carry a begging cup and ring a bell or clappers—as the disease sometimes destroyed their vocal cords—both to warn people they were coming and to attract people to give them alms. Victims were separated from their families and were sometimes blamed for causing plagues, resulting in them being tortured and burned alive. However, beginning in the 1500s, leprosy became rare. Some researchers suggest this may have been due to the rise of a more virulent

In the fourteenth century, priests blessed leprosy victims in order to cure them of the disease as shown in this illustration. (The British Library)

killer—tuberculosis—which killed off leprosy suffers before they could pass the disease to others.

Since early times, those with leprosy have often been isolated, or quarantined, in leper colonies—*lazarettos* or *leprosariums*—often run by monastic orders. Forced institutionalization, more likely among poorer people, occurred throughout the late 1900s in some Western cultures, and colonies are still found in developing regions of the world including parts of Africa, Asia, and South America. Moreover, the American state, Hawaii, did not lift its mandatory leprosy isolation law until 1969. The leper colony at Carville, Louisiana—virtually a prison in its early years—was in operation from 1898 to 1999. Modern treatment of the disease occurred in the early 1940s when the first antibiotic that made leprosy noncontagious was developed. By 1981, the WHO recommended a yearlong multidrug treatment (MDT) regimen of three antibiotics rifampicin, clofazimine, and dapsone that cured patients. MDT also prevented transmission, nerve damage, and disability and dramatically reduced leprosy worldwide.

Leprosy has been featured in the arts. For example, in the film *Ben Hur* (1957), Judah Ben-Hur is an aristocratic Jew living in Judaea. Due to political forces, he is forced into slavery and rescues a Roman aristocrat who adopts him, and later he returns to his home in Judaea. Upon his arrival, he finds that his mother and sister are living in a leprosarium valley as they have contracted the disease. However, they are miraculously cured by believing Christ is the messiah.

UNUSUAL TREATMENTS

Although some healers from earliest times realized that leprosy could not be permanently cured, they attempted remedies to ease the symptoms. In some cases, spontaneous remission may have been viewed as a cure after some treatments. In ancient Egypt, healers attempted to drive away leprous spots on the skin by applying onions cooked with a mixture of sea salt and urine to the lesions. Greek physicians around 200 CE recommended bleeding followed by purgatives to bring the body's humors back in balance. They also administered medications that included a mixture of snake flesh, herbs, honey, and shavings of elephant's teeth mixed in wine. Castration was also suggested as a cure for leprosy. None of these were effective.

Into the Middle Ages, physicians mixed ashes of vine branches with the fat of wild animals and rubbed this mixture on the face to eliminate leprosy tubercles. Since semi-precious jewels were thought to cure or prevent the disease, for those who could afford more expensive treatments, healers spread a paste-like substance made of ground amber and pearls on facial lesions. To prevent the disease in children, red coral was hung around the child's neck.

At various times from antiquity into the 1700s, both human and animal blood, either as a beverage or as a bath, was used to treat leprosy. The blood of a child under two years of age, in particular, was thought to cure the disease. The child's blood was mixed with dog's blood and added to heated water in which the patient bathed. Healers also gave patients the blood of children or virgins to drink as a purification ritual. Blood from dead bodies, menstruation, and the foreskin of a circumcised child was also made into a drink to treat leprosy. These were not effective and could have harmed the donor.

Treatments with snakes and the products of snakes were used from antiquity into the 1800s. Since snakes shed skin, the skin was thought to act as an "anti-venin" and remove the leprosy lesions. Snake skins were dissolved in strong wine, which the patient drank. Folk healers in Brazil supposedly cured leprosy with the bite of a snake, and in Columbia, spiders and scorpions' bites were administered. Repeated doses of bee stings were tried in the 1910s. None of these methods cured the disease and could have harmed the patient.

For centuries in Indian Ayurvedic medicine, healers rubbed chaulmoogra oil from the seeds of the hydnocarpus tree on the skin lesion and gave the oil to leprosy sufferers to drink. These treatments sometimes reduced the tubercles. Supposedly in China into the 1800s, sufferers were enclosed in the carcass of a freshly eviscerated small bull for an hour or more. In another treatment, patients ate snakes, the flesh of a dead child, and a cooked human placenta, which had no effect on the progress of the disease.

In the late 1800s, after the leprosy bacterium was discovered, and throughout the early 1900s, physicians experimented with many substances to kill the microorganism. Syphilis, streptococcus, and other pathogenic bacteria were injected into the skin nodules in the belief that the bacteria would overcome and eliminate the leprosy bacterium. This treatment was quickly abandoned as it only gave the patient the disease. Electricity was applied to the leprosy lesions, and a few patients claimed it restored some feeling. Physicians also rubbed mercury salts, cashew nut oil, and raw petroleum on the nodules. Some physicians found that chaulmoogra oil injected into nodules often reduced them. This was considered the best treatment for leprosy until the 1940s. The oil is still used today in India and some Asian and African nations as part of folk medicine for leprosy and other skin diseases.

Liver and Gallbladder Issues

Liver and gallbladder diseases often cause similar symptoms. These include pain and tenderness under the right ribs that often radiates to the right shoulder blade, nausea and vomiting, fatigue, loss of appetite, and jaundice. Until the early nineteenth century, jaundice was considered a disease in itself, and not a symptom of several illnesses.

Jaundice

In jaundice, the skin and whites of the eyes and mucous membranes turn yellow. Urine may also be dark yellow and the stool gray in color. Jaundice is caused by a buildup of bilirubin in the blood. Bilirubin is formed when red blood cells are broken down that the liver uses to create bile, which is stored in the gallbladder for fat digestion. If anything prevents or blocks bilirubin or bile from freely moving such as inflammation of the liver (hepatitis), cirrhosis (fatty liver), cancer, or gallstones (cholelithiasis), which clogs the common bile duct, jaundice, or icterus, may develop.

The word "jaundice" is from the Middle English *jaunes*, from the Old French *jaunice*, meaning "yellowness." The term "icterus," also meaning "yellow," is from the Latin via Greek. Greek physician Hippocrates (c. 460–370 BCE), for example, described epidemic jaundice (*icterus epidemicus*) and believed it was caused by an imbalance of humors. Later Scottish physician, William Buchan (1729–1805), in the eighteenth century believed that jaundice was caused by "an obstruction of the bile," "bites of vipers and mad dogs," female hysteria, and "violent passions" such as grief and anger. In terms of death rate during 1632, English statistician John Graunt (1620–1674) reported that 43 individuals out of 9,535 deaths had died of "jaundies" in London. In the United States during 1850, the census reported over 500 people had died from jaundice. Now only death rates for diseases in which jaundice is a symptom are reported.

Hepatitis and Cirrhosis

Jaundice is a major symptom of hepatitis. Hepatitis is primarily caused by microorganism, toxic substances, and obstruction of bile ducts. Acute hepatitis usually resolves in a month or so and is caused by the hepatitis A virus, mononucleosis, malaria, yellow fever, and other diseases. Toxic substances, including drugs and medications, also inflame the liver, and the bile ducts can become temporarily blocked. Chronic hepatitis results from hepatitis B or C viruses, blocked bile ducts, and toxic chemicals such as heavy chronic alcohol consumption and can also lead to liver cancer.

By 1942, two forms of hepatitis had been identified. These included "infectious hepatitis" (caused by hepatitis A virus from fecal contamination) and "serum hepatitis" (hepatitis B and C and other viruses transmitted by blood and other bodily fluids). When American recruits during WWII succumbed to the hepatitis from infected serum incorporated in yellow fever vaccines, the two forms were confirmed. The viruses, however, were not identified until the late twentieth century. Infectious hepatitis is more common in developing nations, and most people are infected as children. Individuals with drug abuse disorder who inject drugs are susceptible to serum hepatitis. Worldwide, by 2020 viral hepatitis caused more than 1 million deaths each year, and it was the seventh leading cause of death in the mid-2010s.

The term "hepatitis" is derived from the ancient Greek *hepar*, meaning "liver," and *-itis*, meaning "inflammation," and was first used in English in the early eighteenth century. Viral hepatitis is an ancient disease as the hepatitis B virus has been identified in central European archaeological remains from at least 5,000 BCE. By the mid-eighteenth century, epidemic jaundice, likely resulting from both infectious and serum hepatitis (due to the use of unsterilized surgical instruments), was recorded by military surgeons in European and North American armies. It was so prevalent in militaries that it was sometimes called "campaign jaundice."

Another liver disease that causes jaundice is cirrhosis, which is mostly caused by chronic alcohol dependence or hepatitis B or C viruses. Nonalcoholic fatty liver disease is another form of cirrhosis and is associated with being overweight, diabetes, high blood fats, and high blood pressure. In cirrhosis, fat is deposited in the liver, which becomes enlarged. Cirrhosis produces symptoms of inflammation, fatigue, fever, jaundice, and right upper quadrant abdominal pain. It often leads to mental disorders, fluid within the abdomen (ascites) and extremities, bleeding, confusion, liver cancer, and death. In 2019, cirrhosis was the eleventh most common cause of death globally. Autopsy studies suggest that from 4.5 to 9.5 percent of the population is affected with the condition. It is more prevalent among males and in central Asian nations.

The term "cirrhosis" is from the Greek *kirrhos*, meaning "orange-brown," which was first used in English during the early nineteenth century. Hippocrates in the fifth century BCE gave the earliest description of cirrhosis and the mental health issues resulting from the end stage of the disease. During 1632, the malady was called "livergrown," and Graunt reported that 84 individuals out of 9,535 deaths had died of it in London. Buchan in the eighteenth century believed cirrhosis was caused by eating too much meat and ingesting too much alcohol and recommended patients consume a diet based on fruits, vegetables, and whole grains, which is still recommended today. In 1850, the U.S. census reported over 1,800 deaths from "disease of liver" although it was not considered accurate.

There is no treatment for the hepatitis A infection; however, vaccination against A and B, but not against C, became available in the late twentieth century. Hepatitis B and C infections are now treated with antiviral medications. Another recent treatment is liver transplantation. It was first accomplished in the early 1960s, but it was not until the 1990s that the surgery became safe and common. However, due to an insufficient number of donors compared to the number of potential recipients, this surgery is limited.

Gallstones and Gallbladder Disease

Gallbladder disease (cholecystitis) is generally caused by gallstones. These stones can temporarily block the common bile duct—which joins the liver, gallbladder, and pancreas—particularly after a fatty meal and may result in painful bilious colic. Gallstones are hard masses of bile salts and cholesterol that form in the gallbladder. If they become stuck in the common bile duct, damage to both the liver and pancreas may occur, which is potentially life-threatening. Risk factors for gallstones include being middle aged, female, being overweight, diabetic, taking birth control pills, having rapid weight loss, and leading a sedentary lifestyle. Since the late twentieth century in developed nations, an increased prevalence of gallstones has been seen. About 12 percent of the population has gallstones but up to 80 percent have no symptoms.

The word "gallbladder" was first used in the late eighteenth century in English, and its derivation is not clear but is likely of Indo-European origins. "Cholecystitis" is from New Latin, from the Greek *cholecyst*, meaning "gallbladder," and *-itis*, meaning "inflammation," and was not used in English until the mid-nineteenth century. Gallbladder-related ailments have afflicted humans for millennia. Evidence of gallstones has been found in Egyptian mummies. The Greco-Roman physician Galen (129–c. 210 CE) based on humoral theory believed that the gallbladder was a repository for gall and regulated the emotions. Yellow bile, for example, stored in the gallbladder, made people "choleric," or irritable. In the seventeenth century, English physician Thomas Sydenham (1624–1689) claimed gallbladder pain was from vapors of the blood getting into the intestines. He also thought gall bladder disease in women was caused by hysteria and not real.

Prior to the late 1800s, physicians as a last resort treated gallbladder diseases by cutting the bladder open, removing the stones, and draining the fluid, but many patients died from infection, and it did not cure the problem. By the late nineteenth century when antiseptic conditions and anesthesia were introduced, cholecystectomy (removal of the gallbladder) became successful. Low-fat diets were recommended in the twentieth century, but recent research shows they may increase gallstone formation in some people. By the turn of the twenty-first century, minimal invasive laparoscopic surgery was used to remove the gallbladder rather than through large abdominal incisions.

Gallstones may have affected the course of history. Alexander the Great (356–323 BCE) of Macedonia, for example, died of severe stomach pains that likely resulted from acute cholecystitis fueled by excessive alcohol and food consumption. His early death likely prevented future conquests. French leader Napoleon III (1808–1873), nephew of Napoleon I, surrendered to Germany in 1870 at the battle of Sedan during the Franco-Prussian war. Due to painful bladder and kidney stone and gallstone attacks, he was under the influence of opium, which was reported to make him lethargic, sleepy, apathetic, and indecisive. Hepatitis may even have affected the outcome of wars. For example, in 1944 during WWII (1939–1945), the incidence of hepatitis was around 10, 16, and 50 per 1,000 people in the British, American, and German armies, respectively. The higher rate among the German force may have weakened them and helped in the defeat of Nazi Germany.

UNUSUAL TREATMENTS

Since liver and gallbladder disease largely remained untreatable until the twentieth century, many folk and alternative cures abounded. Healers also relied on various plants and other substances based upon traditions that for the most part were ineffective.

Jaundice

Unaware that jaundice was a symptom of, but not, a disease, many remedies have been tried, and due to the natural waning of jaundice from infectious hepatitis and other acute illnesses, these treatments were thought to be effective. During the Middle Ages, for example, eating the roasted flesh of a "well-nourished" cat was thought to cure jaundice, even swallowing lice mixed with ale each morning for a week. Bezoars, undigested stonelike mass found in animal intestines such as cows, were used as an antidote to poisoning and also as a treatment for jaundice. The patient soaked them in water and drank the fluid or scraped a bit off to consume. None of these were effective for jaundice but were largely harmless to the patient.

Based upon the Law of Similar ("like cures like" still used in some alternative systems), some healers, such as Swiss physician Paracelsus (1493–1541), suggested patients look at yellow birds to cure jaundice. Until the early twentieth century in a few Middle Eastern nations, healers or family members would split open a live pigeon and apply it to the patient's breast so that its "living principle" might enter and sustain the patient. In the twentieth century, Spectro-Chrome Light Therapy suggested patients bathe in lemon, red, yellow, and magenta light. Other than light therapy for neonatal jaundice, these light treatments were ineffective.

Vigorous exercise was also prescribed. Buchan in the eighteenth century ordered horseback, walking, running, and even jumping for jaundice patients. He noticed that after medicines had proved ineffectual, patients on a long journey in a rough coach ride were cured of the condition; exercise may have helped pass gallstones.

Throughout the mid-nineteenth century, physicians, such as American Daniel Drake (1785–1852), practiced the ancient traditional treatments of bloodletting by cutting a vein, cupping (applying hot cups to the skin that had been scratched to draw out blood), leeching, and blistering by caustic materials on the abdomen to draw out the bad humors. They also administered enemas with calomel (poisonous mercury compound). These treatments likely hastened the death of patients.

Various herbs and substances have been tried for jaundice over the centuries. In ancient Egypt and Greece, juniper berries—the berry that gives gin its flavor and has anti-inflammatory properties—was thought to cure the condition along with incantations to deities. In the fourteenth century, English physician John Arderne (1307–1392) still favored juniper berries for jaundice. In addition, he administered crushed earthworms soaked in holy water or wine and considered it a "well-tried remedy." From the thirteenth through the mid-nineteenth century, mumia—ground-up powder from stolen Egyptian mummies—was used as a treatment for jaundice, which were ineffective. Today some alternative practitioners recommend dandelion tea or green melons consumed with parsley tea to reduce jaundice or eating tomatoes, radishes, and colorful vegetables, but the evidence is mixed regarding their effectiveness.

Hepatitis and Cirrhosis

In ancient Egypt, healers gave the plant fenugreek (of the pea family) and licorice "to calm the liver and pancreas," which may have relaxed the common bile duct and reduced pain from stones. Surgery on the liver was a treatment of last resort. Hippocrates suggested that when the condition did not improve with the usual remedies, the abdomen over the liver should be burned with heated "spindles of boxwood," which had been dipped in oil or fungi so pus could run out. He also recommended in the case of severe dropsy (collection of fluid in body), which often accompanied end-stage liver disease, to make many small cuts near the ankle to draw off fluid. These likely hastened death from infection.

Burning into the liver was also tried. For example, Greek physician Aretaeus around the first century suggested pushing red-hot irons into the liver to drain abscesses. Byzantine Greek physician Paulus Aegineta (625–c. 690) recommended using these irons to create an eschar (piece of dead skin that peels off) and ulcers by burning the area over the liver, spleen, or gallbladder in order to reach the liver to drain the pus, which likely hastened death.

English surgeon John Arderne (1307–1392) put a plaster bandage over the liver made of burnt animal remains, juice of nightshade (a poisonous plant), warm vinegar, and rose oil mixed with barley meal. American physician Henry Hartshorne (1823–1897), for nephritis and cirrhosis in the mid-nineteenth century, bled from a vein on the right side and placed leeches over the liver area followed by blistering with Spanish fly (ground-up cantharis beetles), which were not effective and caused weakness in the patient. He also inserted a metal drainage tube into the abdomen. Without sterile conditions, this likely led to infection and death. British physician Edward John Waring (1819–1891) placed leeches around the anus to "unload the portal system" and positioned a plaster bandage of black pepper followed by a heated linseed-meal poultice over the hepatic region, which was not effective. By the early twentieth century, "liver cleanses" and "detoxes" had been advertised to "detoxify the liver" and "rid the body of impurities" and are still advertised today by alternative practitioners. These "detoxification" substances include dietary supplements, special diets, chelation infusions, and enemas. However, no scientific evidence has been found to show that these detox cures are useful. Likewise, folk and alternative systems recommend milk thistle seeds, which contain silymarin to protect the liver from environmental toxins, and CBD (Cannabidiol), a "cure-all" to help liver conditions. However, research about their usefulness is inconclusive.

Gallstones and Gallbladder Disease

Gallbladder disease and gallstones had similar treatments to hepatitis and jaundice as discussed earlier, often based on humor theory. Bleeding, purging, and placing of poultices on the right flank were common from Greco-Roman antiquity into the late nineteenth century. Sydenham, in the seventeenth century, for example, repeatedly bled and purged men at the beginning of a "Bilious Cholick," prescribed opiates, and had them ride horses to break up "particles in the blood." However, for most women, he only administered laudanum (mixture of opium and alcohol). If the woman became jaundiced, he gave a "gentle purge" of Senna (from the dried pods of the cassia tree) and rhubarb leaves and generally considered their extreme pain to be psychosomatic. None of these treatments were effective, but laudanum may have relieved pain.

Buchan in the eighteenth century claimed that copious urination cured colic pain, so he gave patients "sweet spirits of nitre" (distillation of alcohol with caustic nitric and sulfuric acids) mixed with a fluid to drink. The Robbs, a family of several physicians, in their 1890s' medical text *Family Physician*, suggested bathing the patient on the right side with "aqua regia" (mixture of extremely caustic hydrochloric and nitric acid), which likely burned the skin. For severe colic pain, they administered chloroform or ether, which would have helped with pain. Many physicians and alternative practitioners suggested patients drink water from natural sulfur and mineral springs, which were not effective.

In contemporary times, alternative and folk treatment based upon tradition for gallbladder attacks include peppermint, burdock-catnip, and dandelion-root tea along with apple cider vinegar and turmeric. However, the science is inconclusive regarding their effectiveness. Folk remedies include crystal therapy. The patient places yellow jasper on the abdomen to "cleanse and detoxify" the gallbladder and reduce pain. There is no evidence that this is effective.

Malaria and Intermittent Fevers

Malaria, also called "Ague" or "intermittent fever," has three stages that generally come in two- or three-day cycles. In the first stage, the patient shivers and feels cold. This is followed by a high fever for several hours and then profuse sweating and a rapid decrease in the fever. The disease is caused by four forms of the *Plasmodium* protozoa. Of the four, *P. falciparum*, or "malignant malaria," is the deadliest and most common form of the parasite and disease. It can cause severe anemia, cerebral malaria, enlarged spleen, renal failure, and death. The disease is spread from humans to humans by the bite of female *Anopheles* mosquito, which carries the parasites. The protozoa were first identified in a patient's blood in 1880 by a French military physician stationed in Algiers. All ages are susceptible to malaria; however, children have a higher risk of dying. Without treatment, their case fatality rate can be up to 70 percent. Adults in endemic areas of the world generally acquire some immunity to the disease.

The name "malaria" is derived from the Latin *mal aria*, "bad air." The Romans recognized the relationship between foul stagnant swamps surrounding Rome and the presence of fevers. Some physicians until the mid to late 1800s still thought the disease was caused by fetid odors that were carried by the winds—*miasmas* theory. Others believed debauchery, getting wet or chilled, sleeping exposed to the night air, and long rides "through the dews of night" resulted in unbalanced body "humors," which in turn caused the disease.

Malaria is one of the world's oldest infectious diseases and is primarily found in subtropical and tropical areas. The disease became common around 10,000–6,000 years ago with the development of communities and the spread of agriculture. DNA analysis has found the parasite in some Egyptian mummies from over 3,500 years ago—including King Tut. The oldest description of the disease is found in an ancient Chinese medical text *Nei Ching*, written around 2,700 BCE. It was also described in the *Susruta*, an ancient Indian Ayurveda medical text. The disease likely arrived in southern Europe by 400 BCE. Around this time, the Greek physician Hippocrates (460–370 BCE) described it as "the fever."

Intermittent fevers were first described in northern Europe by 1350, where they occurred in the autumn and spring. Later in London during 1632, for example, 43 died of Ague and 1,108 of "fevers" out of 9,535 deaths; many of the fever deaths were likely due to malaria. By the 1600s, the parasite had been transmitted to the Americas by European explorers, colonists, and enslaved Africans. Once the disease was in the southern United States, it mostly affected boatmen who brought it northward along the interior rivers. In other areas of the world, such as British

India, around 824,256 died from malaria in 1869. In Italy at the end of 1800s, malaria deaths ranged from 15,000 to 20,000 per year particularly in rural areas, and in Greece in 1905, almost a half of the population was attacked by the disease. Malaria is still a major global problem. The World Health Organization (WHO) in 2016 estimated that about 216 million cases of malaria occurred with 445,000 deaths—mostly children in WHO's African Region.

The quinine drug was the original effective treatment for malaria and is found in the bark of the Cinchona bush native to the Andes. The bark was used by native peoples to prevent and cure fevers. In the early 1600s, it was transported to, and used on, the continent. However, the English viewed this "Jesuit's powder" with suspicion due to anti-Catholic fears and did not utilize it until the late 1600s. By 1907, many malaria parasites had become resistant to quinine. In China, the *Qinghao* plant that contains artemisinin had been used for centuries for fevers. Quinine today is combined with artemisinin and antibiotics for the treatment of malaria. Insect repellant, synthetic anti-malaria drugs, and environmental measures are now used to prevent the disease.

Malaria has influenced history, warfare, and the arts. Some scholars suggest that malaria was responsible for the decline of many ancient Greek city-states and contributed to the fall of Rome. During the rise of the Roman Republic, engineers drained the swamps, which eliminated malaria in Rome. During the later empire, the infrastructure deteriorated, and the drainage system was not repaired, which led to numerous deaths from the disease and a weakened population. When a Germanic army sacked Rome in 410, the invaders also succumbed to the disease and were forced to abandon the city. Later malaria was depicted by artists, reflecting its occurrence in everyday life, such as English authors Chaucer (1343–1400) and Shakespeare (1564–1616).

In more recent times, workers could not build the Panama Canal until yellow fever (discussed in another entry) and malaria had been controlled, which did not happen until 1906. During WWII (1941–1945), it was the most important health hazard encountered by the Allies and American troops in the South Pacific, where about 500,000 U.S. troops were infected. Malaria also caused severe illness among American soldiers training in southern states. To counter this scourge, an eradication program was established by the Centers for Disease Control and Prevention (CDC). They drained swamps and sprayed houses with insecticides, leading to a malaria-free nation in 1951. Western Europe followed a similar program and became malaria free by 1975.

UNUSUAL TREATMENTS

Various cures for intermittent fevers for centuries have been attempted. These have primarily been bleeding and purging to get rid of "poisons" in the body. Patients likely survived or died because of, or despite, these treatments. Various herbs have also been used for fevers. In England, for example, the noted physician Nicholas Culpeper (1616–1654) recommended mixing the *cinquefoil* herb with wine or white-wine vinegar. He claimed it cooled the patient's blood, body humors, and fever. The renowned English physician, Thomas Sydenham (1624–1689), on the other hand, found that the newly discovered "Peruvian bark" when ground into a powder and mixed with wine and cloves and given to patients for several weeks often cured the disease and was generally harmless to the patient. Many English physicians at first thought this concoction was dangerous as it did not eliminate poisons from the body. By the mid-1700s, the use of this quinine, either ground or as an infusion, became more acceptable, and in malaria-prone tropical areas, British colonists drank "gin and tonics" to prevent and cure the disease. However, the small amount of quinine mixed with water and gin was unlikely to avert the illness.

To reduce the first stage chills of the malarial cycle, by the late 1700s, some physicians administered Fowler's solution (containing highly poisonous arsenic) mixed with quinine, opium, or chloroform to patients. Other physicians recommended firmly rubbing the extremities, stomach, and the back with brushes or woolen cloths moistened with pungent liniments containing ammonia, oil of turpentine, camphor, or mustard to warm the patient. After the fever had abated, physicians gave patients opium mixed with calomel (contained mercury) or quinine to reduce the inflammation. Arsenic and mercury if used for longer periods would cause poisoning and death but may have killed some parasites. The other treatments were not effective.

Besides the Cinchona bark, physicians used other astringent barks in an attempt to decrease fevers. Native peoples in North America, for example, utilized the powdered bark of the dogwood tree. By the 1800s, a few American physicians used dogwood and barks from other native trees, including yellow poplar, oak, and willow, to treat malaria. They were generally mixed with wine or whiskey. Willow bark, which contains salicylate, may have reduced the fever. Likewise, in Australasia, eucalyptus leaves were used, which could have reduced the fever but also did not kill the parasites. Other physicians, during the fever stage of the malarial cycle, used the method of bleeding with leeches, cupping (heated glass cups placed on the skin to draw blood), or blistering (caustic materials that caused a blister) on the abdomen to draw out poisons and "calm the blood" before giving quinine. This could have caused infection and weakness if too much blood was drawn. By the late 1880s, after the parasite had been identified and quinine alone was found to be the best treatment, bloodletting and blistering along with the ancient humoral theory was discredited by physicians.

Menstruation and Gynecological Issues

Gynecological maladies range from menstrual disturbances and infections to prolapsed uteruses. Today females generally begin to menstruate around age 12 (puberty) and stop around age 50 (menopause) except during pregnancy. If an egg from the ovaries is not fertilized by a sperm, the endometrium (lining of the uterus)—which has been growing in preparation for implantation—sluffs off roughly every 28 days. Premenstrually, some women are tired, have mood swings, and are irritable, and during menstruation, they may have cramps. During the flow to collect the discharge, women generally use pads, tampons, or menstrual cups.

Numerous terms are common for menstruation such as "period," "time of the month," "the curse," "Aunt Flo," or "on the rag." The term "menstruation" was first used in 1686, and is derived from the Latin *mensis*, from the Greek *mens*, meaning "month," and the Latin *ātiō*, meaning "action." From ancient Greece into the late eighteenth century, humoral theory taught that menstruation eliminated a woman's extra blood, and if it did not flow, it would result in illness. In addition, exercise was discouraged although poor women continued to work. However, today in modern cultures, most women go about their usual daily activities including exercise.

Although menstruation is a natural cycle, almost all religions have viewed the menstruating woman, and/or the bloody discharge, as impure or unclean. Moreover, similar taboos concerning menstruation are found across religions, cultures, and time. Today some ancient customs are still practiced globally among the orthodox, particularly in Middle Eastern, South Asian, and African nations and among tribal groups. Some of these traditions forbid a menstruating woman to have sexual intercourse, to attend religious services or enter a temple or place of worship, to cook, or to eat with the family, and they require them to remain in an isolated hut or part of a room and to take a purification bath or wash after bleeding has ceased.

Though most women have, more or less, regular periods, problems sometimes result. A few have amenorrhea, the absence of or cessation of menstruation. Amenorrhea is defined as an adult woman missing three menstrual periods in a row or lack of menstruation by age 15 in a girl. Common causes of amenorrhea include pregnancy, lack of hormones such as estrogen and progesterone, some medications, illnesses, low body weight, excessive exercise, or a thick or closed hymen among pubescent girls. Depending upon the cause, it is treated with medications or surgery. The term "amenorrhea" was first coined in 1804 from modern Latin, from the Greek *a* (not), plus *men* (month), plus *rhein* (to flow). The disorder was first described in the ancient *Papyrus Ebers* (c. 1500).

In another common condition, endometriosis, the lining of the uterus grows outside of the uterus resulting in painful cramping that usually occurs near menstruation. Other symptoms include painful sexual intercourse, infertility, low back pain, heavy menstrual bleeding, uterine ulcers, and adhesions. It was first discovered microscopically in 1860, although it had been documented in medical texts since the time of the Greek physician Hippocrates (c. 460–370 BCE). Globally, endometriosis is found in about 10 percent of all women and can cause work and social disruptions. The cause of endometriosis is unknown although it is hypothesized that it has a genetic basis and that increased estrogen levels during puberty may trigger the symptoms. There is no cure for the malady other than over-the-counter pain medication and as the last resort surgery to remove the endometrium growth outside the uterus. The term "endometriosis" was first coined in 1925 and is from the Greek *éndon*, meaning "inner"; *metrium*, "womb"; and *ōsis*, "abnormal condition."

Vaginal infections, "vaginitis," are caused by a variety of microorganisms or even allergic reactions. Some infections are sexually transmitted and are discussed under the "Sexually Transmitted Infections" entry. Several conditions have a whitish discharge (leucorrhea). Candida or "yeast" infections, for example, produces a thick, white "cottage cheese" like discharge from the vagina, along with itching and burning on urination. A yeast infection typically happens when the balance of the vagina changes due to pregnancy, diabetes, weakened immune system, or other factors. It is a common condition with about 75 percent of adult women having at least one candida infection during their lifetime, which most women find distressing. This and other infections are now managed with various medications. The term "leucorrhea" comes from the Greek *leukós*, meaning "white," plus *rhoía*, "flow, flux." Hippocrates in *Diseases of Women* taught that leucorrhea was the "flowing away of the seed of the woman" and attributed it to a disease from the imbalance of the body humors.

Hippocrates also described uterine prolapses. In this disorder, the womb descends into the vagina or protrudes from the vaginal opening. It is caused by weakened pelvic floor muscles and ligaments that are unable to support the womb. Some symptoms of the malady are urinary leakage, painful intercourse, low back pain, and pelvic heaviness. Risk factors include obesity, pregnancy, frequent heavy lifting, being postmenopausal, chronic coughing, constipation, or genetic factors. It is estimated that up to 40 percent of women have varying degrees of this pelvic floor dysfunction, and it can lead to complications such as ulceration of exposed tissue or the prolapse of other pelvic organs. Most women find it a distressing condition. American physician Elijah B. Hammock (1826–1888) thought it resulted from tight Victorian girdles. Treatment today includes Kegel pelvic and other exercises, a vaginal pessary to hold the vagina in place, or surgical repair.

In the arts and works of fiction, until the late twentieth century the topic of menstruation was considered taboo for public discussion. One of the first to explore the topic was British author Doris Lessing's (1919–2013) novel *The Golden Notebook* (1962). This fictional work is based upon four fictional notebooks kept by the female protagonist, a writer. She discusses her emotions, smells, and other aspects of her menstruation.

UNUSUAL TREATMENTS

Since the dawn of humanity, many folk traditions emerged to treat gynecological conditions, and most were handled by midwives. In modern times, male healers attempted to manage "female complaints," which often were not taken seriously based upon the medical theories of the time. Women, over the centuries, to absorb menstrual flow, used various mosses, grasses, hides, and rewashed old rags and sometimes did nothing at all. To regulate or bring on menstruation in ancient Egypt, based upon tradition, women douched with a mixture of garlic and wine, or garlic and cow's horn. If that did not work, the *Papyrus Ebers* recommended douching with honey, wonder fruit, sweet beer, and fennel. The best remedy, however, was a cloth plaster placed on the vulva containing dates, onions, and acanthus fruit crushed with honey. None were effective but generally harmless to the woman. If the hymen was closed in a girl, preventing menstrual discharge, a healer would cut it.

If an adult woman had not menstruated for several years, the papyrus gave a recipe to bring on the flow. The woman recited magical chants and had sexual intercourse. Then she drank a mixture of the berry-of-the-uan-tree, caraway, incense, and uah-grain heated with cow's milk and tallow, which was likely not effective but harmless. In ancient Greece, Hippocrates advocated fumigating the vagina with various aromatic substances and recommended bloodletting to rebalance the humors, which could have caused weakness. In classical Rome, physicians injected a medicated solution using a pair of bellows into the vagina. Scottish physician William Buchan (1729–1805) gave women "Peruvian bark" (quinine), along with other bitter and astringent medicines that caused tissue shrinkage or iron filings infused in wine or ale to bring on menstruation; iron may have helped remedy anemia, which could have prevented the flow, and quinine and other astringents could have caused a miscarriage.

Buchan advised women during menstruation to avoid anything cold or hard to digest such as fish and fruit. When he believed the lack of menstruation was from viscous blood, to rebalance the humors, he also bled from a vein, administered laxatives, and gave foot baths. Hammack, in the nineteenth century, besides laxatives and regular bathing, suggested "good porter beer" and meat. Better nutrition that included iron and protein likely started the flow. He also advised drinking much pennyroyal and tansy tea as both have been found to stimulate menstrual flow, ease pain, or cause an abortion. Some physicians placed leeches in the vagina and on the cervix to encourage menstruation, which would not have been effective.

In the twenty-first century, several stones and crystals were advertised to bring on a period. When carried by the woman, the stones supposedly resonated with the sexual organs. For example, moonstone is supposed to help soothe premenstrual symptoms. Malachite, "the midwife stone" was used to bring on a period and relieve menstrual cramps. Any "cure" was likely caused by the placebo effect. However, sexual activity to orgasm causes uterine contractions that may dilate the cervix and help the uterine lining begin to shed. It may also prevent cramps during a heavy flow as clots may be more readily expelled.

For painful menstruation—often caused by endometriosis—Greek gynecologist Soranus (practicing between 98 and 38 CE) recommended bed rest. If this was not successful, he advised bloodletting for strong healthy patients, which had little effect. If pain continued, he applied linseed meal and turpentine poultices to the abdomen. In obstinate cases, Soranus performed dry cupping (heated cup that produced vacuum) over the belly and groin or, if this failed, wet cupping (cuts made on the skin before cups were applied) and leeching. These were not effective unless by placebo effect.

In the nineteenth century, purging and a poultice of hops or tansy was applied to the lower abdomen, and physicians such as Hammack administered Dover's powders (ipecac and opium). By the 1890s, a variety of advertisements touted cures for painful periods. These included tampons impregnated with opium and tonics such as "Lycia Pinkham's Vegetable Compound," which primarily contained alcohol and herbs. Even the newly invented anesthesia was used for menstrual cramps. These remedies likely helped with pain but could cause addiction.

To prevent and cure infections and leucorrhea, the Egyptian papyri recommended various vaginal douches. Those containing wine may have helped as they could have killed the microorganisms. A last resort recipe was a douche made of cow's bile, cassia, and oil, or of hog's bile and fresh dates. Cautery was also used for sores and ulcers but did not cure infections. Buchan, in the eighteenth century, ordered the patient to exercise and drink red port or claret for yeast infection. If this did not work, he added powdered Peruvian bark and 10 drops of the elixir of vitriol (sulfuric acid) to a glass of red wine. The patient consumed this several times a day, which had no effect as quinine is specific for malarial-type fevers.

Many remedies have been found to return a displaced uterus back into its proper place. The Egyptian papyri, for example, based upon tradition recommended that a woman consume mold from a wooden ship that was added to the froth of fermented beer. Another remedy suggested burning wax, incense, or a foul material and allowing the fumes to penetrate a woman's sex organs. This was because the ancients considered the uterus to have animal characteristics and thought it would be repelled by disagreeable odors. Some cures likely caused infections, rather than putting the uterus back into place. These included a woman rubbing the protruding cervix with dried human excrement or petroleum or "froth-of-beer" mixed with animal manure, which could have caused infections.

Hippocrates recommended that the sides of the uterus be cut and then anointed with cerate (wax or resin) in order for scar tissue to pull it back into the vagina. In another treatment, the physician suspended patients on a ladder head down, manipulated the uterus into its proper place, and then released her to lie flat for a week. A pomegranate, cut into half, sometimes was also put against the cervix to keep the uterus in place. Other physicians in ancient Greece packed the vagina with moistened ashes. These treatments could have been successful.

For a prolapsed uterus, midwives in the early modern period would tightly strap the lower abdomen to keep the "wandering womb" in place. By the nineteenth century, physicians had patients lie down for several days and gave douches of cold water sometimes mixed with astringents including zinc sulfate and toxic "sugar of lead" (lead acetate). These may have helped, but the lead could have poisoned the patient. Pessaries of various substances including ivory, gum elastic, glass, and silver coated with gold were also used to keep the uterus in place.

Mental Disorders ("Neurotic"): Anxiety and Functional Disorders

Anxiety and functional disorders are new names for nonpsychotic or "nervous" disorders and include maladies in the past called "hysteria" (discussed in its own entry) and "neuroses." They include depressive disorders (discussed in its own entry) and anxiety disorders like phobias, panic attacks, posttraumatic stress syndrome (PTSD), and obsessive-compulsive disorder (OCD). The other major category, functional disorders (also called conversion, somatoform, or dissociative disorders), includes identity, multi-personality, or amnesia disorders in addition to functional neurological disorder in which the person experiences paralysis, blindness, tremors, etc. Since the early twentieth century, names for these disorders have been frequently changed, which causes confusion. In addition, different factions of the mental health field sometimes use different terms. A major change occurred when neurosis was placed in the same category as anxiety disorder in the 1980 edition of the *Diagnostic and Statistical Manual of Mental Disorders* (*DSM-3*), which is used by mental health professionals to diagnose mental disorders.

Anxiety and functional disorders are found in roughly 20 percent of the population. It is hypothesized that genetics and stressful life events from childhood or more recent times are risk factors for developing these conditions. In general, people with these maladies can function in society although they may have social and relationship issues. Generalized anxiety disorder (GAD) is characterized by uncontrollable worrying about many circumstances of daily life. The person may feel nervous or irritable and have a fast heart rate, rapid breathing, sweating, a sense of impending danger, panic, or doom without being able to identify the source of the worry. GAD affects about 3 percent of the population, and women are twice as likely to be affected as men; it tends to first occur at about age 30. Women are also more likely to have phobias. A phobia is a strong irrational fear or reaction to something. Some specific phobias include fear of flying, heights, the dentist, elevators, the outdoors, insects, etc. Having a phobia can disrupt personal, family, and work relationships as people with phobias often avoid the object or activity. The term "phobia" comes from the Greek *phóbos*, meaning "morbid fear," and was first used in English in the late eighteenth century. Around 8 percent of the population has a phobia, and they tend to develop in adolescence or early adulthood.

In OCD a person mentally obsesses about something and has a compulsion to eliminate it. A common example is an obsession of something not being clean. A compulsion is the behavior the person carries out to relieve their anxiety such as repetitive and excessive cleaning. It is found in about 1 percent of the population and is equally common among men and women.

Another closely related anxiety disorder is PTSD that used to be known as "shell shock" in WWI or "war neurosis" later. The term PTSD was coined in 1980 in the *DSM-3* after many Vietnam veterans were seen with the issue. However, the condition has been recorded throughout the centuries from antiquity due to fear generated from combat, accidents, assaults, rapes, disasters, childhood abuse, or other harrowing events. Symptoms include irritable or aggressive behavior, self-destructive behavior, hypervigilance, exaggerated startled response, problems with concentration, bad dreams, and sleep disorders. It occurs in the United States in about 3.5 percent of the population. Like other anxiety disorders, it can cause work and relationship issues. Cognitive talking therapy, behavior modification, and in some cases antianxiety medication are commonly used to treat these anxiety disorders. Among soldiers in combat, the best treatment is rest for a few days near the frontline within hearing distance of the battle din.

Among the functional disorders is dissociative identity disorder, or "split personality," in which the person has two or more identities. Each has a different personality. Almost all individuals with a multiple-personality disorder have been victims of childhood abuse or neglect. The person shuts off or dissociates himself/herself from a situation or experience that is too violent or painful. The different personalities are usually inaccessible to each other. Since one personality sometimes exhibits reckless or even hysterical behaviors, it can affect the individual's personal and work life. Less than 1 percent of the population has the disorder. Throughout the end of the eighteenth century, different personalities in dissociative identity disorder were blamed on spirit or demon possession. However, by the mid-nineteenth century, clinicians including French psychiatrist Pierre Janet (1859–1947) thought it had natural causes; he coined the term "dissociation" for this condition; however, no successful treatment exists for the disorder.

Dissociative amnesia is generally caused by recent trauma. In this case, people cannot recall information about themselves. Usually it is fragments of their life, and it is rare to lose their whole identity including name and all memories. In functional sensory disturbance, "somatoform," or conversion reaction, the individual experiences physical symptoms and often goes from physician to physician to find a cause. Symptoms include numbness, tingling, paralysis, or even loss of vision as the result of severe psychological trauma or anxiety. These symptoms often fluctuate and may vary from day to day or be present all the time. Management is with talking therapy and sometimes medications.

Anxiety issues have appeared in fiction and memoirs. For example, *The Strange Case of Dr. Jekyll and Mr. Hyde* (1886) is a novella by Scottish author Robert Louis Stevenson (1850–1894). This work alludes to dissociative identity disorder. Dr. Jekyll is a friendly researcher with several friends who has developed a serum that turns him into Mr. Hyde who is an evil reclusive serial killer. The serum can cause him to transform into one character or the other. American author and successful businessman Clifford Beers (1876–1943) suffered severe episodes of generalized anxiety and depression. During his confinement at various private and state mental institutions, he was seriously abused. His book *A Mind That Found Itself* (1908) spearheaded a successful mental health movement in the United States and Canada for more humane treatment of those with mental disorders. OCD is featured in the romantic comedy film *As Good as It Gets* (1997). The main character suffers from OCD and displays obsessive throughs and repetitive behaviors such as not stepping on sidewalk cracks. Through various situations including a romantic interest, his anxiety and obsessions wane.

UNUSUAL TREATMENTS

Many treatments for anxiety and functional disorders have been tried over the centuries. Egyptian priests in antiquity, for example, recited incantations to banish evil spirits thought to be causing the problem. Greek physician Hippocrates (c. 460–370 BCE) advocated trepanation, or scraping or drilling a hole in the head, to allow demonic spirits, pain, or air to depart to relieve mental disorders. Some patients did survive the operation, but it was only useful for relieving pressure from blood clots or swelling in the brain. This treatment was used from antiquity into the eighteenth century for mental illness issues. Another treatment that began in the Middle Ages and lasted into the eighteenth century was the exorcism of demons as an accepted remedy for multi-personality disorders, which was likely not effective.

Throughout much of Western history, for other nervous and anxiety issues—based upon humoral theory—bloodletting by leeches, venipuncture, or cupping was carried out to drain out "bad blood" and balance the humors. Eliminating bad humors also included drugs to induce vomiting and purging. To force healthier blood to the patient's brain, American physician Benjamin Rush (1746–1813) strapped patients in a chair and then whirled it around. None of these treatments were successful and likely caused some conditions to get worse and may have caused further anxiety.

For nervous disorders, Scottish physician William Buchan (1729–1805) recommended riding horseback as it gave "motion to the whole body, without fatiguing it." He primarily favored long sea voyages or journeys for those who could afford them as the patient could focus on things other than what was causing anxiety, fear, and nervousness. Some studies have found that vacations are helpful in reducing anxiety as it gets the person out of the environment that is causing the issue.

From at least the ancient Greece era, spas and baths for hygiene, social engagement, and for treating nervous conditions were common. However, baths and bathing went out of favor after the fall of the Roman Empire in the fifth century. By the turn of the seventeenth century, spas as therapy were rediscovered, and numerous "baths" were established throughout Europe and North America over the next two centuries to treat nervous disorders. Many were hot mineral springs, where people "took the waters" by both drinking and soaking in the water. One of the most famous spas in the United States was the Battle Creek Sanitarium, "the San," headed by American physician John Harvey Kellogg (1852–1943). Patients suffering from "nervous breakdowns," anxiety disorders, and other ailments went for treatment at his sanitorium, which included, baths, showers, and enemas. Many patients recovered from their illnesses due to this hydrotherapy. Kellogg also advocated light baths, in which individuals sat in cubicles of incandescent or ultraviolet (UV) light, which could also have been helpful in cases of depression. Light therapy was also advocated by alternative healer Indian American physician Dinshah P. Ghadiali (1873–1966), who developed the "Spectro-Chrome Therapy" (method of healing using different color light waves), which was popular in the mid-twentieth century. He recommended yellow, lemon, purple, scarlet, and magenta light for phobias and emotional disorders. Various colored "light saunas," are still advocated by some alternative practitioners in the early twenty-first century. These treatments may, or may not, have been successful but were largely harmless to the patient.

Another harmless treatment is an air bath. It has been advocated for general well-being and mental health from the eighteenth century into current times. The air bath often is accomplished in the morning upon rising when the person sits or walks around a cool room without clothing for a half hour or so. Kellogg recommended a hot-air bath in a closed chamber.

To treat nervousness and other mental conditions of women during the nineteenth century, physicians sometimes performed hysterectomies, including removal of their healthy ovaries, as it was still accepted by some that the uterus and even ovaries were the cause of female nervous problems. It was not effective, and some women died of infection.

From the 1927 into the 1960s, a "modified insulin shock therapy" was used to treat nervous disorders. Patients were given a dose of insulin that put them into a "subcoma," which appeared to alleviate anxiety. However, it could be a frightening experience for the patient.

In the eighteenth and early nineteenth centuries, German physician Franz Anton Mesmer (1734–1815) suggested that the body contained an invisible natural force called "animal magnetism." He claimed that the magnetic action of his hands could cure general anxiety and phobia along with other illnesses. It became known as hypnosis or hypnotherapy and is based on the hypothesis that anxieties and phobias reside in the unconscious mind. When a person is hypnotized, they are in a trance-like state, have heightened concentration, and are more open to suggestions. The therapist can suggest that their fear or obsession will disappear. It has been found to help some patients manage anxiety, phobias, and PTSD. However, some in the field of psychology and medicine are suspicious of the therapy as it has resulted in more serious mental disorders in some clients.

In the late twentieth century, as part of the new age movement, various new therapies emerged. Psychiatrist Wilhelm Reich (1897–1957), for example, believed that a life force he called the "orgone" was present in all living matter and that all mental and physical illnesses were the result of the orgone flow being restricted. Therefore, to decrease anxiety and other nervous illnesses, the person sat inside an "orgone box" for a few hours to unblock these energies. No research suggests that this is an effective treatment, and any positive results were likely due to the placebo effect.

Based upon behavior theory that suggest that anxiety disorders have been learned, "implosion," "flooding," or "exposure" therapy is used for phobias and other anxiety disorders such as posttraumatic stress disorder (PTSD). Implosion therapy works by exposing the patient directly to their phobia. For example, a person who has claustrophobia will be closed in a closet for several hours until the fear response is extinguished since nothing negative befalls them. For some anxiety or phobia disorders and PTSD, the patient is exposed to repeated imagined images of the trauma or feared objects until it no longer causes a severe anxiety reaction. These methods are accomplished in a safe and controlled environment and in many cases is helpful.

In the late twentieth century, several fad therapies emerged for "neurosis" and anxiety disorders. For example, American chemist Ida Rolf (1896–1979) developed a painful deep muscle massage called "Rolfing" or "structural integration" that claimed to rid the person of traumatic memories stored in muscles and in turn eliminate the anxiety disorder. However, there is no evidence that the technique is an effective treatment for any condition and may cause bruising to the patient. Another technique was "primal scream" therapy introduced by American psychologist Arthur Janov (1924–2017) in his *Primal Therapy: The Cure for Neurosis* (1970). He argued that neurosis was caused by the repressed pain of childhood trauma, and if it was brought to consciousness and recreated, the psychological issue will go away. As part of the therapy, the patient is encouraged to scream and express repressed anger. There is no evidence that this is a successful therapy other than placebo effect, and it has no scientific basis.

A few folk remedies were developed in the twenty-first century. Some individuals who had anxiety disorders and phobias of electromagnetic fields or potential "mind control" made "tin foil" hats or lined a regular hat with this aluminum foil.

It had little effect on anything. A "cure-all" for mental disorders popularized by the 2010s was CBD, which is a derivative of the hemp plant. Proponents claim that it helps to alleviate symptoms of anxiety disorders and PTSD. However, little research has been done on this drug to determine its effectiveness.

Mental Disorders ("Psychotic"): Bipolar Disorder and Schizophrenia

Until the late eighteenth century, mental disorders called "madness," "insanity," "craziness," or "lunacy" were considered one condition with different symptoms. However, by the late eighteenth century when people, usually women, could function, they were diagnosed as being "neurotic" or "nervous" or "hysterical." On the other hand, if people had difficulties in functioning; exhibited extreme moods; and had a loss of contact with reality, delusions, and/or hallucinations, they were considered psychotic. Until the turn of the twentieth century, psychotic behaviors were considered different forms of the same madness. However, German psychiatrist, Emile Kraepelin (1856–1926) divided psychotic symptoms into two different maladies, namely, "dementia praecox" and "manic depression."

The term "manic" is from the Greek *mainomai*, meaning "to rage," and "depression," first used in the fourteenth century, is from the Latin *deprimere*, meaning "to press down." The term "manic depression" was used throughout most of the twentieth century. In the 1950s, German psychiatrist Karl Leonhard (1904–1988) introduced the term "bipolar" mood disorder to differentiate manic depression from unipolar and other depressive disorders discussed in the "Depression and Melancholia" entry. Bipolar means "two poles," signifying the polar opposites of mania and depression. In the 1980 edition of the *Diagnostic and Statistical Manual of Mental Disorders* (DSM-3), which classifies various mental afflictions, bipolar was entered as the official term.

Bipolar mood disorders, sometimes referred to as bipolar spectrum disorders, includes bipolar I, bipolar II, and cyclothymic disorder. The disorders range from mild to severe. During an episode, elevated energy level, racing thoughts, decreases sleep, and irrational behavior often alternated with severe depression. The illness usually begins in the late teens or early adult years and lasts a lifetime. About 3 percent of the population has bipolar disorder, and it is found equally among males and females. Episodic cycle can last from a few days to years and alternate with normal periods of mood, thought, and behavior. Bipolar I is marked by severe depressive and manic episodes and is more common among males. Extreme mania can become psychotic. The depressed state is like the older descriptions of melancholy in which the person has a depressed mood and sometimes loses contact with reality. Individuals with bipolar I find it difficult to function in society without treatment, which can include mood stabilizers such as lithium, anticonvulsants, and antipsychotics, along with psychotherapy and sometimes magnetic or electroconvulsive therapy (ECT) discussed under the "Unusual Treatments" section.

In bipolar II, a less severe form of mania may alternate with depressive episodes. The person often makes poor choices and has work, social, and relationship issues. It is more common among females. Episode cycles are often more rapid.

Cyclothymia has similar characteristics to the bipolar disorders but is much milder. The person experience cyclic highs and lows, and in between the episodes, their mood is normal. However, like the other bipolar disorders, their behavior can be detrimental to work, social, and family.

The other major psychotic disorder is schizophrenia originally termed "dementia praecox," meaning "precocious madness." In 1911, Swiss psychiatrist Eugen Bleuler (1857–1939) renamed the affliction "schizophrenia." The term "schizophrenia" is from the Greek *skhizein*, meaning "to split," plus *phrēn*, meaning "mind." Schizophrenia can lead to faulty perceptions, inappropriate actions and feelings, and a withdrawal from reality into fantasy and delusion to the point where the individual cannot function in society. Common symptoms include auditory hallucinations, paranoia, delusions, and/or disorganized thinking. But these symptoms can also occur with brain tumors or alcohol withdrawal. A little less than 1 percent of the population worldwide is schizophrenic, but this lifetime illness can severely disrupt families and the patient's social and work life even with the use of antipsychotic medications. Another serious mental health issue is schizoaffective disorder. It combines the characteristics of both mood disorders and schizophrenia, and treatment is similar to those disorders.

A few individuals with these illnesses might become catatonic. In the most common type of catatonia, the patient often stares blankly and does not respond when spoken to or will sit or lie in an unusual position for a long time. In some cases, they exhibit "waxy inflexibility," when their limbs can be placed at odd angles and they will not move. Although the cause of these psychotic disorders is unknown, it is hypothesized that they have a genetic component or are the result of childhood trauma or abnormalities in the brain chemistry or circuitry.

Mental disorders have been found since ancient times and were thought to be caused by spirit or demon possession. However, Greek physician Hippocrates (c. 460–370 BCE) believed they had natural causes resulting from an imbalance of the four humors and was a dysfunction in the brain. But a few ancient Greek and Roman physicians still believed these illnesses were the result of spirit possession, and this belief was carried into Christianity. After the fall of the Roman Empire in the fifth century and under the control of the growing power of the Church, these illnesses were blamed on demonic possession or sinful behavior. If the person had mild symptoms, they were usually cared for by their family and often hidden. In Western cultures into the 1960s, stigma surrounded mental illness, and people were sent off to asylums as the condition was thought to be incurable. With the development of antipsychotic medications in the 1950s, many people with these psychological disorders can now live successfully in society. However, those who are rejected by their families, who are not compliant with treatment, or despite the best treatment slowly deteriorate often become homeless, and institutionalization is inevitable.

Humanity's fear and ignorance about mental illness have often been the subjects in works of art. For example, *Extraction of the Stone of Madness* (1494) by Dutch painter by Hieronymus Bosch (1450–1516) shows a man with a funnel hat drilling into the head of a patient (trepanation) and removing a flower bulb while a man and a woman view the procedure. American novelist Ken Kesey's

(1935–2001) *One Flew Over the Cuckoo's Nest* (1962), and later film (1975), follows various characters who were committed to a mental institution exhibiting different mental disorders. The protagonist, with an antisocial personality disorder, disobeys the rules and entertains his fellow patients. However, he is punished for his behavior by shock treatment and finally a lobotomy. The film helped to form negative popular opinions about these treatments and psychiatric institutions. The biography, *A Beautiful Mind* (1998), by German American Sylvia Nasar (1947–) and film (2001) is loosely based upon the life of the young mathematical genius and Nobel laurate John Forbes Nash Jr. (1928–2015), who developed a mathematical model of game theory in the early 1950s. He marries and soon plunges into schizophrenia and is hospitalized several times for the disorder, and he struggles with his illness, career, and personal life.

UNUSUAL TREATMENTS

Treatment for people with mental disorders has often been brutal and harsh. For example, for lethargic or depressed patients or those with catatonic episodes, some ancient Greek healers pulled the person's hair, applied hot mustard to the groin, and blew mustard and vinegar up the nostrils. This may sometimes have stimulated them. Greek physician Hippocrates (c. 460–370 BCE), based upon humoral theory, advocated trepanation—scraping or drilling a hole in the head—to allow evil spirits, pain, or air to depart to relieve many mental disorders. This treatment was used from antiquity into the eighteenth century. Some patients survived the surgery, but it did not prevent or cure anything. Roman physician Celsus (25 BCE-50 CE) recommended starvation, being shackled, and beating as a treatment. Other Roman physicians tickled patients, squeezed their feet, pulled their hair, placed mustard plasters on their heads, shouted in their ears, or made them sneeze with pepper, which was likely not very effective. Physicians also bled patients to balance their body humors to eliminate symptoms, which did not reduce symptoms and led to weakness.

During the Middle Ages when a person was thought to be possessed, or to be a witch, they were referred to the clergy. Physicians were only called in to confirm or reject a diagnosis. Since it was believed that people exhibiting abnormal behaviors could infect others, they were sometimes executed by drowning or burning at the stake. Some were tied up, shackled with chains, and locked away in "lunatic asylums" that were established in the fifteenth century to isolate them from society. In many cases, family members and villagers tolerated mild eccentricities and only locked up more dangerous people. It was believed that the insane were insensitive to most sources of discomfort and that those with mania possessed unusual strength. This resulted in few comforts for the institutionalized patients. Locked doors and window bars to keep patients from escaping the asylums were also common.

One of the most notorious and filthy asylums was Bethlem Royal Hospital (Bedlam) in London that was operated from 1247 through 1948. During its earlier years, patients received inadequate meals and often lay in their own excrement while being chained to posts. During the seventeenth into the mid-eighteenth

People in the eighteeth and nineteenth centuries with mental disorders were often locked up in asylums. The middle class would visit the asylums to watch patient's antics as entertainment. This is illustrated in British artist William Hogarth's (1697–1764) engravings. (National Library of Medicine)

century, visitors could visit and watch patients' antics as entertainment. Patients were often beaten and confined to cages to maintain order. By the late eighteenth century, the asylum became a place of therapy as well as human storage. For people exhibiting agitation, such as that found in patients of bipolar or even schizophrenia, opium was used to calm them. Sometimes it was mixed with cannabis. Although it was calming, patients often became addicted to the drugs and suffered withdrawal symptoms when it was discontinued. For patients who bit off their fingers or lips, the front teeth were removed. In the eighteenth and into the late nineteenth centuries, bloodletting, purging and emetic medications, starvation, and very hot and cold baths were used to reduce agitation from manic depression or lift mood in depressed patients. Patients were confined to the tub with only their head sticking out to receive food and water. Often, they were left in the baths for days to wallow in their own urine and feces. These treatments were used to "bring them to their senses." Into the mid-twentieth century, asylum staff used fear to keep their charges in line such as threats of lobotomies, scolding, confinement, strait jackets, padded cells, or loss of privileges. These were terrifying to patients.

In the early twentieth century, insulin shock or "coma therapy" was developed and mostly used for schizophrenia. For several weeks, patients were injected daily with enough insulin to put them into a coma. They were revived with intravenous or nasal-gastric glucose. Sometimes it was helpful, but the treatment could cause convulsions, obesity, brain damage, and death.

In the 1930s, frontal lobotomy was developed for the treatment of schizophrenia and bipolar disorders. It was popularized by American physician Walter Freeman (1895–1972) who would insert sterilized icepick-like instruments through the eye orbit into the prefrontal area of the brain to destroy the tissue. It was extensively used in the 1940s and 1950s and fell into disrepute by the early 1960s. Lobotomies sometimes were successful but often led to serious personality changes, resulting in apathy, lack of initiative, social disinhibition, poor judgment, incontinence, and sometimes coma or even death.

Another treatment developed in the 1930s was "shock therapy" or electroconvulsive therapy (ECT) to produce seizures, which sometimes was effective for schizophrenia and affective disorders. Although electric fish and eels had been placed on the head in antiquity and electricity from batteries was used from the eighteenth century to cure mental disorders, ECT was not commonly used until the 1940s. Severe convulsions were common in earlier procedures that could result in broken limbs or spine, but by the 1970s, the patient was given muscle relaxants and sedation to prevent this. Approximately 70 percent of ECT patients are women, and more than a third are 65 and older. Stigma attached to the use of ECT led to a decline in its use during the 1960s along with the emergence of effective psychiatric medications. Although it can be helpful, variable memory loss, confusion, and high blood pressure can occur in some patients around the time of treatment. In the mid-1980s, transcranial magnetic stimulation (TMS) was developed. It creates magnetic fields to stimulate nerve cells in the brain to improve depressive and psychotic symptoms and has fewer side effects.

In the early twenty-first century, some people who were paranoid believed that a tin hat, in other words a hat that was lined with several layers of aluminum foil, would shield the brain from threats such as government mind control, electromagnetic fields, mind reading, and other conspiracy theories. The hat did nothing and was harmless to the patient.

Neurological and Dementia Disorders

The major neurological disorders include epilepsy (discussed in its own entry), dementia, which used to be called "senility," and Parkinson, which used to be called "shaking palsy." Common symptoms of dementia include impaired memory and judgment, confusion, personality and behavior changes, impaired communication, and language deterioration that are severe enough to interfere with the tasks of daily living. Dementia is a collective term used to describe the symptoms of progressive cognitive decline and is not a normal part of aging. A high proportion of Parkinson disease patients also exhibit dementia over the course of their illness.

The most common causes of dementias are Alzheimer's disease, followed by vascular dementia, Lewy body dementia, and frontotemporal dementia. Several rare causes of dementia are also found. For instance, Creutzfeldt-Jakob disease is thought to be caused by prions acquired from infected meat, and Huntington's disease (chorea) is caused by a dominant gene. Two infectious diseases—HIV and tertiary syphilis—can also lead to dementia. Some other neurological diseases found in adults include ALS (amyotrophic lateral sclerosis) or Lou Gehrig disease, which affects nerve cells in the brain and spinal cord, causing loss of muscle control. There is no cure for any of these neurological conditions. Some medications may reduce the symptoms.

In the past, from an inadequate intake of niacin (vitamin B3) from protein, pellagra was a common nutritional disease that caused dementia but could be reversed with sufficient protein intake. Chronic exposure to lead, other heavy metals, and pesticides can cause dementia symptoms as can some psychotropic medications, but the disorder can often be reversed when the toxic substance is removed. Years of heavy drinking can result in Wernicke–Korsakoff syndrome with symptoms of dementia; however, it cannot be reversed.

Alzheimer's disease accounts for 60–80 percent of all cases of dementia. Around 10 percent of people over the age 65 have Alzheimer's disease with most being over the age of 75. Among those with Down syndrome, at least 25 percent of those older than 40, and more than 50 percent of those over 60, develop Alzheimer's disease. This disease is characterized by abnormal clumps called amyloid plaques and jumbled fiber bundles called tau tangles in the brain. German psychiatrist Alois Alzheimer (1864–1915) in 1906 discovered these microscopic brain changes, and the disease was named after him. About two-thirds of patients with Alzheimer's are women.

Similar to Alzheimer's is Lewy body dementia (LBD) that is frequently misdiagnosed as Alzheimer's or Parkinson's disease in its early stage. LBD accounts for about 5 percent of all dementia cases. It is caused by abnormal deposits of alpha-synuclein, or Lewy bodies, in the brain that were discovered by German American Frederick Lewy (1885–1950) in 1912 while doing autopsies on brains of people with Parkinson's disease. He found, depending upon where the abnormal deposits were located, that the person would either develop dementia with Lewy bodies or Parkinson disease. In LBD, as opposed to other conditions, it may begin with REM sleep behavior disorder where the person talks or acts out their dream in their sleep, has visual hallucinations, or their attention and alertness varies widely from day to day, or even during a single day. Other patients will start out with a movement disorder, leading to a diagnosis of Parkinson's disease, and later develop dementia and other symptoms such as REM sleep disorder.

Many people with both LBD and Parkinson's dementia also have plaques and tangles, which is a characteristic of the brain changes found in Alzheimer's disease. Lewy bodies in the brain stem cause a disruption in the production of the brain neurotransmitter dopamine. Too little dopamine can cause Parkinson disease (PD), which is characterized by tremor, slow movement, rigidity, and balance issues. It is estimated that PD affects 1 percent of the population over the age of 60. In 1817, English physician James Parkinson (1755–1824) was the first to diagnose the disease as a neurological condition in his *An Essay on the Shaking Palsy*. Although there is no cure for this degenerative illness, in 1961 L-dopa was first given to patients and was found to help reverse some symptoms. Likewise, deep brain stimulation surgery, where electrodes are implanted into a specific part of the brain, was first approved in 1997 to treat PD tremors.

Frontotemporal dementia is found in around 1–5 percent of all dementias and has similar characteristics to the other conditions. Vascular dementia is caused by a series of "little strokes" that deprive the brain of oxygen. The person may fall; exhibit shuffling steps; suffer from incontinence and memory loss; and have inappropriate behavior and other symptoms of dementia. Vascular dementia is considered the second most common cause of dementia and accounts for 5–10 percent of all cases. These conditions can be extremely stressful to family and loved ones, and to patients in the early stages of the disease who may be confused of what is happening around them.

The first-known use of the term "dementia" was in the late eighteenth century from the Latin *demens*, *dement*, meaning "out of one's mind." Likewise, the first-known use in English of the term "senility" from the Old French *senile*, from the Latin *senīlis*, meaning "old age," was in the late eighteenth century. Senile or senile dementia tended to be most common words for these neurological conditions until the late twentieth century. From ancient Greece and Rome into the twentieth century, senility was considered a normal and inescapable part of aging. In addition, Greco-Roman physician Galen (129–210) noted that "shaking palsy" was also a disease of aging. Galen's writings influenced medicine into the nineteenth century. In the Middle Ages, science and research in the West was almost nonexistent, due to religious beliefs and the power of the Christian Church, which sometimes considered senility and palsy as signs of possession of the devil even in aged individuals. Autopsies to determine causes of death were generally forbidden. Therefore, it was not until the late seventeenth century that widespread dissection was acceptable, and scientists were able to observe brain anomalies in these patients.

Dementia has been depicted in some art forms. For instance, in William Shakespeare's (c. 1564–1616) *King Lear*—one of Shakespeare's most tragic plays—the king progresses into what has been interpreted as Lewy body dementia. The king knows something is cognitively wrong and spirals down through the steps of dementia throughout the acts. Another disturbing play that shows dementia is the 2012 production, *Le Père* (*The Father*), by French playwright Florian Zeller (1979–). It was produced as a film in 2020. In this one-act play, the audience look through the eyes of a proud man experiencing rapidly progressing dementia. At the end of the act, he is unable to recognize where he is and who is his real family, and the audience is also confused, not knowing realty from delusion. American artist William Utermohlen (1933–2007) was diagnosed with Alzheimer's disease at the age of 61. From the time of his diagnosis until he was admitted to a nursing home, he began to paint self-portraits as he progressed through the different stages of Alzheimer's. His first portrait shows details of his face, and his last painting was an abstract scribble in the vague shape of a head.

UNUSUAL TREATMENTS

Since dementia was thought to be a normal part of aging for most of history, few treatments were offered, and families took care of their aged relative. However, in some periods of time, such as the Middle Ages into the seventeenth century, elderly women with dementia were often thought to be witches or possessed by the devil and were burned at the stake. They were also locked away in "mental institutions" into the 1960s and by the 1980s into "nursing homes" now called long-term care or memory-care facilities.

In the mid-twentieth century, sleeping pills and the newly developed tranquilizers were administered to dementia patients. However, they often made the patient more confused, agitated, and anxious. Agitated patients or those who wandered were often restrained in chairs so they could not move around, resulting in further confusion and paranoia among patients. Since there is no cure for these diseases, folk and alternative remedies were also tried. By the twenty-first century, soothing music, ice cream, and over-the-counter pain medicine was found to calm agitated patients due to relief from minor pain and the introduction of a pleasure into their life. Even a glass of wine was found to result in happier patients. Alternative remedies were also tried. Aroma therapy was promoted as a way of helping those with any type of dementia. Aroma therapy proponents claim that fresh citrus oils like sweet orange in the morning helps people perk up. They propose that woody oils like cedarwood or sandalwood in the afternoon have a calming effect and that lavender at night promotes sleepiness and reduces agitation. Some evidence suggests that smelling these essential oils may have some effect on the person. Lavender has been found to damp down the glutamate NMDA-receptor in the brain, resulting in decreased agitation. The oils are most often put into a diffuser or rubbed on the skin. In some cases, however, they can cause skin irritations. Turning bright lights on in the early evening also prevents agitation, which is sometimes referred to as the "sundowner syndrome."

By the 2010s, advertisers marketed brain and memory supplements, some of which contained jellyfish compounds. However, research has not found these supplements to slow, reverse, or stop cognitive decline or dementia. In the 2010s, CBD (Cannabidiol) from the *Cannabis sativa* plant was advertised as helping to ease anxiety in dementia. A few studies have shown that CBD and cannabis may help to manage symptoms of agitation and aggression.

For "shaking tremors," Parkinson in 1817, based upon ancient humoral theory, recommended bloodletting from the neck and blistering. He inserted small pieces of cork into the blisters to let out a purulent discharge to divert blood and inflammation from the brain and spinal cord, which Parkinson believed was the seat of the problem. Scottish physician Alexander Macaulay (1783–1868) in his *A Dictionary of Medicine* (1831) likewise suggested these treatments. After these therapies, Macaulay applied stimulants to the limbs, including ammoniated oil, camphorated oil, turpentine or even stinging nettles, mustard, and electric sparks. These treatments were not effective, and the patient could have developed infections and become weak from loss of blood. In the mid-nineteenth century, some physicians used anticholinergic agents such as hyoscyamine from the deadly nightshade (Atropa belladonna) plant for Parkinson's disease, which helped reduce tremors. However, with high dosages, poisoning or even death could occur. It is still used today to control tremors and spasms.

French neurologist Jean-Martin Charcot (1825–1893) developed treatments for Parkinson's disease. He had observed that after bumpy carriage, train, or horseback rides Parkinson's patients had decreased symptoms. Therefore, to replicate this bouncy rhythmic movement, he developed a "shaking chair" powered by electricity. One of his students then invented a shaking helmet to vibrate just the brain. Other physicians administered hemlock and marijuana, which were found to be effective in temporarily reducing tremors. Cannabis and opium in combination were found to be particularly useful. However, their use could lead to addiction, and too much hemlock could lead to poisoning.

In the late nineteenth century, a Russians scientist developed a system to stretch the spinal cord. A halter was attached under the chin, the patient was hoisted off the ground, and gravity put traction on the spinal cord. After this treatment, muscle rigidity and some sensory symptoms improved, but tremor were not reduced. However, the treatment was dangerous as it could lead to a spinal cord injury and death.

Obesity and Weight Loss

Obesity is abnormal or excessive fat accumulation, particularly around the abdomen, and is a major risk factor for several chronic conditions including diabetes, cardiovascular diseases, and cancer discussed in other entries. People with all these conditions together are sometimes considered to have "metabolic syndrome," which is associated with higher mortality rates. Some people can be overweight with few risk factors, as is found in weightlifters and Suma wrestlers who do not carry visceral fat. A crude measure of overweight and obesity is the body mass index (BMI). To derive this, a person's weight (in kilograms) is divided by the square of his or her height (in meters). A person with a BMI from 25 to 29 is defined as overweight. A person with a BMI of 30 or more is usually considered obese, and a BMI over 40 indicates extreme obesity.

Overweight and obesity at one time were thought to be a problem for the rich due to an abundance of food. But by the twenty-first century, this chronic health issue began to increase in low- and middle-income people and nations. Globally, by 2020, about 40 percent of adults aged 18 years and over were overweight, and 13 percent were obese. Risk factors include both hereditary and environmental factors in addition to high caloric intake and lack of physical activity.

The term "obesity" first appeared in the English language in the early seventeenth century for excessive fatness or corpulence. It is borrowed from the Middle French *obesité*, from the Latin *obēsitāt-*, from *obēsus*, meaning "fat or stout." The attitudes toward obesity and overweight have waxed and waned over the centuries. Paleolithic hunter-gathers societies often had little food and were emaciated. Fatter people, due to more access to food, were associated with wealth and power. But as agriculture developed and more food was available, obesity began to be seen as unnatural. For example, Greek physician Hippocrates (c. 460–c. 375 BCE) believed that all diseases began in the gut and that everything in excess was opposed by nature. Christianity, by the fourth century, considered gluttony one of the seven deadly sins. This led to the Christian tradition of considering obesity to be unattractive. Slight plumpness among women was still acceptable as extreme thinness often meant ill health and lack of fertility. By the seventeenth century, difficult childbirth among corpulent women was noted. However, obesity as a chronic condition with serious health consequences was not perceived until the mid-nineteenth century.

Concern about dieting and weight loss began in the early nineteenth century as Victorian women wished to have small waists and wanted to look pale and frail as a new beauty ideal. Rebellion against this frailness emerged in the late nineteenth century. As part of a clean living movement, in which healthy living was emphasized and fit and natural figures were considered beautiful rather than corseted waifs, exercise and better diets were emphasized. However, due to societal changes in the post–WWI years, many weight-loss schemes to become slim were perpetuated during the flapper era when thinness was desired. After this trend, less emphasis was placed on diet and weight during the depression and WWII years. In the postwar era, beauty standards changed again to extreme thinness, which emerged out of the modeling industry. But in a world obsessed with overweight and weight reduction, a fat acceptance movement to promote dignity among "people of size" for equal access to opportunities emerged by the end of the century. Also, by the late twentieth century, for the morbid obese, medical management offered the new bariatric surgeries to reduce weight and prevent other chronic conditions.

By the twenty-first century, beauty was viewed as a well-toned body among both men and women. In addition, concerns of a growing worldwide obesity crisis emerged. Numerous diet and weight-loss fads arose, but most were only temporarily successful, and some had serious health consequences. Clinicians recommended that weight loss for overweight or obese individuals was best achieved by reducing caloric consumption and increasing exercise. Self-help programs and support groups also developed to give support.

Corpulence and overweight in the past are reflected in the arts and literature. The earliest carving, the Venus of Willendorf, is of a woman. This limestone figure from around 25,000 BCE has huge pendulous breasts and a large belly and may have been a fertility goddess. In the early modern period, obese men were often portrayed as comic characters. An example is Falstaff, companion to Prince Hal, in William Shakespeare's (1564–1616) *Henry IV*, who is depicted as a heavy eater and drinker. However, in this era the standard of beauty was a full figure for women, but not obesity. In modern times, few TV series have depicted fat individuals as protagonists. One exception was actor William Conrad (1920–1994), the central character in several TV drama or detective series in the late twentieth century. Some TV reality shows emerged in which obese people attempted to lose massive amounts of weight. For example, in *My 600-lb Life* (2012–) reality show, people who weight over 600 pounds must first demonstrate they can lose a certain amount of weight through strict dieting and exercise before they are eligible to receive bariatric surgery.

UNUSUAL TREATMENTS

Of all medical and health conditions, more methods have been concocted to reduce weight than any other health condition. However, many were, and are, unhealthy, dangerous, or ineffective.

Nostrums and Drugs for Weight Loss

In the nineteenth and early twentieth centuries, substances to boost metabolism such as thyroid extracts became popular. Dried and powdered thyroid glands from cows and pigs were advertised such as "Dr. Newman's Obesity Pills." They did cause the person to lose weight but also caused hyperthyroidism that led to rapid or irregular heartbeat, excessive sweating, and even death. A tapeworm weight-loss fad also started in the nineteenth century. The person bought and ingested beef tapeworm eggs, which grew in the intestines. The worm consumed much of the food the person ate, causing weight loss, but it could grow to more than 50 feet long and live for years. Although finally banned in developed nations, the tapeworm for weight loss is still advertised on the internet, and people can also go to Mexico to obtain the treatment. After the individual has lost the desired weight, the tapeworm is killed by medication. This folk remedy is dangerous as it can cause anemia and loss of essential nutrients and disrupt the functioning of body organs including lungs, liver, and brain, leading to seizures and death.

In the 1920s, *Lucky Strike* cigarette advertisements recommended that women smoke a cigarette instead of eating food to keep slim. Since cigarettes contain nicotine, which is a stimulant that increased their metabolic rate and is an appetite suppressant, people were often able to keep their weight down. However, smoking was later found to lead to cardiovascular disease and cancers. This nostrum is still used in the twenty-first century by some people. A compound called dinitrophenol (DNP) entered the market as a weight-loss remedy in the mid-1930s. It rapidly increased metabolism and did result in rapid weight loss. However, the substance caused a high body temperature and sometimes led to blindness and death. Due to these effects, the fad only lasted a few years, but it is still advertised on the internet under different names for weight loss.

In the immediate post–WWII era, when amphetamines were found to decrease appetite, many women took over-the-counter Benzedrine diet pills, "bennies," to lose weight. In 1959, the U.S. Food and Drug Administration (FDA) made it a prescription drug. However, besides weight loss, they could cause psychotic symptoms, and in the 1970s, they were put under strict restrictions in Western nations. Physicians decreased prescribing them for weight loss. Additionally, an even more dangerous obesity control drug, fenfluramine, was prescribed in Europe in the 1960s and in the United States in the 1970s. Sometimes it was combined with the drug phentermine and sold under the name of Fen-Pen. But fenfluramine was found to cause serious health issues such as heart valve damage and was taken off the market in 1997.

In the 1950s, the human choriogonadotropin (HCG), which is produced during early pregnancy and primarily used to treat female fertility problem, was also used to reduced obesity. It was taken daily with a very low-fat, extreme weight-loss diet of 500 calories per day and is still advertised on the internet. Proponents claim that HCG resets the metabolism and prevents weight gain. However, studies have shown it is ineffective, and it is the limited calories that cause weight loss. Headaches, depression, and fatigue are common results from this fad. Sleeping for a long period of time for weight loss emerged in the 1970s. The person takes sleeping pills to sleep all day for several days and does not eat. People may lose weight, but the oversleeping can result in blood clots and muscle loss from lack of exercise.

In the 2010s, a folk remedy, which may have arisen from the modeling industry, suggests dipping cotton balls into orange juice or smoothies and then swallowing them. The balls are intended to make the stomach feel full without weight gain. Weight loss might be achieved, but the cotton polyester mix is not digestible. People following this dangerous folk fad risked choking or obstructing their intestinal tract, potentially leading to death.

Fasting and Diets

Several individuals in the first decade of the twentieth century recommended fasting, not only for weight loss but also for a healthy life, including American physician John Harvey Kellogg (1852–1943) who ran the Battle Creek Sanitarium. He used water therapy as the primary treatment, which consisted of enemas, baths, saunas, and drinking a lot of fluids along with vegetarian meals. People did lose weight but often lost muscle mass. Along with Kellogg's regimen, Horace Fletcher (1849–1919) launched a popular fad called "Fletcherism," of chewing each bit of food numerous times to the point where it liquefied and became tasteless. Since people were not eating as much, weight loss occurred.

Starvation diets became popular in the mid-twentieth century. In the "Prolinn Diet," the only thing consumed was a low-calorie liquid drink derived from collagen found in the hooves of slaughterhouses animals such as cattle and pigs. Other starvation diets included the consumption of just lemonade, maple syrup, condensed milk, or cayenne pepper mixed with water for about a week to 10 days. These diets caused weight loss but also led to health problems. From the 1950s into the 2020s the "cabbage soup diet" has been advertised. It is a short-term weight-loss program that is low in calories, fat, and protein. It can reduce weight but lacks nutrients needed by the body. Another twenty-first century diet advertised on the internet is the "baby food diet." One or two adult meals a day are replaced with baby food. It can cut calories but is not sustainable and lacks fiber. Weight loss using apple cider vinegar has been advertised for decades, and research is mixed as to whether it is safe.

By the twenty-first century, the most popular fad diets included the ketogenic (high fat, low carbohydrate), the carnivore (only meat and other animal products), and paleo (high protein, low carbohydrate). They were advertised through books, television, and the internet. One of the first high-protein, low-carb diets created in the 1970s was the Scarsdale diet developed by a Scarsdale New York physician. This 7- or 14-day regime was designed for rapid weight loss by drastically reducing calories. The "grapefruit diet" is similar to this diet with a 12-day plan. Grapefruit or grapefruit juice is consumed at every meal. Proponents claim that grapefruit juice can burn fat, but there is no scientific evidence for this. Ketogenic diets include the Atkins or South Beach diets, in which much fat and protein are consumed. Weight is lost and blood cholesterol often declines, but it is debated as to its safety over a long period of time.

Another diet regimen that began in the 1970s and became popular by the twenty-first century was the "paleo," "hunter-gather," or "caveman" diet. It is based on foods that might have been eaten during the Paleolithic era before the emergence of farming and is devoid of grains. This high-protein, low-carbohydrate diet consists of meat, wild game, fish, fruits, vegetables, nuts, and seeds. Similar to this diet is the "clean eating" diet that had emerged by the mid-2010s. Proponents believed that only eating "natural foods" and avoiding processed foods would result in better health and weight loss, but little evidence for this is found. Some variations of this diet are the gluten-free, grain-free, dairy-free, carnivore diet that excludes all foods except meat, eggs, and low-lactose dairy products.

Gadgets and Equipment

In the nineteenth century, sweating became fashionable to reduce fat. Rubber corsets were invented as a result of the recent vulcanization of rubber. Wearing one caused sweating and temporary weight loss by dehydration. These along with vapor baths and dry heat and light therapy to aid sweating were also used. Heat stroke could have been a side effect of this remedy, and these regimens did not produce long-lasting weight reduction. Variations of belts to reduce fat around the abdomen were advertised into the twenty-first century. These gadgets caused sweating but were not effective in eliminating fat. A similar method is a slimming belt that compresses the fat cells. The waist may appear smaller once the belt is removed, but the fat cells quickly return to normal shape.

By the early twentieth century, mechanical belts were advertised as a method for losing weight and were popular in gyms in mid-century. The belts massaged the waist and abdomen and supposedly removed fat but were not effective. In the early twenty-first century, shaking tables, sometimes called vibrating machines, upon which a person stood several times a week were advertised as an aid to reduce fat. Some evidence exists that users may lose a few pounds using this mechanical device several times a week.

Another remedy of the early twenty-first century for weight loss was ear stapling based upon the principles of acupuncture. A staple is put into the ear that supposedly stimulates the pressure point that decreased the appetite. Mixed results were found from this remedy. In the 2010s, a Japanese company suggested that the color blue suppressed the appetite as red and yellow are often used for food advertisements. Therefore, they created blue-tinted glass lenses to make the food look unappetizing to suppress the appetite. It was not effective.

By the twenty-first century, crystal therapy became popular. Proponents believed that vibrations from crystals could influence different organs of the body. The royal blue sodalite crystal, for example, was thought to boost metabolism, reduce fat, and encourage people to make healthy food choices. This folk remedy has little scientific evidence. Another folk treatment is ingesting or smelling essential oils. Proponents of the use of these oils claim they can help in weight reduction. Some oils that are purported to decrease appetite, or even burn fat, include peppermint and lemon oil and spices such as ginger and cinnamon. Any weight loss with the use of these folk remedies is likely from the placebo effect and lower caloric consumption.

In the 2010s, cryotherapy or freezing fat cells to destroy them became popular. In one method, a person stood in a chamber that covered all the body except the head for a few minutes in a temperature of around −200 degrees. The theory behind cryotherapy for weight loss is that extreme cold temperatures force the body to work hard and burn more calories. Studies suggest that some fat cells may be destroyed using this method, but it can be dangerous for people with severe hypertension and heart problems. Another form is called cryolipolysis, or "fat freezing." In this treatment, the operator places paddles on the fat area, such as the abdomen, to freeze and break down fat cells, which was often effective. Smaller devices that are belted around the waist and are advertised for home use had minimal effect.

P

Parasites and Worms

A parasite is an organism that lives on, or within, another organism called a host. The parasite uses the host's cells and fluids as a resource to maintain and often propagate itself. Some parasites carry serious diseases, and others, by their activity, cause sickness or death. Parasites include *ectoparasites* that live on, rather than in, their hosts and include lice and fleas discussed under the "Bites and Stings: Insects and Arachnids" entry. *Protozoa* include the single-celled organisms known as plasmodium, such as the malaria protozoa, which is discussed in the "Malaria and Intermittent Fevers" entry, or *Giardia lamblia* discussed in the "Bowel and Rectal Issues" entry. Finally, *helminths* or worms live inside the host. Worm parasites include roundworms, hookworms, pinworm, trichina spiralis, and tapeworms. They can infect both humans and animals. About 10 percent of the developing world is infected with intestinal worms, but parasitic worm infections are not common in developed nations.

Numerous parasitic worms exist. Some can be deadly, and others have few symptoms. Many are transmitted through soil or by unwashed hands soiled with fecal material that in turn contaminates food or drink with the worm's eggs. Parasites until the turn of the twentieth century were thought of as normal pests. Due to the developments of sanitation and clean water, their prevalence has decreased worldwide, especially in developed nations, and they became vilified.

Several roundworms infect humans. For example, the roundworm *Ascariasis lumbricoid* infects the small intestine and usually does not cause symptoms but may be seen in feces. Large infestations may cause lung damage and intestinal issues. It is the most common roundworm infection and enters the body through consuming food or drink contaminated with the round worm's eggs. About 25 percent of the world population is infested with this parasite, mostly in developing nations.

Hookworm infections used to be common in the southeastern United States into the 1930s when people defecated on the ground, where the eggs were deposited in the soil, and not in outhouses or indoor toilets. The larva of this roundworm bores into the skin when walking barefoot on contaminated ground. Early symptoms include itching and a rash ("dew itch") usually on the foot. The worms live in the small intestines and cause blood loss leading to anemia, malnutrition, and intestinal disease. Another common intestinal round worm is the whipworm (*Trichuris trichiura*), whose eggs also live in the soil and cause similar problems and is transmitted by poor hygiene.

A rare roundworm condition that may lead to blindness, liver problems, and death is the raccoon roundworm, *baylisascaris*, that is passed through raccoon feces. It is found in North America where racoons are present. Young children are most at risk when they play in soil contaminated with the worm's eggs and put their fingers in their mouths. The most common worm infection in developed nations is the pinworm (*Enterobius vermicularis*). It is an extremely small roundworm and most commonly occurs among children, people who take care of infected children, or are institutionalized. In these groups, the infection can be up to 50 percent. When the eggs of the worms are ingested, infection is established in the colon. At night, the female lays eggs around the anus causing an itch and can contaminate many surfaces, thus spreading the infection. They cause concern among most parents but do not lead to serious issues.

A roundworm from the trichinella family causes trichinosis. It is generally contracted by eating undercooked meats including pork or wild game such as bear and cougars, which contains larval cysts. During the initial infection, invasion of the intestines often results in diarrhea, abdominal pain, and vomiting. However, when the larva invades the tissues and forms cysts in the muscles, it may lead to muscle aches. A common symptom is swelling around the eyes. The larva may cause central nervous system damage and inflammation of heart muscle or the lungs. The disease was common in North America, but by the late twentieth century, with better feeding of pigs and thorough cooking of meat, it became rare. Most cases in the 2020s are from wild game.

In addition to roundworms, flat or tapeworms can affect the intestines, liver, or lungs. These flat worms are contracted by eating undercooked beef or pork contaminated with larvae cysts. One tapeworm that infests humans is from the *taenia* family. Its ingested cysts develop into adult tapeworms and attach to the small intestines. It develops segments called proglottids, which produce eggs. Adult tapeworms can measure more than 80 feet (25 meters) in length and can survive for years in a host and can cause weight loss. Most of these parasitic worms can be eliminated by over-the-counter anthelmintics or prescription medications.

The term "parasite" was first used in English in the mid-sixteenth century. It is from Old French, from the Latin *parasitus*, from the Greek *parasitos*, meaning "eating at another's table." The term "worm" was first used before the twelfth century and is from Middle English, from the Old English *wyrm*, meaning "serpent," and the Latin *vermis*, meaning "worm." Worms and other parasites have always been a part of the human condition. In antiquity, tapeworms and round worms were recognized. The *Papyrus Ebers* (c. 1500 BCE), for example, notes that "AAA" disease was likely the intestinal parasite *Ankylostomum duodenale*, or Egyptian hookworm. Greek physician Hippocrates (460–375 BCE) identified worms from domesticated animals, fish, and humans. Roman physicians, such as Celsus (c. 25 BCE–c. 50 CE) and Galen (129–200), were familiar with the human roundworms, pinworms, and tapeworms. Based upon humoral theory, intestinal worms were thought to arise in the gut due to an excess of "phlegmatic humors," which were produced by overeating or by eating the wrong food. Spontaneous generation in the gut was another theory. Archaeological evidence from the Middle Ages has found high levels of roundworms and tapeworms in latrines. After the fall of the Western Roman Empire in the fifth century, studies in the West on worms did not begin until the seventeenth century when numerous new species were identified. Their relationship to human infection and disease, however, was not discovered until the early nineteenth century. For example, the association between trichina infections and pigs had long been recognized. But the encysted larvae in pig muscles was not noted until 1821, and they were not seen in humans until 1835.

Some science fiction novels and TV series have depicted parasitic worms. In the film *The Rath of Kahn* (1982), a part of the *Star Trek* science fiction series, the evil Kahn out for revenge against Captain Kirk, commander of the Enterprise, places a "Ceti eel larva" into two captured Enterprise officer's space helmets. These wormy eels slither into the ear canal and brain and take control of the mind. As they grow, they cause "madness and death." After a convoluted plot, one officer kills himself rather than assassinate Kirk, and the wormy eel oozes out of the other officer's ear, thus saving his life.

UNUSUAL TREATMENTS

At least from antiquity, humans have attempted to eliminate worm infestations. The *Papyrus Ebers* lists several recipes to eliminate worms. The healer administered an infusion of pomegranate root in a large jug of water. The alkaloid in the root may have paralyzed the worm, which would have let go of the intestinal wall, and been expelled in the feces. Another concoction was the "Inner-of-the-Fruit-of-the-Caster-oil-tree" added to yeast and water. Castor oil has been used as a purgative for centuries and would have caused diarrhea but may not have expelled or killed any worms. Another recipe called for "Flesh-of-a-Live-Cow" combined with freshly baked bread, incense, lettuce, and sweet beer. This was drunk for four days as was a mixture of "Ass's Milk with Dough, Wine, and Bitter Beer." These would have been harmless to the patient but not effective in removing worms.

Tapeworms were often difficult to expel, so the healer first administered worm cakes. The first cake was composed of field herbs and natron (the salt used for embalming) and baked into a cake with cow's bile. The other cake, in addition to field herbs, was composed of "Heart-of-the-mesa-Bird," along with honey, wine, and sweet beer. Neither of these would have been effective in removing tapeworms and would have been relatively harmless to the patient.

One recipe that may have helped to expel tapeworms included acanthus resin, peppermint, and wild lettuce. It was used as both a poultice applied to the body or strained and drunk. Since it contained flavonoids, alkaloids, and tannins, it may have in some cases helped to loosen the hold of the tapeworm in the intestine and caused minimal harm to the patient.

In ancient Greece, Hippocrates treated worms by administering the crushed root of the wild herb *seseli* mixed with water and honey, which was drunk. The oil from the *seseli* plant has anti-inflammatory and antifungal properties but likely did not expel worms although it was harmless to the patient. Galen in the first century advised bloodletting and a modified diet to balance the humors, which was ineffective.

In the Middle Ages, religious cures were tried to cure worms and other parasites. Wisewomen or priests would give blessings over the belly of a sick children to rid worms, which was unsuccessful. In the late Middle Ages, physicians treated worms with bitter, purgative plants such as wormwood or gentian. Wormwood has been shown to kill the parasites and causes severe diarrhea, which would have passed dead worms out of the body. However, it could cause dehydration. In the fourteenth century, several treatments for worms were proposed. These included eating garlic, vinegar, and peach tree leaves. Peach tree leaves contain cyanide, which may have killed worms. But it could also have killed the patient or resulted in breathing difficulties or shock if too much had been ingested. In the fifteenth century, based upon humoral theory, bloodletting by cutting a vein or leaches were tried, which did not eliminate worms or other parasites.

In the sixteenth century, Swiss physician Paracelsus (1493–1541) and other physicians administered calomel (mercurous chloride) for many illnesses including worms. Large doses of mercury chloride could have killed the worms, but frequent or long-term use could have resulted in toxic effects that included excessive salivation, gum inflammation, loosening of the teeth, gastrointestinal upset, and arm and facial tremors or even dementia. Calomel was recommended for elimination of worms into the late nineteenth century.

Scottish physician William Buchan (1729–1805), in his *Domestic Medicine* (1774), also recommended the purgative calomel along with jalap (root of the Mexican climbing plant) every other day. On the odd days, he administered powdered tin mixed with honey. The tin could cause nausea, vomiting, and diarrhea. This toxic regime may have worked to eliminate worms but likely harmed the patient. Sulfur mixed with honey was also used to expel worms. It caused diarrhea but was not toxic to the patient.

In the nineteenth century, American physician Henry Hartshorn (1823–1897) in his *Essentials of the Principles and Practice of Medicine* (1867) recommended ingesting an ounce of turpentine or petroleum to eliminate worms, which could be toxic to the patient. He also administered carbolic acid (phenol) mixed with licorice extract. This may or may not have worked but could have caused severe corrosive injury to the mouth, throat, esophagus along with nausea and vomiting in high doses. Coma and seizures and death could also occur. Hartshorn also suggested that the patient eat pumpkin seeds on an empty stomach, which was harmless but likely did not get rid of worms. Physicians in this era also gave arsenic compounds to eliminate roundworms and tapeworms, which may have poisoned both the worms and the patient.

In the Victorian era, tapeworms were advertised for weight loss. Women who wanted to be extremely thin to achieve what society viewed as beautiful would swallow a tonic that contained a tapeworm egg. When the egg hatched, the worm grew, usually inside the intestine, and used nutrients from whatever the person was eating. This did cause weight loss, and when the person wanted to get rid of the tapeworm, they used various substances such as gentian root to kill the worms. However, the worm could attach itself to other organs and cause serious issues or even death. Tapeworm eggs in the 2020s are still sold via the internet as a folk treatment. To eliminate these worms, "parasite cleanses" were advertised. These concoctions contain wormwood, oregano, black walnut, and clove oil along with other plant extracts. Alternative medicine proponents—in addition to these substances—suggest that raw garlic, pumpkin seeds, papaya seeds, pomegranates, beets, and carrots can help eliminate parasitic worms. Research is mixed as to their effectiveness. These plants and foods, for the most part, are harmless if consumed in directed doses.

Plague

Plague is caused by the *Yesenia pestis* bacterium. Signs of the disease include pains in the chest, cough, high fever, chills, vomiting of blood, red splotches on the skin, and swollen lymph nodes, called "buboes," which over time turn black. In addition, gangrene sometimes occurs in fingers, toes, and the nose also turning them black, thus the term Black Death. As the disease progresses, most patients experience delirium, coma, and death without treatment. The bacterium is generally transmitted by the bite of a flea that has fed on infected rats—the primary host—or other rodents. In 1894, two bacteriologists, French Alexandre Yersin (1863–1943), for whom the bacterium was named, and Japanese Kitasato Shibasaburō (1853–1931) independently identified the bacillus, but its transmission was not known until 1905.

Three forms of the disease are found. "Bubonic plague" is the most common with about a 50 percent mortality rate. In "pneumonic plague," the lungs are infected, and bacteria are transmitted by droplets from sneezing and coughing. This form has a 100 percent mortality rate. "Systemic plague" involves most organ systems and results in death within a couple of days. All ages and levels of society are susceptible to *Y. pestis*, but if they survive the infection, they become immune to the disease.

Both the terms "plague" and Black Death were first used in the mid-1300s with the arrival of the disease in Europe. However, this scourge is ancient, and its DNA has been found in the skeletal remains from a nearly 5,000-year-old tomb in Sweden. Since classical antiquity, three major pandemics of the disease have been recorded.

The first pandemic with symptoms of bubonic plague was the Plague of Justinian (541–542 CE) in the Eastern Roman Empire. Physicians of the era thought it was caused by atmospheric contamination. DNA analysis has suggested that the disease was from Central Asia and was carried to Constantinople by infected rats on grain ships from Egypt. The scourge brought numerous deaths in some areas. It caused a disruption of agriculture, trade, and society and in its wake allowed Germanic and eventually Arab invaders to conquer the region.

The second European pandemic of *Y. pestis* raged from around 1347 to 1350. The disease was first recorded in Central Asia around 1338. It reached China and India by 1346 and Western Europe in 1347 when ships with infected people, and likely rats, arrived in Mediterranean ports from Asia. The disease initially affected the poor living in filthy environments but then rapidly spread throughout the whole European population as both bubonic and pneumonic forms. When plague emerged, parents abandoned sick children, and the rich fled the cities and towns but often carried the infection with them to other areas. In some communities, a few people remained to bury the dead, and bodies lay in the streets rotting. Several communities instituted "pest houses" to quarantine patients, and seaports prevented ships suspected of carrying plague from entering harbors. This pandemic killed between 30 and 50 percent of the European population. Conversely, the dramatic decrease in population and increase in inherited wealth was a factor in the European Renaissance. After the initial onslaught, the disease became endemic with outbreaks occurring about every 10 years until the mid-1750.

Some believed the plague was caused by God's wrath and a punishment for sinful behavior, but physicians offered other theories. Most blamed it on foul-smelling air (miasmas theory). Some contended that it was caused by contagion from the plague victim's clothing. Astrologers suggested the disease resulted from a conjunction of Saturn, Jupiter, and Mars in Aquarius. Most thought all these factors caused an imbalance in the body humors leading to the disease.

However, by the 1860s the disease had become so rare in Western nations that it was omitted from many medical texts. Ominously, a third pandemic emerged in 1894. It originated in China and spread to Hong Kong and India and then to the rest of the world. In developed nations with good waste management systems, the disease did not gain a foothold. For example, it arrived in San Francisco's China Town in 1901 on immigration ships, but it was contained due to enforced sanitation. Although this epidemic ended in 1904, the bacillus spread to wild rodents in the western region of the United States, resulting in a few cases every year among outdoor enthusiasts and workers.

By the mid-twentieth century, outbreaks of the plague were only found in Asia, Africa, and some Latin American nations in rural mountainous areas. By the 1950s, antibiotics cured the disease although the *Y. pestis* bacillus has become resistant to some. The World Health Organization (WHO) now estimates there are about 3,000 cases with 500 deaths per year worldwide. Plague immunization had become available in the late nineteenth century against the bubonic, but not against the pneumonic form, and is now used by plague researchers, the military, and workers in plague-infested areas. The WHO considers plague a reemerging disease that could develop in conflict-torn regions and refugee camps with poor sanitation and rat infestations.

Black Death changed European history. Fear of the plague became embedded into European culture and was found in artistic and literature works. For example, it was mentioned in Italian writer Giovanni Boccaccio's (1313–1375) *The Decameron* and Geoffrey Chaucer's (1343–1400) *The Canterbury Tales*. In Chaucer's the "Pardoner's Tale," the disease is connected to immorality, in particular, excess food, drink, gambling, or sex. Numerous woodcuts, paintings, and frescos illustrated pictures of Death, often presented as a skeleton, with a scythe mowing down sinners. People offered prayers and fasted, and some fervently religious flagellants roamed the countryside dragging crosses and publicly whipped themselves as penance to prevent the scourge. These were not effective and led to the decline in the reputation of the Church and was a factor in the Protestant Reformation.

UNUSUAL TREATMENTS

During the medieval pandemic, physicians tried in vain to prevent or cure the disease. Special plague doctors were hired by communities to treat everyone. Some examined patients with a long pole so they would not be contaminated by the patient's odor. They also wore a beak-like mask, which was filled with straw and aromatic herbs, and covered their bodies with a long gown. These efforts may have minimally helped to prevent inhaling *Y. pestis* bacillus from the pneumonic form.

Doctors recommended that healthy people bathe in rosewater or vinegar to drive away any contagion. They also suggested burning aromatic herbs and fumigating rooms to eliminate noxious vapors. In addition, they suggested using sweet-smelling substances in nose masks and gave patients vinegar and Armenian bole (red clay) to treat the disease, which did not prevent it.

In the fourteenth century, plague doctors wore long beaked masks that they stuffed with herbs and flowers in the belief that it kept the deadly disease away, which they thought was carried by bad odors or "miasma." (Courtesy of the National Library of Medicine)

To rebalance body humors and to purge the blood of putrid matter, physicians bled patients. They placed leeches on the buboes and cut veins. Cupping (hot cups placed on the skin to draw blood) was used for those under 14 years of age. Doctors also lanced the buboes or burned them with red-hot irons and administered drugs to cause purging and vomiting, but these treatments were ineffective and painful to the patient. Most physicians emphasized the dangers of miasma fumes, which led some communities to eliminate the sources of foul-smelling air. These included wastes from overflowing latrines, butcher shops, and tanners. These efforts likely decreased the rat population.

By the mid-sixteenth century, people smoked tobacco as the smoke was thought to counteract the plague poison in the air. Physicians recommended ingesting a "philosopher's egg" composed of an egg, saffron, mustard seed, and ground unicorn horn. Where the horn was obtained was not mentioned. This was also a treatment for influenza, colds, and poisons. Several cure-alls were used. Theriac containing viper's flesh, snake venom, and opium was a standard remedy for both the prevention and cure of plague into the eighteenth century, but it was only affordable by the rich. Mumia (powdered black substance from Egyptian mummies) was also used into the eighteenth century. By the mid-nineteenth century, although plague was now rare, opiates, drugs to cause sweating, small doses of carbolic acid, and sulfites were thought to be helpful in some cases of plague. None of these nostrums were effective.

Pneumonia and Lung Conditions

There are several diseases of the lower respiratory systems. Pneumonia is the most prevalent with symptoms that generally include a cough, fever, chills, trouble breathing, loss of appetite, bluish color in lips and fingertips, chest pain, rusty-colored sputum, and malaise. The *Streptococcus pneumoniae* bacterium, also called pneumococcus, is the most frequent cause of the disease. The bacterium inflames the alveoli (air sacs that exchange oxygen and carbon dioxide) and was first isolated in 1881 independently by French microbiologist Louis Pasteur (1822–1895) and American army officer George Sternberg (1838–1915). Pneumonia can also result from a virus or fungus infection. Two types of pneumonia are found: "bronchial" and "lobar." In bronchial pneumonia, pus is found in many alveoli in one or both lungs while lobar pneumonia is characterized by an acute inflammation of the entire lobe or lung and tends to be more lethal.

Pneumonia has been greatly feared throughout history as it was a major killer of young children and the elderly. It has been known as the "old man's friend," as it allowed the patient to slip away and die peacefully. The noted Canadian physician William Osler (1849–1919) termed it "captain of the men of death" in the 1920s. It was a leading cause of death throughout the early twentieth century, where the mortality rate was estimated to be between 30 and 40 percent. The rate today among the elderly is around 8 percent even with antibiotic treatments. It is often a sequela of influenza (discussed in another entry). In the United States, for example, about 50,000 die from the disease each year. Worldwide, pneumonia accounts for 16 percent of all deaths of children under 5 years old.

A condition, which was often confused with pneumonia in the past, was pleurisy. Symptoms were similar other than a hacking nonproductive cough, lower fever, and a severe pain in the side of the chest when breathing. The pain was due to irritation or inflammation between the two pleural membranes surrounding the lungs that normally slide over each other when a person breathes. Caused by an infection, injury, or other factors, pleurisy was generally self-limiting and had a lower death rate compared to pneumonia. In the past, both pneumonia and pleurisy were thought to be caused by cold northerly winds, drinking cold liquids when the body was hot, sleeping in a draft or on wet ground, and other environmental factors that cause an imbalance in the body "humors."

Some other ancient lung disorders include emphysema and asthma. Emphysema causes shortness of breath and coughing. In this disease, the alveoli of the lungs are enlarged and lose their elasticity. Chronically inhaling smoke, mineral, and other dusts contribute to the condition. The death rate for emphysema in 2021 was 2.2 per 100,000 people in the United States. In asthma, bronchial spasms cause the patient to find breathing difficult. Wheezing, coughing, and shortness of breath are also common. Around 8 percent of adults and 7 percent of children have asthma, and it has been increasing since the 1980s.

Pneumonia and other lung conditions have been discussed by medical personnel for centuries. Greek physician Hippocrates (c. 460–370 BCE), for example, described various diseases of the lower respiratory system, which had a high mortality rate. Into the eighteenth century, some medical books classified lung diseases as pleurisy, asthma, and tuberculosis (discussed in another entry). English physician William Bullein (c. 1515–1576) was the first to write about "pleurisie" as a separate condition from pneumonia. The term "pneumonia" is from the Greek meaning "lung" and "pleurisy" from the Greek meaning "side."

In addition, until the mid-1600s, asthma (from the Greek to pant or breath hard) was a description for any condition with dyspnea (shortness of breath). However, Swiss physician Théophile Bonet (1620–1689) described emphysema—from Greek meaning "to inflate"—as a separate disease, and by 1819, stethoscope inventor French physician René Laennec (1781–1826) had accurately described various lung conditions. In 1965, a combination of emphysema, asthma, and chronic bronchitis was termed COPD (chronic obstructive pulmonary disease), now considered a major chronic health condition. The United States in 2021 had a COPD death rate of about 36.9/100,000 population.

The death rate from all these conditions, however, has largely decreased since the mid-twentieth century. This is due to the widespread use of antibiotics beginning in the late 1940s and immunizations in the early 2000s in developed nations, which has decreased the lethality of pneumonia and other bacterial lung infections. Bronchodilators, inhaled steroids, and other drugs have helped quell asthma attacks and eased breathing issues in emphysema and COPD.

However, in December 2019 a new virus (SARS-CoV-2) called COVID-19 began to rampage the world, killing hundreds of thousands of people mostly 65 years and older until a vaccine was developed in early 2021. They died from pneumonia, blood clots, and organ failure. The pandemic caused many aspects of modern life to slow down as businesses, schools, bars, churches, restaurants, gyms, theaters, and other venues were closed or were required to have limited capacity. People were advised to stay at least 6 feet apart—"social distancing"—from each other, wear masks, and avoid crowds to prevent its spread. New variants of the virus that were more easily transmitted and harmful arose. Immunization and drugs for treatment were available for this scourge by mid-2021, and nations slowly began to get back to pre-pandemic conditions, but the delta and other variants caused further surges. Like other pandemics, it had worldwide social, political, and economic effects and the disease became endemic.

Pneumonia and asthma have been depicted in fiction. American author O. Henry—pseudonym of William Sydney Porter (1862–1910)—in a short story, "The Last Leaf" (1907), depicts a poor young Greenwich village artist with pneumonia. She believes that when the last leaf on the ivy vine on the wall outside her window falls off, she will die. Her old neighbor artist, who has never created a masterpiece, goes out in the night during a blustery storm and paints a leaf on the wall near the vine. The young woman sees it and recovers. However, the old neighbor catches pneumonia while painting the leaf and dies. In the film *Toy Story 2*, Wheezy the asthmatic penguin has a broken squeaker and is destined for a yard sale. Woody the cowboy protagonist toy tries to save him from the sale. Through a lot of adventures, he rescues the penguin, and the toys are all back together. Another toy restores Wheezy's voice.

UNUSUAL TREATMENTS

Healers, from at least ancient Egypt (c. 1500 BCE) based upon humoral theory, recommended bloodletting as the first treatment for lung infections. It was used throughout the early twentieth century and was accomplished by cutting a vein and using the method of placing several leeches on the chest or by wet cupping, in which the skin was scratched and a warm cup that created suction was placed over the incision to draw out blood. As late as 1942, one of the most popular medical textbooks, Osler's *The Principles and Practice of Medicine*, 14th edition, still recommended bleeding for pneumonia and in the case of acute congestive heart failure to reduce fluid volume. The treatment may have helped in some cases. He also suggested patients sleep out of doors and wear woolen pajamas. Cupping is still used today in folk and traditional systems and is thought to remove "toxins," which is unproven.

In addition to bloodletting, Hippocrates recommended treating pneumonia with a "cerate," which was a salve made of wax or resin mixed with oil, lard, and herbs and rubbed on the chest and sides. The patient would also be purged and given a concoction of pepper and black hellebore, a dangerous herb that causes heart irregularities, along with opium to relieve the pain of pleurisy.

British physician Nicholas Culpeper (1616–1654) suggested chewing on woodbine, or honey-suckle, to ease breathing difficulties, which may have helped. Thomas Sydenham (1624–1689) endorsed milk and sugar enemas but not immediately after bleeding, which were not effective. In the following century, Scottish physician William Buchan (1729–1805) thought bleeding had a good effect on pneumonia and pleurisy with either leeching or cupping and recommended an application of warm cabbage leaves for side pain. Physician also suggested drinking milk, eating figs as a purgative, and applying blisters to remove poisons. Powder of pearl (ground-up pearls) was used to reduce inflammation and is still utilized today in Chinese medicine. Most of these remedies were likely ineffective other than by placebo effect, and bleeding could have caused weakness.

In the nineteenth century, for pneumonia and pleurisy, many physicians, such as the American Henry Hartshorne (1823–1897), purged the patient to loosen the bowels to eliminate excess fluid. He used several concoctions including "citratized magnesia," still used today to clean the bowels before a medical procedure. For nourishment, brandy and beef broth were freely offered. Elijah B. Hammack (1826–1888) recommended a mixture of "poppy heads or hops" used alternately with a turpentine "stupe" (cloth dipped into hot water with a mixture of turpentine and oil that was wrung out) and placed on the painful area, which may have reduced pain. Physicians also often ordered Dover's powder that contained opiates to relieve pain and ipecacuanha that causes vomiting. In the early twentieth century rather than sponging a patient with ice water to reduce pneumonia fever, one physician, John Haddon (1845–1924) suggested throwing a pail of cold water suddenly over the patient in a tub to "shock the nervous system" and reduce the fever, which likely hastened death.

For treating shortness of breath and asthma, Buchan, based upon humor theory, recommended a "seton" (an incision in which a thick thread was pulled through the skin and a bean placed in it to cause pus formation to "draw out poisons") be placed on the back or side. To relieve an acute asthma attack, Hartshorne in the nineteenth century instructed the patient to inhale the fumes of a piece of paper that had been soaked in silver nitrate, dried in an oven, and ignited, which would have likely made the attack worse due to irritation. Physicians also used ether and nitrous oxide to quell an attack, which temporarily helped by relaxing the bronchial tubes. However, other gasses such as hydrogen were sold by "miracle cure" peddlers, and even smoking tobacco was recommended for breathing difficulties, which likely made the conditions worse due to lung irritation.

Polio

Poliomyelitis, shortened to polio, produces three forms of the disease with increasingly severe symptoms that often occur in a matter of hours. "Abortive polio" begins with a sudden onset of a slight fever, malaise, headache, sore throat, and vomiting. "Non-paralytic polio," besides these symptoms, also causes deep muscle pain and tenderness and stiffness of the neck and back. Complete recovery occurs in these forms of poliomyelitis. However, in "paralytic polio," about 50 percent recover without paralysis, 25 percent have mild disabilities, and the remaining patients experience severe and permanent disability or death. The most severe is bulbar in which the patient is unable to breathe, or even swallow, for weeks or permanently. About 20–30 years after the original onset of the disease, many former victims develop "post-polio syndrome," with a loss of muscle strength.

The disease is caused by three types of *enteroviruses* that can damage the spinal cord or brain stem. They are transmitted by close person-to-person contact and by objects contaminated with fecal material. Up to 72 percent of all polio infections in children are asymptomatic who shed the virus in their stool. Throughout the nineteenth century, polio was a rare disease. In 1843, it was first termed "infantile paralysis," a term used into the mid-twentieth century, as it mainly affected children under five years old. In 1874, the word "poliomyelitis" was coined by German physician Adolph Kussmaul (1822–1902) from the Greek *polios* (gray), *myelos* (marrow), and *-itis* (inflammation), as gray matter in the spinal cord led to paralysis when inflamed.

It was not until 1908 that Viennese physicians Karl Landsteiner (1868–1943) and Erwin Popper (1879–1955) discovered the infectious agent for poliomyelitis was a "filterable virus," but the virus was not cultivated in tissue culture until 1949. Humans were found to be the only natural reservoir for the illness, and lifetime immunity generally occurred once a person had been infected.

Polio became increasingly common from the late nineteenth through the mid-twentieth century in industrialized nations and was typically found in the summer months. At its peak in the 1940s and 1950s, it paralyzed or killed over half a million people worldwide each year, and during 1952, a record 57,628 cases were reported in the United States. It is hypothesized that better hygiene, sanitation, and water purification meant that infants and young children were less exposed to the disease, so they did not develop immunity. Exposure to the polio virus was therefore delayed until late childhood or adulthood, when it was more likely to take the paralytic form.

The disease has likely existed since prehistory. The earliest probable evidence of polio is from ancient Egypt found in a mummy's skeletal remains (3700 BCE) and on a stela (1580–1350 BCE) that shows a man with a withered leg. However, throughout history polio was a rare disease. It was not until the late eighteenth century that English physician Michael Underwood (1736–1820) first described debility of the lower extremities in children—now recognized as poliomyelitis. During the nineteenth century, a few localized outbreaks occurred in North America and Europe, but it was not until the early twentieth century when major epidemics were seen in industrialized nations. For example, a massive epidemic occurred in New York City in 1916 that overwhelmed hospitals and caused fear among the middle class whose children were more prone to the disease than poorer children living in unsanitary environments. In this epidemic, of more than 9000 cases, 20–25 percent died. The outbreak led to widespread panic, causing thousands to flee the city to nearby mountain resorts. Children with the disease were quarantined at home, and parents forbade their healthy children to go to public places such as movies.

Since transmission of the virus was not known until the 1940s, physicians in 1916 formed many theories regarding its spread. Some thought the virus was transmitted by contaminated water and advised people against drinking from public fountains or swimming in public pools and beaches. Others thought it was carried by dogs and cats, so many pets and strays were slaughtered. Still others suggested it was carried by stable flies or was due to lack of personal hygiene or mold. The public also came up with theories including the summer use of feather pillows or even poison injected into food by German saboteurs as the epidemic occurred during WWI in Europe.

Although outbreaks occurred every summer in Canada, the United States, the United Kingdom, and other high-income nations, the most serious epidemics occurred in the 1940s and 1950s that resulted in polio during these decades to be feared as much as the atomic bomb. But dread of the disease spurred the development of immunizations, rehabilitation therapy, and the rise of disability rights movements worldwide. One individual who fostered these causes was American President Franklin D. Roosevelt (1882–1945) who acquired polio at age 39 and was paralyzed for life.

The development of an injectable inactivated virus vaccine in 1955 by Jonas Salk (1914–1995), and a live oral vaccine in 1962 by Albert Sabin (1906–1993), dramatically reduced polio in higher-income nations within the space of a decade. In 1988, the World Health Organization (WHO) proposed a global polio eradication program. By 2019, polio had been eradicated from all nations with the exception Afghanistan and Pakistan due to political issues.

Polio has been featured in the arts. For instance, American writer Philip Roth's (1933–2018) novel *Nemesis* (2010) describes a 1944 polio epidemic in a small Jewish community. As children die, panic spreads, and fear of the disease emerges. It is blamed on various ethnic groups, and libraries, pools, playgrounds for children are closed, and there is talk of quarantining the whole community. A teacher flees the town and joins his fiancé at a summer camp that is polio free. However, he is stricken with the disease that disables him, which he likely brought with him. It emerges in the camp, and children are sickened, some seriously. One child dies, and several are seriously ill.

UNUSUAL TREATMENTS

Since polio was a rare condition until the mid-nineteenth century, only a few treatments for its prevention or cure were used. In the eighteenth century, one English physician advised wrapping a child in the skin of a freshly killed sheep. In the 1860s, American physician Henry Hartshorne (1823–1897) prescribed heat, hot sand, and sunbaths and soaking the limbs in hot water to relax the spasms. Cod liver oil, cold ocean bathing, passive exercise, tonics such as strychnine, galvanism (electric current), and blistering plasters to the spine were also thought to reduce spasms and paralysis.

During the early-twentieth-century epidemics, most physicians rejected these methods and tried other treatments although the rationale for their use was often

The iron lung (c. 1933) used to "breathe" for polio victims into the late 1950s. Many were children. (The George F. Landegger Collection of Alabama Photographs in Carol M. Highsmith's America, Library of Congress, Prints and Photographs Division.)

not given. Based upon very old methods, they tried cupping (hot cups placed on skin that had been scratched) to draw out blood and their supposed poisons. Other treatments included radium water, gold chloride, quinine, lime (calcium) water, and wine of pepsin (a digestive enzyme). They also prescribed hydrogen peroxide as a nasal spray, strychnine, and other tonics to do something, which likely caused more issues for patients. Physicians such as American Hardee Johnston (1873–1950) also recommended tetanus, diphtheria, or rabies antitoxin as they thought these would block the polio virus from attacking nerve cells; it was not effective. A standard procedure was lumbar punctures to relieve pressure on the spinal cord often followed by an injection of adrenalin, quinine, disinfectants, urea, or strychnine. By the 1950s, the lumbar puncture method was found to make paralysis more likely, and many of these treatments led to the deaths of numerous children.

The primary treatment during the early twentieth century was placing paralyzed limbs, or even the whole body, in ridged casts or splints, sometimes for months, so as not to further damage weak muscles. However, lack of passive exercising led to atrophy of all muscles and increased permanent paralysis. Most physicians rejected heat treatment and exercise during the initial phase of the disease, which was the basis of Australian nurse Elizabeth Kenny's (1880–1952) treatment. Since the muscles were often only temporarily paralyzed, her method was found to be effective. Although opposed by many physicians, Sr. Kenny used and taught her method from the 1930s through 1950s until polio dramatically decreased. In cases of bulbar polio, health workers placed patients with breathing difficulties into an "iron lungs"—a body size respirator—sometimes for life.

Since physicians were unable to cure or prevent the disease, ineffectual folk methods emerged. These included bathing in ox blood or almond meal; plasters (poultices) of mustard or Spanish fly; and a diet of rice, almond meal, and oxygen water. One inventor marketed his Spectro-Chrome light box and claimed that various colored lights cured illnesses. A light was shown through green and then blue glass to cure polio. The device was marketed into the 1950s but was not effective.

Pregnancy: Common Issues and Potential Complications

A pregnancy is divided into three stages of about 12 weeks each, called trimesters. The first sign of pregnancy in the first trimester is generally a missed menstrual period, sore breasts, and nausea or vomiting called "morning sickness." Moodiness and constipation are also common. In Western cultures, some women experience an unusual appetite (pica) for special foods. Women with poor nutrition in developing nations sometimes will crave dirt, chalk, clay, peeling paint, or animal feces. At about 12 weeks, the end of the first and the beginning of the second trimester, the embryo becomes a fetus. During this trimester, nausea and fatigue usually decreases. After 19 weeks, the woman usually feels the fetus moving called "quickening." As the abdomen stretches and the fetus grows, aches in back and other places are common. Stretch marks form, and skin darkens around the nipples. In the third trimester, the growing fetus can put pressure on the lungs and bladder, causing difficulty in breathing and frequent urination. At roughly 40 weeks, contractions, which can be a sign of real or false labor, begin and the infant is born. (See the two "Childbirth" entries for more details on the birth process.)

The first-known use of the term "pregnancy" or "pregnant" was in the fourteenth century and is derived from Middle English, from the Latin *praegnant*, "carrying a fetus." Pregnancy for women throughout the centuries has usually been a time of joy to fulfill the desire of having a child. In most cultures, pregnant women are given special privileges, attention, and even food in times of hardship. In some cultures and time periods, having a male child, for the perpetuation of the male family line and property, was most important. Pregnancy has also been a time of distress for many women if it was an unplanned or an unwanted condition.

Pregnancy can also have complications. One of the most traumatic events of early pregnancy is an abortion often called a miscarriage (for elective abortion, see the "Pregnancy: Prevention and Termination" entry). Unless bleeding stops, it can lead to death. Although it is often a taboo subject for discussion, studies estimated that up to 30 percent of all pregnancies may end in miscarriage due to chromosomal abnormalities in the fertilized egg, an embryo with structural defects, or other physiological factors. Over 80 percent occur during the first trimester when some women do not know they are even pregnant. A miscarriage can lead to self-blame, depression, and anxiety of ever getting pregnant again, along with grief and deep mourning. Numerous factors cause spontaneous abortions including older maternal age, obesity, diabetes and other chronic diseases, polycystic ovary syndrome, and medications. In Mesopotamia (c. 3100–539 BCE), healers believed that miscarriages were caused by evil magic. In addition, from antiquity into the late nineteenth century, heavy lifting, jumping up and down, swearing, being frightened, vigorous exercise, weak sperm, and even tight corsets were thought to cause miscarriages.

Another serious pregnancy disorder is preeclampsia, formerly called toxemia. It is pregnancy-related hypertension and is most common during the second half of pregnancy. It occurs in about 3–5 percent of pregnant women. Symptoms include a sudden rise in blood pressure; protein in urine; swelling in legs, feet, and hands; headaches; abdominal pain; and other symptoms. Preeclampsia can lead to eclampsia with convulsions and death. Scottish physician William Buchan (1729–1805) described it in the seventeenth century as "convulsion fits." Treatment by the mid-twentieth century included newly developed antihypertensive drugs and magnesium sulfate to prevent seizures, so few women die of this complication today.

Gestational diabetes is a malady that sometimes occurs during pregnancy. It causes high blood sugar resulting in health complications for both the mother and fetus, and it was not recognized until the 1950s. From 6 to 9 percent of pregnant women develop this condition. It can lead to excessive birth weight, stillbirths, and early labor. Risk factors are obesity, lack of physical activity, polycystic ovary syndrome, and diabetes in an immediate family member. Controlling blood sugar with diet, exercising, and sometimes medication prevents complications.

The most dangerous complication is an ectopic pregnancy in which the fertilized egg implants and grows outside of the uterus, usually in a fallopian tube. Pain is usually felt on one side of the abdomen, and it can be a life-threatening event as the growing embryo can cause the tube to tear, resulting in fatal bleeding. This dangerous condition occurs in around 1 out of 50 pregnancies and causes 10 percent of all maternal deaths. Risk factors include age over 30, smoking, fertility treatment, and pelvic infection. This condition like other pregnancy losses can be emotionally devastating to a woman. The term "ectopic" is from the Greek *ektopos*, meaning "out of place," from *ex*, meaning "out," and *topos*, meaning "place," and it was first used in English in the late nineteenth century. In the 1880s, due to safer surgery from better asepsis and anesthesia, the fatality rate decreased from around 80 percent in 1880 to less than 0.1 percent in the late twentieth century, particularly in developed nations. Ultrasound and laparoscopic surgery can now diagnose and remove the embryo, and drugs such as Methotrexate can stop its growth.

Another serious but rare pregnancy malady is a hydatidiform mole or molar pregnancy. An abnormal fertilized egg causes tissue overgrowth. The tissue from the placenta or egg begins to proliferate into fluid-filled grape-like structures, and the lesion can become cancerous. Approximately 1 in every 1,000 pregnancies is diagnosed with this condition primarily among very young or premenopausal older women. The term "hydatid" was first used in the mid-seventeenth century from the Greek *hydatis*, meaning "cyst." "Mole" is from the Latin *mola*, meaning "mill" or "millstone." In the past, this uncontrolled growth could kill the woman. As with other pregnancy losses, women are usually devastated by the loss and fear of having another molar pregnancy. Today it is easily diagnosed by ultrasound and can be surgically removed.

In another rare condition pseudocyesis, or false pregnancy, a woman believes she is pregnant. It is most common between ages 20 and 39 years. Other than negative pregnancy and ultrasound tests, the woman has the signs of a pregnancy. The cause is unknown, but one hypothesis suggests that psychological factors may trick the endocrine system into believing that she is pregnant. However, a phantom pregnancy can cause a woman anxiety, despair, and low self-esteem. The term "pseudocyesis" was first introduced by English writer and physician John Mason Good (1764–1827) in 1823 from the Greek *pseudo*, meaning "false," plus *kuēsis*, meaning "conception." The condition was first noted by Greek physician Hippocrates (c. 460–370 BCE) who thought it was caused by air and retained menstrual or humoral fluids.

Pregnancies, or lack of, have been represented in art forms since early times from Venus figurines in antiquity to numerous paintings, films, and works of fictions. They have also changed the course of history. For example, King Henry VIII (1491–1547) divorced two wives as they did not give him a male heir, which precipitated the English Protestant Reformation. His daughter Queen Mary Tudor (1516–1558), a devout Catholic, to atone for her lack of pregnancy, as she believed God was punishing her, persecuted and killed many Protestants. After this bloodshed, she felt she was pregnant, but it was pseudocyesis. In the arts, for example, in American playwright Edward Albee's (1928–2016) convoluted play *Whose Afraid of Virginia Wolf* (1962), pseudocyesis is implied when the new faculty member's wife hints this may have occurred and traps her husband into marrying her. In the 2016 BBC drama *Call the Midwife* Christmas special, the nurses travel to South Africa to help in a mission hospital. One patient has a phantom pregnancy because she has nothing else in her life to look forward to.

UNUSUAL TREATMENTS

Many folk remedies have been offered over the centuries for minor issues of pregnancy such as morning sickness and pica. In addition, treatments have also focused on more serious problems. Many remedies were based upon tradition and humoral theory. To ensure a healthy pregnancy, for example, Swiss physician Jakob Rueff (1500–1558) advised women to wear gems and stones around their neck such as sapphire and coral as good luck charms to ward away evil influences. In the twenty-first century, some women still used stones, such as quartz, for anxiety reduction. They placed the stone over the baby bump while lying down, which may have worked by placebo effect.

In ancient Egypt, to quell nausea, mint was a popular remedy and has been shown to have anti-nausea and vomiting properties. Rueff gave patients a syrup of pomegranates, musk, lignum, aloes, and cinnamon. Due to the placebo effect and anti-nausea properties of some of these spices, this concoction may have been effective. To relieve nausea and vomiting in the seventeenth century, Buchan recommended "keeping the body gently open" by giving the patient prunes, figs, and roasted apples. He also bled the woman if she had much pain to rebalance the humors, which would have weakened her and perhaps caused anemia.

American physician Elijah B. Hammack (1826–1888), for severe morning sickness, gave the patient morphine or opium, which often quelled the nausea but could have caused drug dependence. He also recommended a piece of ice be laid on the stomach and the patient be administered blue mass pills to relieve both nausea and constipation. However, this medication contains toxic mercury that could have poisoned the woman and caused birth defects or miscarriage. For constipation, he recommended an enema of cold water mixed with a well-beaten fresh egg and sugar, which likely was effective and caused little harm to the woman or fetus.

Remedies for serious complications were also prescribed. In Mesopotamia, to prevent a miscarriage, healers wrote out magical incantations and recipes to prevent evil spell. One magic recipe included crushed magnetite, antimony, dust, agate (considered a powerful magic talisman stone), and dried "fox" grapes mixed with the blood of a female shelduck and cypress oil. This mixture was rubbed on the woman's heart, stomach, and vulva, but it did not prevent the miscarriage. Since spontaneous abortions were also thought to be caused by evil spells in ancient Egypt (c. 1500 BCE), women used a plug-like amulet named for the goddess Isis called the "Isis knot" or "Isis blood." The pregnant woman placed this object that looked somewhat like an ankh in her vagina and recited magic spells, but it likely did not stop the miscarriage.

Based upon Hippocrates, Greek gynecologist Soranus who practiced in Alexandra (c. 98–138) suggested procedures to stop spontaneous abortions. When a pregnant woman was bleeding and in pain, he bathed her in hot water. In addition, a mixture of lead oxide, gall that grows on oak trees, myrrh, and Egyptian alum was injected into the vagina. The gall and alum have astringent properties, and the alum has blood coagulant property, which may have sometimes stopped the bleeding, but hot water would likely have not been effective. Also, lead is toxic, and myrrh is known to cause uterine bleeding, so they likely were counterproductive.

The Roman naturalist Pliny the Elder (23–79), to stop miscarriage, recommended a mixture of cut leeks, mint, heracleum (hogweed), coriander, and other herbs. This plant mixture may have been somewhat effective as some of the plants including coriander and leeks help quell bleeding, but hogweed can cause serious irritation. In the sixteenth century, Rueff suggested that a pregnant woman should wear a bear claw around her neck to prevent miscarriage, but no reason for this was given. He also recommended that gall, nutmeg, cloves, and musk be ground and mixed with red wine and put into the vagina. Like the ancient Egyptians, he recommended a gall and alum pessary to stop the discharge in addition to bloodletting near the ankle to get rid of bad humors. The pessary might have helped. Rueff also recommended a drink that contained pearl powder, ivory scrapings, and powder of precious stones, which would have had no effect.

Buchan in the seventeenth century advised fragile women to avoid tea and other "weak watery liquids" and to not go outside in damp foggy weather or sleep in damp places as they could catch a cold, which could put them in danger of losing their pregnancy. On the other hand, he claimed that overweight women should avoid anything that would heat the body as he believed that heat caused a miscarriage. By the twenty-first century, fever and soaking in hot tubs had been found to cause birth defects in early pregnancy.

In the nineteenth century to prevent a threatening miscarriage, Hammack administered a saline cathartic, potassium nitrate, and digitalis (used to slow heart rate) until the pulse dropped. If the patient was nervous, he gave a laudanum (opiate and alcohol) enema. Laudanum might have relieved pain, but nitrates have been found to induce spontaneous abortions, so it could have been counterproductive. If the uterus was dilated, he put his finger in it and pressed it "backward and forward," which he claimed expelled the ovum but which likely also led to infection. For spontaneous abortions, he also put ice water on the abdomen, which may or may not have stopped heavy bleeding.

When a woman was having convulsions ("fits"), due to eclampsia, Rueff in the sixteenth century rubbed a mixture of salt and vinegar into her feet and hands, which likely did nothing. He then scarified the skin and applied cupping glasses (heated cups that cause a vacuum) to remove bad humors in the blood. Then smelly and stinking items such as castoreum (fluid from the sacs of the North American Beaver), asafetida herb, feathers, and animal horns were burned, and the fumes applied to her nostrils as bad smells were thought to expel the corrupt humors in the womb and to prevent death. Since he had no concept of hypertension, only the bloodletting would have sometimes worked to reduce blood pressure and in turn convulsions. Some English physicians by the eighteenth century administered alkanet root, which has anticonvulsant and sedative properties, and may have helped quell these "fits."

In the nineteenth century, clinicians, to prevent convulsions, also bled the women if she had a bounding pulse as they noticed it foreshadowed seizures. Others recommended an enema of castor oil, molasses, and warm water, which was harmless but ineffective. Physicians also gave Dover's powder (ipecac and opium), an emetic, along with brandy and quinine. Opium and alcohol did act as sedatives and may have helped. Some clinicians, however, administered potassium bromide, which has been found to prevent seizures.

For either a hydatidiform mole or a false pregnancy, Rueff devised several treatments to rebalance the humors and draw the supposed water and air out of the uterus. He bled from a vein in the ankle and had the woman bathe twice a day in a bath with a bag containing various herbs. This bag was also laid on her vulva. He also pushed a suppository into the uterus made of castoreum and various waxes and herbs. In addition, the woman was not allowed food and drink that caused "air to form," such as beans that would "puff up the body." Treatments for a false pregnancy may have worked by placebo effect but would not have worked for the mole that often killed the woman.

For an ectopic pregnancy from the seventeenth century through the late nineteenth century, since the ectopic embryo was thought to be responsible for killing the mother, treatments attempted to kill the embryo. The arsenal included starving, purging, and bleeding the woman; administration of strychnine; passage of electromagnetic currents through the ectopic mass; and injecting morphine into the fetal sac. Whatever the treatment, the prognosis was not good, and most women died.

Pregnancy: Infertility, Fostering Conception, and Pregnancy Tests

Infertility in women is caused by several factors. (Male reproductive issues are discussed in another entry.) They include endometriosis or growth of the uterine lining outside the uterus; damage to the fallopian tubes from infections where the egg is fertilized; uterine polyps and fibroids that prevent the egg from implanting; abnormalities of the cervix or thick mucus that plugs the os (opening into the uterus); hormonal dysfunctions; undesirable weight; older age, stress, and in about 20 percent of infertile couples unknown factors. About 10 percent of women in the United States, ages 15–44, have difficulty becoming pregnant or staying pregnant after trying for a year. The infertility rate is higher in developing nations. Attempting to conceive and finding failure year after year can cause some women to feel guilty, ashamed, frustrated, and a failure.

In the past, women who were barren were often looked down upon with pity or sometimes even scorn and disapproval. Some were condemned as witches and killed. In addition, until the nineteenth century the woman was usually blamed for the lack of pregnancy. It was not until the eighteenth century that in America a minister, Benjamin Coleman (death c. 1747), wrote that a woman who did good works even if childless should not be considered barren. It was viewed as a sad but not a moral defect in the woman. The first use of the term "barren" is from thirteenth century from the Middle English *bareine*, from the Anglo-French *barain*, meaning "not producing" or "not fruitful." Infertile was first used in medical texts in the late seventeenth century and is from Middle French, from the Late Latin *infertilis*, meaning "not fertile or productive."

Infertility was mentioned in the Christian Old and New Testaments and Hebrew and Islamic scriptures that described God's intercession, which enabled a few women to become pregnant. Reasons for infertility were offered in different time periods. In ancient Greece, it was thought to be from wandering womb. In the sixteenth century, some English physicians held that infertility was caused by the women being too fat or thin or the womb being too hot or cold. In the seventeenth century, the noted Swiss physician Jacob Rueff (1500–1558) in *The Expert Midwife* suggested that barrenness was caused by "enchantments of evil arts." Scottish physician William Buchan (1729–1805), in the next century, on the other hand, wrote that high living, sudden fear, anxiety, and grief resulted in barrenness. By the late 1700s, infertility was thought to be "a mechanical problem."

By the mid-twentieth century, tests had been developed to determine reasons for infertility in women. These included testing for adequate hormone levels in the blood, biopsy of the endometrium, and dye test using a fluoroscope to examine the fallopian tubes. By late century, ultrasound or laparoscopic procedures were used to check for abnormal growths on the ovaries or uterus. Treatment included fertility drugs and hormones to aid ovulation, surgery to open blocked fallopian tubes, or assisted reproductive technology (ART). Several types of ART by the early twenty-first century had been developed. These included intrauterine insemination in which sperm is placed inside the uterus when the woman was ovulating, in vitro fertilization in which the egg and sperm are mixed together and the resulting embryo is placed into the uterus, transferring both sperm and egg into a fallopian tube, or placing a fertilized egg—sometimes from a donor—into a fallopian tube. The success rates ranged from 20 to 60 percent.

References to infertility and desire for pregnancy are found in the arts and literature. One of the earliest prehistoric works of art is of a figurine from about 26,000 years ago. About 200 similar ones have been found in Europe with the youngest being about 11,000 thousand years old. Most of them have wide hips, large breasts and stomachs, small heads, and legs and arms that taper to a point. They are thought to have been used for fertility rituals and were carved out of various materials. In many cultures, statues of pregnant or breastfeeding women, or women holding small children, were used as charms to foster pregnancy. Infertility themes are also found in literature. In William Shakespeare's (1564–1616) play *Julius Cesar*, Caesar asks Mark Antony to "touch"—likely spank—his barren wife Calpurnia so that she will become pregnant. *The Handmaid's Tale* (1985), a sci-fi novel by Canadian Margaret Eleanor Atwood (1939–), film (1990), and television series (2017) by the same name depict a dystopian society where few women are fertile. Elite married men are assigned fertile young women, called handmaids, to bear children.

UNUSUAL TREATMENTS

Since the ability to conceive was important in most cultures, women and healers of the time developed tests to determine if a woman was fertile or pregnant, the gender of the fetus, and actions to help a woman become pregnant based upon religious beliefs and trial and error. However, most of them were ineffective. One Mesopotamian (c. 4500 BCE) prescription for helping a barren woman conceive, for example, was to string 21 stones with linen thread and put them around her neck along with incantations to the gods.

Numerous remedies were given in the Egyptian *Kahun Papyrus* (c. 1825 BCE) to test for fertility and pregnancy. To see if a woman was fertile, a spell for bringing on nosebleeds was prescribed as it was believed that if a young woman's nose bled, she would get pregnant; if it did not, she would remain childless. Another indication of fertility was for the priest or healer to touch the lip of a woman with a finger, and if it twitched, she was fertile; if it did not, she was infertile. In another test, the practitioner placed an onion on her abdomen overnight, and if the odor appeared in her mouth the next day, she would bear children.

Other folk remedies were suggested in ancient Egypt. To conceive a boy, a woman drank breast milk from a woman who had recently delivered a male child. The milk was sometimes stored in a jar that was shaped like a mother nursing her infant. To determine if a woman was pregnant, she drank a pounded watermelon mixed with the milk of a nursing mother who had recently given birth to a son. If she burped, the test was considered negative for pregnancy. But if she vomited, she was pregnant, likely due to nausea of pregnancy. However, the papyrus did describe a possible accurate test for pregnancy. A woman was asked to urinate on wheat and spelt daily. If both grew, she was pregnant. This test may have been somewhat successful as a pregnant women's urine contains hormones that have been found to influence the germination of seeds. On the other hand, it was believed that if the wheat grew, she would likely have a boy, and if the spelt grew, it would be a girl. If neither grew, she was not pregnant and likely infertile. Secret incantations and rituals in a sacred temple to the god Bes, who protected pregnant women and children, were also used to foster fertility.

In ancient Hebrew culture, a test for pregnancy, after the fourth month, was for the woman to tread on soft ground. If her footprints sank more deeply than at a previous time, she was pregnant as she likely gained weight. The ancient Greeks also administered fertility and pregnancy tests. The Greek physician Hippocrates (c. 460–370 BCE) suggested that a woman drink mead (fermented honey and water) upon going to sleep. If she developed abdominal distention and cramps during the night, she was pregnant. This method was a common pregnancy test for centuries but was not accurate. To conceive a boy, the *Hippocrates Corpus* advised that a man should tie his left testicle for a boy and his right for a girl. In order to conceive, women were advised to consume a mixture of men and bulls' urine, tar water, chaste tree, pomegranates (which might stimulate progesterone), cantharides (Spanish fly), or castor oil. Spanish fly and pomegranates were thought to be aphrodisiacs. These likely caused nausea and vomiting but not pregnancy. To cure sterility, healers recommended fumigation of the uterus. A lead or wooden tube was introduced into the os that was connected to a vessel that emitted aromatic fumes, but it was ineffective other than placebo effect.

In ancient Rome, various amulets or charms to help a woman conceive, or prevent conception, were inscribed with magical formulas. One charm box represented a uterus that could be opened and closed with a key. The woman opened it when she wanted to become pregnant and kept it locked when she did not. The Romans also believed that a man spanking a barren woman on the buttocks would help her conceive. This practice was also carried out by the mother of the bride at a temple in secret rituals. A husband spanking his wife over his knees to increase fertility was also practiced in the nineteenth century. If she did become pregnant, it was likely placebo effect.

By the thirteenth century, physicians also recommended dubious pregnancy tests and methods for conception. For example, Rueff suggested that an iron needle placed in urine would be covered with black spots if the woman was pregnant. To cure infertility, he suggested that the woman should eat food that "does not cause wind." Rueff also suggested that the vulva should be anointed with salves containing ingredients such as spices, brains of roasted hares and duck, gees, and hen grease. He also recommended consuming a cooked mixture of pigeon and other bird brains, bull, "lecherous goat," or boar's testicles mixed with nuts, dates, bull urine, herbs, honey, and other ingredients. These remedies were largely harmless but did not likely result in pregnancy. Buchan in the seventeenth century for barrenness suggested a milk and vegetable diet, astringent medicines, such as alum, dragon's blood (tree resin), elixir of vitriol (sulfuric acid), Peruvian bark (quinine), and a cold bath. These remedies were largely harmless other than sulfuric acid, which could cause burns.

A folk remedy for infertility from an unknown time until the present is found in Dorset, England. Ancient tradition suggests that sleeping a night on the *Cerne Abbas*, or chalk giant, is supposed to aid conception. The giant is an outline of a man drawn into the chalk hillside who has a massive club in his right hand and a large erect phallus. In some Eastern European countries, on the Monday after Easter, young ladies dress in ritual outfits, and young men pour water on them to increase their fertility, but no reason is given for why this might work.

In the early twentieth century, researchers developed the mouse and then "rabbit test" to detect pregnancy. Rabbits were injected with a woman's urine or blood sera, which contains pregnancy hormones. The animal was then killed, and if the ovaries indicated ovulation, the woman was pregnant. However, a better method, the "frog test," was developed that did not kill the animal. Medical personnel injected a female frog with the material, and if the frog produced eggs within the next 24 hours, the test was positive, and the frog could be used again. Devices were also developed that did not work to detect pregnancy. For example, American physician Albert Abrams (1863–1924) invented a device that he claimed diagnosed and cured everything, but it was just a jumble of wires and radio tubes. One used for pregnancy tests was the "electrobioscope" based on "the sexuality of numbers and sounds;" it was not effective.

In the early twentieth century, lack of libido was thought to produce infertility. Therefore, some folk cures advertised radium suppositories, which only resulted in cancers of the reproductions organs. By the early twenty-first century, folk remedies were still forthcoming to foster conception. For example, honey and spices had been utilized for centuries to promote fertility. Some suggest that cinnamon mixed with the honey helps with blood flow in the ovaries as they contain amino acids helpful for ovarian function. Others recommend cough syrup that contains guaifenesin. This substance thins mucus not only in chest but also in the cervix, which may allow sperm to enter the uterus more readily. Folk and alternative medicine has also recommended ingesting vitamin D supplements and Irish moss, a form of red seaweed found mainly in the North Atlantic, which is rich in fiber, vitamin B, magnesium, and potassium, in addition to avocados, brazil nuts, and walnuts to foster conception. Whether or not these remedies are effective is debated. Some health-care providers also recommend a stress-free vacation in the sun to promote vitamin D. It has also been observed that reducing stress such as going on a vacation can result in pregnancy.

Pregnancy: Prevention and Termination

Since the beginning of civilization, women have attempted to prevent or terminate a pregnancy. In hunter-gatherer and agricultural communities, numerous children were generally needed to work. However, when cultures became more complex and evolved into urban city-states and crowded metropolitan areas, smaller families became desirable. Additionally, in tribes or communities with limited resources, children were sometimes seen as a liability to the family or community, resulting in newborns being killed and/or sacrificed to the gods.

The live birth rate was largely unknown before the nineteenth century when it began to be recorded and a decline was seen in industrialized nations. For example, in the United States the birth rate fell from at least 43 in 1850 to about 30 in 1900 and to 12 per 1,000 people in 2020. In the United Kingdom (UK), the birth rate fell from almost 36 in the 1870s to about 29 per 1,000 people by 1900. By 2020 in primarily English-speaking developed nations, the birth rate further declined to 12 in Australia and New Zealand, 11 in the United Kingdom and Ireland, and 10 per 1,000 people in Canada. This decline was due to advancements in public health and hygiene, technology, middle class's desire to educate their children, increased education of women, and cultural changes along with the availability of inexpensive rubber condoms and diaphragms. Although coitus interruptus ("withdrawal"), crude condoms, the rhythm method, and breastfeeding had been used for centuries, they often were not effective. In the 1960s, a hormone-based contraceptive pill was developed along with other hormone procedures, various intrauterine devices, and surgical sterilization that safely and effectively prevented conception.

Another factor for the decline in birth rate was the availability of safe pregnancy termination through surgery and, by the mid-twentieth century, hormone therapy. Induced abortions have been carried out for thousands of years at some point in most cultures and were generally acceptable before the time of quickening (when a woman first notices the fetus is moving, which is around four to five months). On the other hand, pregnancy termination has been, and is still, controversial and prohibited by some religions and nations. Nevertheless, in some countries induced abortions have been an aspect of family planning, and by the late twentieth century, childbirth was more dangerous than an induced abortion under the care of a health professional. Also, if safe pregnancy terminations were difficult or impossible to obtain, women often went to dangerous and desperate lengths to terminate a pregnancy, leaving some injured or dead from unsafe procedures or substances. This is reflected today in the higher maternal morbidity (sickness) and mortality rates in regions where abortions are difficult to obtain or illegal.

Induced abortion rates have often been an estimate, as women who obtain pregnancy terminations in regions where it is illegal are often not counted unless they need medical attention. By the late twentieth century, abortion rates began to decrease in many nations likely from increased family planning and a desire for fewer children as more infants survived into adulthood. Between 1990–1994 and 2010–2014, the induced abortion rate steeply fell in developed regions from 46 to 27 per 1,000 women but remained about the same in developing regions. In the United States and Canada, for example, the abortion rate declined from 25 to 17 per 1,000 people over this time.

The first use of the term "contraception" in English was in late nineteenth century from the Latin *contra*, meaning "against," and *conception*, meaning "becoming pregnant." The term "abortion" in the medical literature is still used to refer to a "miscarriage." The first use of the term "abortion" is from mid-sixteenth century Latin *abortio*, from *aboriri*, "to miscarry." In the mid-twentieth century, the term "abortion" began to be used for induced pregnancy termination. The legal, religious, and social status of contraception and pregnancy termination has waxed and waned over the course of history in different cultures, time periods, and nations. In the United States, for example, in the 1870s due to the federal Comstock Act and various states' Comstock laws, distribution of contraceptives or family planning information through the postal services along with the practice of pregnancy termination—which had been indirectly openly advertised and commonly performed since the beginning of the nation—was outlawed. Through various legal challenges in the Supreme Court, it was not until 1936 (*United States v. One Package*) when contraceptives and 1973 (*Roe v. Wade*) when medically safe induced abortions became available and legal again in the nation. By the 1960s, most non-Catholic Western nations had made abortion legal in some circumstances.

Prior to the nineteenth century, the infant mortality rate was high, and it is estimated that at least 1 of every 10 babies was stillborn or killed at birth. Although neonatalcide (newborn killing) has been illegal in Western cultures for centuries, it was sometimes used as a form of birth control. It occurred because a woman already had too many children, was poor or unmarried, had delivered multiple births, the infant was deformed, or a desired family size had been achieved. Ancient religious groups sometimes saw the practice essential for survival. Neonatalcide has been found in all societies at some point, and rather than being an exception, it has been the rule.

The topic of pregnancy prevention, termination, and even infanticide has been found in the arts and literature for centuries. Infanticide, for instance, was a common plot found in Greek mythology. An illustration of Giacomo Casanova (1725–1798) shows Casanova inflating a condom to test it while visiting prostitutes. In 1845, a drawing by French artist Charles-Michel Geoffroy (1819–1882) shows a woman abandoning her unwanted infant in the winter to wolves. Numerous works of fiction in the twentieth century began to mention contraception and abortion, which were then considered taboo for public discussion. For example, in British author George Orwell's (1903–1950) *Keep the Aspidistra Flying* (1936), a woman backs off from having sex when her lover refuses to use a condom. Abortion has also been depicted in American TV sitcom series. In *Maude* (November 1972), for example, the 47-year-old protagonist, Maude, has an abortion when she finds herself unexpectedly pregnant. By the twenty-first century, birth control or abortion decisions were part of the plot line of many series.

UNUSUAL TREATMENTS

Over the centuries, numerous folk remedies to prevent or end pregnancies have emerged, but some were toxic or not very effective. A few are still used by those who cannot obtain reliable contraceptives. These include cutting an orange or lemon into half and placing it over the cervix, which may act as a barrier to sperm. The acid in lemon juice may also act as a spermicide. Douching with a vinegar solution or soapsuds post intercourse to act as a spermicide is not very effective as sperm likely may have already traveled into the uterus. Extreme exercise has also been tried and sometimes can cause a miscarriage.

The first evidence of contraception is from Mesopotamia, where women placed small circular stones deep into their vagina to prevent pregnancy. To block and/or kill sperm, the *Papyrus Ebers* (c. 1500 BCE) recommends placing over the cervix lint (cotton) soaked in honey, acacia leaves, or acacia gum or even a pessary (a solid material) made from crocodile dung and honey. Acacia gum does have some spermicidal action. Sea sponges drenched in lemon juice and vinegar and placed high in the vagina was another recipe to block sperm from entering the uterus. Some of these may have been marginally effective.

The classical Greeks and Romans relied upon various plants for contraception or early pregnancy termination such as the herb silphium, which was so popular that it was harvested to extinction by the end of the first century. Greek physician Hippocrates (c. 460–370 BCE), along with others into the nineteenth century, advised women to jump up and down and kick their buttocks with their heels to prevent conception or cause an early abortion. The Hippocrates works also describes a drink made of black hellebore—a poisonous plant that was known to cause miscarriage—myrrh, spikenard (an herb), pine resin, and saltpeter. The Greek playwright Aristophanes (c. 446–c. 386 BCE) in some of his comedies noted that the herb pennyroyal was useful for early abortions. Later it was found to be toxic to the kidneys and liver.

Likely based upon Hippocrates' writings, Greek gynecologist Soranus, who practiced in Alexandra (c. 98–138), recommended that prior to intercourse, a woman should smear her cervix with rancid oil, or honey, or cedar oil or push into the opening of the cervix a thin strip of lint. Soranus also suggested women put a pessary made of a rolled ball of wool or cotton soaked in various substances into the vagina before coitus; these may sometimes have been successful. A pessary to cause an abortion was made of cantharides (Spanish fly from the blister beetle), elaterium (exploding cucumber), and colocynth (an herb), which probably only caused irritation in the vagina. He also recommended that women drink the lead-laden water that blacksmiths used to cool metal, but lead is highly toxic. Hellebore and ergot (fungus on rye) was also used for abortions at least throughout the eighteenth century, which sometimes worked. Women have also consumed the seeds of Queen Anne's lace—also called wild carrot—to prevent or abort an early pregnancy, but it can be poisonous.

In the first and second centuries, other herbs and plants were recommended to prevent or terminate pregnancy in classical Rome. For example, Pliny the Elder (23–79), along with Soranus, suggested the oil of the "common rue plant" was a powerful abortifacient. Studies have confirmed that rue does contain some abortive compounds. Greco/Roman physician Galen (129–c. 200 or 216) suggested the birthwort plant (flowers resembling the uterus), used to ease childbirth, could also be used to induce a miscarriage. The practitioner administered it by mouth or a vaginal pessary, but it was toxic to the kidneys. The plant tansy has been taken in early pregnancy to restore menstruation. Many of these concoctions and remedies were used throughout the early twentieth century as herbalists, apothecaries, and midwives knew of the supposed contraceptive and abortion properties of certain plants. This information was also passed among women. Some of these plants are still used today in developing nations or by alternative medical systems.

The most common practice of neonaticide has been by exposure of newborn to the elements, in particular, girl babies. In ancient Rome, for example, the *paterfamilias* (male head of the household) would determine whether the infant lived. In some Western cultures, at least from the 1600s, if a woman did not want a child, she paid the midwife more for a stillborn than a live birth. Couples even secretly practiced neonatalcide and eliminated female fetuses in China during the "one-child" policy years (1980–2015) as boy babies were desired.

In contemporary times, various folk remedies have been used by young or poorly informed women concerning birth control. These include inserting a piece of laundry soap in the vagina before intercourse or the use of disinfectant douches that can cause irritation and chemical burns. In the 1980s and 1990s, women drank milk with toxic iodine after sex to prevent pregnancy. Folk remedies that were not effective but causes little harm include women using Coca-Cola as a douche and drinking Coca-Cola with several aspirins and men using plastic or sandwich bags or balloons as condoms. Women have also attempted to promote an early abortion by placing onions in their underwear, sitting in hot water, or plunging into cold water; these do not work.

Rheumatic and Scarlet Fever

Rheumatic fever (RF) is a sequela of streptococci pharyngitis ("strep throat") or of scarlet fever. Both are caused by the bacterium *Streptococcus pyogenes* (Group A beta-hemolytic Streptococcus). In addition to these diseases, the pathogen causes impetigo, toxic shock syndrome, and streptococcus necrotizing fasciitis—the "flesh-eating" disease discussed in another entry. In 1874, Austrian surgeon Theodor Billroth (1829–1894) first described the streptococcal bacteria, and in the 1920s, American physicians George (1881–1967) and Gladys Dick (1881–1963) showed that streptococci were the cause of strep sore throat (discussed in another entry) and scarlet fever. Until the late nineteenth century, healers thought the causes of rheumatic and scarlet fevers, depending upon the culture, were bad air, an imbalance of the body humors, a contagion, or even a change in wind direction. However, the pathogen is spread by infected droplets in the air as a result of coughing and sneezing, close contact, and objects contaminated with the pathogen.

Rheumatic Fever

RF occurs from 1 to 5 weeks after a strep infection and can damage several organs of the body. The patient experiences high fever, shivering, and severe inflammation of joints, which become swollen, hot, tender, and painful. The pain often moves from one joint to another. The attack can last several weeks. Involuntary jerky movements (Sydenham chorea or "St. Vitus's dance") occur in 20–30 percent of people—primarily children—with acute RF. Other symptoms include nodules under the skin and a flat red rash. These symptoms result from the body's immune response to the *S. pyogenes* Group A pathogen. The most serious consequence of RF is rheumatic heart disease (RHD), which results in damaged heart valves, heart failure, strokes, endocarditis, and death. The signs and symptoms of RHD may not develop until years after the initial infection.

Acute RF and RHD in developed nations have been decreasing since the early 1900s. This trend is believed to be the result of better sanitation, healthier living conditions, and availability of antibiotics, in particular penicillin. In the 1920s in the United States, for example, RF and RHD were the leading causes of death in youth between 5 and 20 years of age, but in the early twenty-first century, the mortality rate was less than 0.01 per 1,000. However, it is still a major cause of illness and death among youth in developing nations. In the late 2010s, it was estimated that there were over 15 million cases of RHD worldwide, with 282,000 new cases and 233,000 deaths annually.

RF is likely an ancient condition. The Greek physician Hippocrates (c. 460–370 BCE) described the symptoms of the disease in detail in the fourth century. The word "rheumatic" was first mentioned in English literature in early fourteenth century and is derived from the Greek *rheum*, which means "to flow" as the disease was seen to go from one part of the body to another. The noted English physician Thomas Sydenham (1624–1689) described an "acute, febrile polyarthritis" in the late seventeenth century. The word "rheumatic fever" in English was first used in the early eighteenth century.

The illness may have affected the fitness of soldiers or even the outcome of war in the past. For example, in 1943 during WWII, problems with RF in training camps in the United States were prevented using the newly developed sulfonamides antibiotics. In epidemics of RF, more than 25 per 1,000 troops at some air bases had the condition. This changed when the use of sulfa drugs protected about 85 percent of the sick troops. Prevention of RF and RHD today is based upon fever-reducing drugs and treatment of strep throat with antibiotics, especially penicillin.

Scarlet Fever

Symptoms of scarlet fever, sometimes called scarlatina, usually includes a "strep" sore throat and fever. Chills, vomiting, abdominal pain, and a strawberry like bumpy tongue may also occur. The *S. pyogenes* can produce a toxin that causes a bright red rash that usually starts in the groin or armpits and covers the whole body. The rash usually begins as small, flat blotches that slowly become fine bumps that feel like sandpaper. The face may be bright red. As the rash fades, the skin may peel. Sequela of the disease can include nephritis (kidney disease), RHD, and RF. It is most common in children from age 5 to 15.

Although descriptions that may have been of scarlet fever, measles, or diphtheria are found in classical Greek and Roman writings, the first definitive description of scarlet fever was written by Italian physician John Philip Ingrassias (1510–1580). He named it *rossalia* and described the rash as being different from measles. Sydenham labeled it *febris scarlatina* around 1676 and described it as being a different disease.

Scarlet fever was viewed as a mild disease until the early nineteenth century. However, between 1820 and 1880, a world pandemic emerged, likely caused by overcrowded cities and poor hygiene. In the mid-nineteenth century, it had a death rate as high as 150 per 100,000. At this point, scarlatina was greatly feared by parents. In the state of Massachusetts during the epidemic of 1868–1869, 42 percent of all deaths in children were from the disease. Although cases of scarlet fever decreased over the twentieth century into the twenty-first century, in the mid-2010s an increased incidence was seen in some Asian nations and the United Kingdom for unknown reasons.

As with RF today, medical management consists of antibiotics for strep throat or tonsillitis. Among those who have had RF or RHD, penicillin injections are given frequently until the child reaches the late teens to prevent further damage.

Scarlet fever has been a plot in several works of fiction in the nineteenth and early twentieth centuries. As it was a frightening disease for parents and children, it was found in young adult and children's literature. For example, in Louisa May Alcott's (1832–1888) *Little Women* (1869), the third daughter is infected with SF. In its aftermath, she becomes progressively weaker and eventually dies of RHD. English American author Margery Winifred W. Bianco's (1881–1944) *The Velveteen Rabbit* (1922) tells the tale of a young boy with the fever who loves his toy so much it becomes real. However, the adults want to destroy it to prevent transmission of the infection.

UNUSUAL TREATMENTS

Various treatments for rheumatic and scarlet fever have been tried over the centuries until the development of antibiotics in the 1950s. Some may have temporarily relieved discomfort, but none cured the condition, and others were harmful.

Rheumatic Fever

Hippocrates suggested that patients should avoid extremely hot and cold temperatures. Syndenham in the seventeenth century recommended applying a mixture of "white bread and milk" to the affected joints to cool them. He also bled 10 ounces of blood on several days and gave an enema of milk and sugar on days when the patient was not bled. This was followed by "gentle purges" in the morning and the following evening with a large dose of diacodium (an opium herbal concoction) in cowslip water. He suggested that patients get out of bed every day and drink cool juleps (sugar syrup drink) and barley and oatmeal broths to quell the illness. None of these had any effect on the disease. Opium took away any pain, and bleeding and purges could have caused weakness or dehydration.

Throughout the late nineteenth century, physicians administered calomel (mercury chloride) and opium based on the tradition that mercury was a panacea. Calomel is still sometimes found in traditional and folk medicine, but this toxic substance is not available in most developed nations. Dover's powder, containing opium and ipecac (to cause vomiting), was thought an effective way to get rid of so-called body poisons as were "flying blisters"—creating small blisters with ground-up cantharis beetles on different parts of the body. Quinine was also tried but did little for the fever as it is specific for malaria discussed in another entry. A common practice as mentioned by American physician Henry Hartshorn (1823–1897) was the "alkaline treatment." Patients drank a concoction of bicarbonate of potassium along with Rochelle salt (potassium sodium tartrate), which acted as a laxative. Laudanum (an opiate) was applied to the painful joint and covered with oiled silk. The opiates may have dulled the soreness, but the purgatives could cause dehydration.

Scarlet Fever

Sydenham treated scarlet fever with bloodletting, enemas, and sweet drinks. He did not allow the patient meat and fermented liquors. After the skin had peeled off, he purged the patient with a mild laxative. If a child at the beginning of the disease was in a coma or had convulsions, Sydenham ordered a blister on the back of the neck followed by "syrup of white poppies," an opiate. Milk boiled with water for drinking was ordered, and patients were told to stay in their room, but they did not need to stay in bed. These treatments, other than pain relief, could have caused dehydration unless adequate fluids were consumed.

For scarlet fever in the mid-nineteenth century, although a few physicians still practiced bloodletting, it was going out of favor for this disease. However, based upon traditional treatments for a sore throat, several leeches were used externally on the throat. Physicians recommended gargling with hot red peppers, vinegar, and water. Silver nitrate was applied to the pharynx with a "large hair pencil" (brush). If there was ulceration in the throat, muriatic acid with honey, nitric acid, and other toxic chemicals were applied. For severe cases of scarlet fever, mustard plasters, hot water bottles, and fresh nettles were placed on the skin. Cupping (heated glass cups that cause a vacuum) was employed over the kidneys if the patient had swelling due to kidney issues. The poisonous herb belladonna was used as a folk cure and is still used today in alternative medical practices. However, these treatments were ineffective and often caused harm.

S

Seizures and Epilepsy

Seizures in the past were called epileptic seizure, convulsions, the "falling illness," or fits and occur from epilepsy and other conditions. During a seizure, a person may suddenly fall to the ground and have convulsions with jerking or shaking movements, loss of consciousness, and sometimes foaming at the mouth and loss of bladder control. The attack generally lasts for under two minutes and stops on its own. This "tonic-clonic"—previously known as "grand mal"—seizure is the type most often identified with epilepsy and is found in about 25 percent of epileptic patients. Another type of seizure is the "absence seizures" known in the past as a "petti-mal" seizure. In this type, the person often exhibits a vacant stare and fluttering eyelids and may be confused for a few seconds; it is found in less than 5 percent of cases. A "focal aware" seizure, previously called "temporal lobe" or partial seizure, makes up about 60 percent of all cases of epileptic seizures. During the attack, the person is conscious and remembers it. In some cases, it is an aura to a tonic-clonic seizure. Another type of seizure is the "atonic," or "drop attack," seizure in which the person's muscles suddenly relax and become floppy, and the patient falls usually forward. They quickly recover but can injure their face or head. These seizures are caused by abnormal brain activity, and in between seizures, the brain functions normally.

Epilepsy is a chronic neurological condition that involves recurring seizures from abnormal electrical brain that are "unprovoked"—not caused by a known condition. It affects about 1 percent of the population and is partly genetic. It is most common in children and in those over the age of 65. One-third of people with epilepsy experience depression and anxiety, and they are sometimes discriminated against. In developed nations, epileptic patients may be prevented from driving a vehicle until they have been seizure free for a certain time. Job opportunities can sometime be limited leading some people with epilepsy to keep it hidden from friends or coworkers.

Non-epileptic seizures, sometimes called "provoked" seizures, are caused by a brain injury or infectious diseases, tumor, stroke, high fever in infants, low blood sugar, high blood pressure in pregnancy, anesthesia, and alcohol withdrawal. When these medical issues are managed, the seizures generally go away. Another type of non-epileptic seizure is a "psychogenic seizure" that is primarily found in young females and is related to psychological issues. During the attack, the person stiffens and jerks.

The term "seizure" is from in the late eighteenth century. It is based upon "seize" plus "ure," meaning a "sudden attack of illness." The term "epilepsy" is derived from the Greek word *epilambanein*, meaning "to seize or attack," and was first used in the sixteenth century. In ancient Babylonian and Egyptian civilizations, individuals with the falling illness were thought to be possessed by spirits. The ancient Greeks thought the disease was a spiritual condition associated with the divine and called it the "sacred disease." However, Greek philosopher Hippocrates (c. 460–370 BCE) was the first person to propose that it had a natural and not spiritual causes. He believed epilepsy began in the brain and was caused by an imbalance in the body's humors. Greco-Roman physician Galen (130–200) taught that epilepsy was caused from the excess humors of phlegm or black bile in the brain.

However, from the fifth through the sixteenth century, as Christianity gained influence in the West, the scientific views of epilepsy experienced a setback, and the idea of demonic possession or God's punishment as the result of sin was accepted as the cause. Seizures were also considered a sign of being a witch, resulting in tens of thousands of people in Europe and in New England, mostly women, being tortured and killed, and anyone with seizures was stigmatized.

It was not until the eighteenth century that the view of epilepsy changed. Various types of epileptic attacks were also described, and the condition was viewed as a disease originating from the brain or other organs. However, patients were often institutionalized in special epilepsy hospitals. By the late nineteenth century, epilepsy was accepted by most physicians as a disease, but some patients were still institutionalized into the mid-twentieth century.

In the mid-1800s, the first effective anti-seizure medication, bromide, was introduced, and by the late twentieth century, many medications to halt seizures had been developed. Today depending upon the type of seizures, medications, brain surgery, and deep brain or vagus nerve stimulation with implants help most, but not all, patients.

Seizures and epilepsy may have changed the course of history and have been depicted in paintings and fiction. For example, French military leader and emperor Napoleon Bonaparte (1769–1821), who conquered much of Europe in the early nineteenth century, had several health problems and experienced a few seizures that resembled epilepsy. These issues may have affected his thinking, where he made poor decisions for some battles including defeat in the Battle of Waterloo. In the Christian New Testament, the gospels report that Jesus cured a boy who had fallen to the ground with seizures and banished the evil spirits possessing the boy. This is illustrated in Italian painter Raphael's (1483–1520) *The Transfiguration* (1500). A supposed link between epilepsy and evil is found in the film *The Exorcist* (1973). Physicians initially diagnose temporal lobe epilepsy in a young girl but soon realize she is possessed by an evil spirit that a priest attempts to eliminate by exorcism.

UNUSUAL TREATMENTS

Historically, trepanation, or scraping or drilling a hole in the head, to allow evil spirits, pain, or air to depart, thereby relieving seizures, was used from antiquity into the eighteenth century. Some patients did survive, but it did not prevent seizures. Hippocrates was one of the first to recommend the procedure. Likewise, various herbs have been and still are used, such as kava, mistletoe, and cannabis, to prevent seizures as part of folk medicine. However, the research is mixed as to their effectiveness, and many have toxic side effects.

The ancient Egyptians may have employed the electric torpedo fish as an early form of electrotherapy to treat epilepsy. The Greeks attempted to cure it by ritual purification and by reciting healing incantations. It is debated whether or not Julius Cesar had epilepsy or mini strokes, but folklore suggests he touched electric fish as a cure for his disease. On the other hand, Galen in the second century recommended that people with seizures eat the torpedo fish as it was easy on the digestion. In the eleventh century, Persian physician Avicenna or Ibn Sina (980–1037) recommended placing a live electric catfish on the brow of a person suffering from a seizure. Experiments of electricity to the head were also done in the eighteenth and nineteenth centuries, but they had no effect on halting seizures or curing epilepsy.

To prevent "catching" epilepsy, ancient Romans did not eat or drink out of the same cups and dishes as someone who had seizures. In addition, Romans would spit on their chest as a preventive measure against getting the falling illness. Roman philosopher Pliny the Elder (23–79) noted that epileptics drank the blood of dead gladiators to cure the disease and ate pieces of liver from these gladiators. Roman physician Celsus (c. 25 BCE–c. 50 CE) in his *Of Medicine* only allowed the patient to drink gladiator blood as a last resort when the condition had become chronic, but this did not prevent seizures. Celsus also bled the patient immediately after a seizure had ended to balance the humors. He also applied a cautery iron to the back of the head and other parts of the body to cause a blister to remove the bad humor in the head. It did not cure the malady and could have resulted in infection and death from the burns. Celsus also administered an enema or a purge of black hellebore to remove bad humors. This herb can cause dangerously irregular heartbeat, dizziness, and problems in breathing and did nothing for seizures. He also clipped the victim's hair short and placed oil and vinegar on it, although no explanation was given for this harmless treatment.

During the Middle Ages and Late Middle Ages, exorcism to drive out demons was a common treatment. For example, the noted English physician John Arderne (1307–1392) used both religious incantations and concoctions. He advised patients to write the names of the Three Kings from the Christmas story in blood on paper, sometimes this was hung around the patient's neck, and to pray for the souls of the parents of these kings for a month along with drinking peony juice (it acts as an antispasmodic) added to beer or wine as a toast to the Three Kings. A recipe to prevent seizures included a decoction of the "bone of a stag's heart" mixed in wine and powdered pearls and burnt human bone consumed with licorice. The licorice could cause heart arrythmias but did not stop seizures.

Folk medicine in the sixteenth century suggested that to prevent epilepsy in a child, red corral and "pionie" (peony) seeds should be hung around a child's neck. Various cooking spices were also used to prevent and cure the malady. These generally had no effect upon seizures, but large amounts of some, such as nutmeg, could provoke seizures and hallucinations.

In the seventeenth century, clinicians derived several remedies for epilepsy from the human body as it was believed that the body held magical properties. One physician administered dried human heart. The "essence of man's brain" to cure many diseases, including epilepsy, required taking the brain of a young man who had died a violent death, adding wine to it, and letting it digest by placing it for a year in warm horse manure. Another remedy for epilepsy was the "spirit of skull" also called "Goddard's drops" used into the nineteenth century. This was made by distilling dried human skulls in a glass retort and then adding spirit of nitre (mixture of nitric acid and water) to the oily mixture that was created. "Spirit of human blood" was made by distilling blood and mixing it with a tincture of peony flowers and angelica water. Mumia or ground-up dried Egyptian mummies was considered a "cure-all" including epilepsy. None were effective but were largely harmless unless too much nitric acid was used, which could scar the esophagus.

Metals were used against seizures. For example, German physician Paracelsus (c. 1494–1541) dissolved gold in a toxic mixture of nitric and hydrochloric acid, newly discovered by alchemists, in the belief that gold would make the body indestructible. He administered the resulting salt, gold chloride, to patients to treat epilepsy. However, the salt was toxic to the kidneys, caused massive salivation, and was not effective for seizures.

In the eighteenth century, Scottish physician William Buchan (1729–1805) in his *Domestic Medicine* (1774) bled patients immediately after a seizure if they had an outgoing temperament to balance the humors. If the attack was caused by general debility, he gave the patient Peruvian bark (quinine used for malaria) to strengthen the nerves. These treatments did not work but were largely harmless other than bleeding, which could have caused weakness.

Into the late nineteenth century, other treatments were administered still based upon humoral theory. As a last resort, for women suffering seizures, surgeons removed the ovaries and bled them for seizures correlated with menstruation. Physicians bled the patient with leeching, cupping, or cutting a vein. Hartshorn applied cupping or a seton (a tread passed under a cut to keep it open to allow drainage) on the neck to remove bad humors from the brain. Clinicians tried anesthesia, but it could cause further seizures. None of these remedies were effective and could cause infection or death. In the nineteenth century, hypnotism became popular. Although physicians gradually accepted this method, it was not effective for epilepsy but was for some psychological issues.

From the late twentieth century, the use of crystals to prevent seizures has been advertised. Proponents believe that carrying or wearing a combination of amethyst, jasper, lapis lazuli, and black tourmaline might prevent and help reduce epileptic seizures due to their "electric vibrations." This is not supported by research but is a harmless remedy. Although controversial, light therapy with exposure to high-intensity light or blue light implanted into the brain might be a marginally effective treatment for temporal lobe epilepsy. Other remedies with mixed results for epilepsy include a ketogenic (low-carb, moderate-protein, and high-fat) diet.

Sexually Transmitted Infections

Some sexually transmitted infections (STIs), also called sexually transmitted diseases (STDs), have been noted since antiquity. The microorganisms causing these infections are generally transmitted in semen, vaginal, and other body fluids through sexual activity. They can also be transmitted non-sexually to an infant during pregnancy and childbirth or from transfusions of blood products or shared needles.

Physicians in antiquity—and up to 60 years after the introduction of syphilis into Europe—realized that gonorrhea, chancroid, and genital herpes were separate conditions. However, in 1587, one physician claimed gonorrhea and these other STIs were symptoms of syphilis, and this error was carried into the nineteenth century. STIs for centuries have created psychological and social issues for the infected person including discrimination, fear, and a lack of a sense of security. Scottish physician William Buchan (1729–1805), for example, using the term "venereal disease," noted that it was considered a shameful disorder.

Gonorrhea, Chlamydia, and the "Whites"

The term "gonorrhea" for centuries has been classified as any whitish discharge from the genitals. It was known in the past also as "whites" in women, "gleets" in men, or "running of the reins" in both. Physicians until the early twentieth century believed this discharge might be caused by masturbation and accidents to the lower back, and not just sexual intercourse. Gonorrhea (also called the "clap") and chlamydia, the two most reported STDs in North America and the United Kingdom have similar symptoms. Both can cause burning on urination and discharge from the penis and sometimes the vagina. In some cases, they do not cause any symptoms. In women, they can spread to the reproductive organs and cause pelvic inflammatory disease (PID), which if untreated with antibiotics can lead to long-term pelvic pain, ectopic pregnancy, and infertility.

During pregnancy, these infections can lead to miscarriage and premature labor. During delivery, the bacteria can be transferred to the infant's eyes and result in an eye infection and, without treatment, blindness. In men, both bacteria can also cause sterility. Gonorrhea is caused by the *Neisseria gonorrhoeae* discovered by German physician Albert Neisser (1855–1916) in 1879. Chlamydia is caused by the *Chlamydia trachomatis* bacterium discovered in 1907. The term "gonorrhea" is from the Greek *gonos*, meaning "semen," and *rhoia*, meaning "flux" or flow, and the term "chlamydia" was first coined in the 1960s from the Greek *khlamus*, meaning "cloak."

Besides these infections, other common STDs were also classified as "gonorrhea" until the twentieth century. They often cause vulva itching, soreness, odor, and more rarely in men a discharge from the penis or inflamed testicles. Trichomoiasis is caused by the parasite *Trichomonas vaginalis*, and from 2 to 20 percent of the population are infected with it. Although considered a nuisance by most women, it is a risk factor for prostate and cervical cancer and is medically managed with antiprotozoal drugs. Vaginal candidiasis, a yeast infection caused by the *Candida albican*, affects three out of four women at some point in their lifetimes. It can be transmitted sexually or arise from an imbalance of microorganisms in the vagina. It is common with pregnancy, antibiotic therapy, and diabetes and managed by antifungal medication. The term "candida" is derived from Latin from the white togas, *candida*, worn by Roman senators.

Genital discharges were first recorded in Mesopotamian and in the Egyptian *Papyrus Ebers* (c. 1500). They were also mentioned in the scriptures of the Abrahamic religions—Christianity, Judaism, and Islam. In London by 1161, it had become evident that some diseases were transmitted sexually resulting in a law that forbade prostitutes from working in brothels if they had "burning in the genitals" in an effort to halt the diseases.

Syphilis and Chancroid

Syphilis, until the mid-twentieth century, was greatly feared. When it first arose, it created severe disfigurement from ulcers, which ate away faces and resulted in an early death. As it became endemic, outward deformities disappeared. By the late nineteenth century, it was found to have different stages. In the primary stage, a painless chancre (ulcer) on the genitals develops several weeks after sexual activity with an infected person. It then disappears. In the secondary stage, a rash over the body and/or swollen glands may occur. After this, it goes into a latent stage that can last for years. In the late or tertiary stage, in untreated people, about 15–30 percent have organ damage. In this stage, serious heart problems or dementia called "general paresis" may develop followed by death. In pregnant women, it can also lead to spontaneous abortions, stillbirths, and congenital syphilis in infants.

Syphilis is caused by the spirochete bacterium, *Treponema pallidum*, first identified in 1905. In the sixteenth century, astrological misalignments were believed to be a cause. In addition, based upon humoral theory, into the nineteenth century it was thought to be caused by poisonous matter introduced into the blood. After the introduction of penicillin in 1943, syphilis dramatically decreased worldwide. However, since 2000 it has been rising in Western nations primarily among young men who have sex with men.

The term "syphilis" was coined by Italian physician Girolamo Fracastoro (1478–1553) in 1530 after a shepherd in his poem, who had been given the disease as a punishment from the gods. The disease was likely brought to the Old World by Columbus's sailors, although a few scholars suggest it may have been present in the Old World and became more virulent. The first recorded outbreak in Europe occurred in 1494–1495 in Naples during a French invasion and was brought back to France. When it spread from France to other nations, it was called the "French pox" or the "great pox" to distinguish it from smallpox. Within a decade, syphilis had spread around the world. Many Europeans believed it was a punishment for illicit sex and kept the disease secret and hidden. However, in the eighteenth century Buchan, seeing that venereal diseases could be transmitted to "innocent infants, midwives, and married women whose husbands lead dissolute lives," recommended education so they could be treated.

Sometimes in the past, physicians confused syphilis with chancroid—also called "soft chancer"—an ancient STI. Chancroid causes painful ulcerations usually in the genital area with ragged edges. It can also cause the lymph nodes in the groin to harden and swell. Chancroid is caused by the *Haemophilus ducreyi* bacterium. It is now common in tropical countries but rare in North America and Europe. Without medical management with antibiotics, the ulcers can sometimes lead to gangrene and the loss of the penis. The term "chancroid" was first used in 1858 and is from the French *chancre*, from the Latin *cancer*, meaning "crab" or later "tumor." In ancient Greece, Hippocrates (c. 460–370 BCE) noted ulcers on both male and female genitals who were very sexually active.

Genital Herpes and Warts

Two other STIs that cause lesions on the genitalia arise from viruses. These included genital herpes and genital warts. Genital herpes, from the herpes simplex virus 2 (HSV-2), typically causes blisters on the genitals or groin area. They break open and cause pain, itching, and burning and can manifest as a recurrent infection. About 12 percent of the world's population has the infection. There is no cure although antiviral medications may prevent or shorten outbreaks. The term "herpes" is from Latin, from the Greek *herpein*, meaning "to creep," and was first used in the fourteenth century.

Genital warts (*condyloma acuminata*) are caused by the human papilloma virus (HPV) and form small, flesh-colored, or brown swellings in the genital area. When several warts grow together, they have a cauliflower shape. Around 75 percent of all sexually active people will become infected with HPV at some point during their lifetime. By the late eighteenth century, physicians noted that few celibate nuns had warts or died from cervical cancer compared to married or widowed women. In 1985, an HPV virus infection was discovered to be associated with cervical cancer. In 2006, an HPV vaccine was introduced to prevent the infection. These warts were known at least from the time of Hippocrates. They were also described by Roman physician Aulus Cornelius Celsus (c. 25 BCE–c. 50), who noted they were transmitted by promiscuous sex and were more frequent in male-to-male sex, which was common in ancient Rome.

HIV/AIDS

In the late twentieth century, a new STI, acquired immune deficiency disease (AIDS), caused by the human immunodeficiency virus (HIV) emerged. It was first identified in 1982 among men who had sex with men and resulted in a high death rate from opportunistic infections. Since their lifestyle was blamed for the disease, only minimal research was initially focused on this "gay plague," and some people saw the disease as a punishment from God for sin. However, by 1990, AIDS had emerged among heterosexuals and children who had acquired the disease via blood-transfusion products, sexual contact, or IV drug use. Those with hemophilia were in particular danger until the blood supply was screened. Intensive research increased to determine the nature of the infection, its prevention, and treatment. By 1997, HIV was considered a chronic condition as drugs for its treatment were developed to keep the viral load down. However, it is incurable. By 2013, AIDS-related deaths had fallen 30 percent since their peak in 2005 due to a cocktail of various drugs.

STIs have posed a threat to military strength and efficiency throughout history. In ancient Rome, for example, civilians including sex workers, common-law wives, children, and slaves often followed the army into battle. These "camp followers," who often established communities around the military post, supplied needed services such as mending and washing clothes, nursing the injured, and cooking along with taverns and sex, which often resulted in the transmission of STDs. These infections followed the Crusades and various European conflicts into the nineteenth century. To reduce the transmission of STIs during the American Civil War, the Confederate military established brothels where sex workers were medically inspected, which successfully reduced STI rates.

Combatants of WWI (1914–1918) dealt with sexual encounters and preventing STIs in different ways. American troops were ordered to be chaste and were punished if they did not comply. If personnel had any evidence of gonorrhea or syphilis, they were hospitalized and/or confined to the stockade, and their pay was suspended. On the other hand, the British and French passed out condoms, and the French and Germans established brothels where the women were medically inspected. In the American military, about 15 percent of troops were unable to fight on the front line due to STIs. Because of this disaster by 1931, condoms became standard issue in the American military although they often were not used. In WWII (1939–1945) and subsequent conflicts in Southeast Asia, STDs were a serious problem for the U.S. military. With antibiotics, they generally could be cured other than HIV. However, chlamydia, gonorrhea, and syphilis are still prevalent in the military due to lack of condom use, multiple sexual partners, and the influence of heavy alcohol consumption.

UNUSUAL TREATMENTS

Treatments of various STIs have existed since antiquity. However, most healers did not give a reason for the remedy, and only a few were effective.

Gonorrhea, Chlamydia, and the "Whites"

For the treatment of ureteral or vaginal whitish discharges, the *Papyrus Ebers* suggested that a man chew manna—resin from the tamarisk tree—and apply a duck's egg yolk to the penis to draw out the poison, which was ineffective. Healers also inserted aromatic sandalwood oil, which has antimicrobial properties, into the urethra, which may have sometimes cured the condition. Roman encyclopedist Pliny the Elder (23–79) recommended consuming chicory and aster flowers (a "sovereign remedy" for diseases of the groin), which have antibacterial properties, to treat discharges in both men and women. Greek gynecologist Soranus, who practiced in Alexandra (c. 98–138), advised a hard bed for the treatment of gonorrhea although no reason was given for this remedy, and it was not effective.

In the thirteenth century, physicians treated a discharge from the penis with leeches applied to a large leg vein to remove "poisonous blood" and injected goat or breast milk into the penis, but these were ineffective. Swiss physician, Jakob Rueff (1500–1558) gave purgatives for gonorrhea including rhubarb and jalap until the "weeping from the penis is gone." He also administered ineffective pills that were a mixture of turpentine, coral, and terra sigillata (stamped medicinal clay that supposedly cured anything). Turpentine may have led to kidney damage and death. By the seventeenth century, camphor, an extract from the camphoratus leaf, was claimed to help cure the "whites." Moreover, in the late twentieth-century researchers found that camphor did have some anti-trichomonas activity and may have been marginally effective for managing this infection.

By the eighteenth century, mercury was used indiscriminately for gonorrhea, syphilis, and soft chancre. Trial and error over the centuries had shown that mercury appeared to cure these infections but was highly toxic to the person. Likewise, Buchan injected lead acetate, which also has antimicrobial properties, into the urethra, which may have in some cases cured the disease. However, these metals were also highly toxic. Clinicians also treated gonorrhea by injecting metal salts, including silver nitrate, potassium permanganate, gold chloride, and zinc sulfate, into the urethra or vagina. These daily injections for several weeks were often successful and used until the 1940s when antibiotics were developed. Scottish physician Alexander Macaulay (1783–1868), in the early nineteenth century, to rebalance the humors, unsuccessfully treated inflamed testicles from gonorrhea by bleeding and purging and placed poultices on the scrotum.

By 1912, gentian violet, a purple dye, was successfully used for vaginal infections, which become popular again in the twenty-first century due to drug-resistant microorganisms. Some current folk remedies that may be marginally effective as they have antibiotic properties include garlic, apple cider vinegar, and golden seal. Gargling with Listerine for oral STI infections has been found only to slightly reduce the bacteria.

Syphilis and Chancroid

For painful chancroid ulcers, Hippocrates, and Greek physician Galen of Pergamon (129–c. 210), rubbed arsenic disulfide (which has antibiotic properties) on the sores, while some physicians cauterized the lesion, which may have killed the bacteria. With the emergence of syphilis in the early sixteenth century, management of this new disease was like that of chancroid. Physicians bled the patient and administered mercuric chloride in the form of calomel to purge the patient and to cause massive salivation, which were both thought to eliminate the disease poisons and balance the humors. Patients were also instructed to anoint the body from the knees to the groin with a salve made with mercuric chloride for several days. The mercury may have helped to destroy these bacteria but likely caused mercury toxicity leading to kidney damage, neurological issues, and death.

In the sixteenth century, several remedies for syphilis were tried. The resin from the guiacum bush brought back from the New World became a popular but useless therapy. It was made into a drink to produce copious sweating, which was believed to cleanse the body of the syphilitic poison but had no effect. In addition, the noted Swiss physician Paracelsus (1494–1541) disputed the effectiveness of this treatment and instead used mercury, which became the "gold standard" of treatment for syphilis into the early twentieth century as it sometimes cured the disease. In the eighteenth century, physicians administered the newly compounded "Fowler's solution"—that contains poisonous potassium arsenite—to syphilis patients, and by the Victorian age of the nineteenth century, a mercury steam room or cabinet to cure syphilis became popular. Nitric acid was also used on chancroid buboes and syphilis chancres. Toxic salts of lead, silver, and gold were also employed, both externally and internally, but it is unknown how effective these heavy metals were in destroying the spirochete. In addition, they could cause liver and kidney damage.

Many worthless devices based upon radio waves, light, and electronics in addition to alcohol-based tonics were advertised to cure the syphilis in the early twentieth century. However, some partially effective remedies did emerge. In the 1910s, an early sulfa drug salvarsan (arsphenamine) was found to cure syphilis, but it could also cause serious side effects. Other remedies were also developed. People had observed that patients who developed high fevers were often cured of syphilis. Therefore, physicians infected patients with malaria. It was reasoned that later the malaria could be treated with quinine; however, many people died of this treatment. Sweat-boxes were also developed for the same purpose. These treatments finally became obsolete by the discovery of penicillin in the 1940s, which cured most bacterial STIs.

Genital Herpes and Warts

Until the late twentieth century, treatments for viral STIs were generally ineffective. Some treatments were tried based upon tradition. Greece physician Aretaeus (130–140 CE), for example, placed a mixture of honey, water, and milk in the urethra and on the penis to treat genital warts. Honey has been used for centuries to treat many infections, but it was unlikely that it was effective on warts. Paulus placed thyme on herpes ulcers and warts. For herpes, and likely chancroid, clinicians advised patients to "avoid drinking wine and strong drinks" and to apply a poultice of herbs on the lesion to relieve the pain. Buchan in the eighteenth century placed asters and dry cotton in the ulcers to absorb the pustular matter. At the same time, he dissolved toxic mercuric chloride in brandy and added it to sarsaparilla tea and had the patient drink it once a day for six weeks, which likely only poisoned the patient.

From the Middle Ages onward, the tansy plant has been used as a folk remedy for various conditions. Recent research has shown that tansy has antiviral properties and may be a potential treatment for herpes. Several folk remedies have also been used to treat genital warts. These include swabbing the wart with tea tree oil, apple cider vinegar, garlic and onions, other herbs and in recent years ozone treatments. However, their effectiveness has not been thoroughly researched, but they generally do not harm the patient.

HIV/AIDS

In the late 1980s, desperation for a cure for HIV/AIDS resulted in activists demanding that an anticancer drug AZT be used as a treatment. This drug was rumored to be successful in lab tests against HIV. The FDA's review of AZT was fast-tracked, but the clinical trials were not rigorous. It was later found that the drug caused anemia for which blood transfusions were required, along with liver damage and other problems. AIDS patients had grown so desperate that they also tried many folk remedies. Some of these cures to relieve symptoms of HIV included milk thistle to improve liver function and medical marijuana. Garlic, St. John's wort, echinacea, and ginseng were also touted but were found to interact with antiviral medications, and there was little evidence that these substances could treat AIDS.

Crystal therapy that claims that vibration of certain stones neutralizes a disease was attempted but has no scientific basis—bloodstones were advertised on the internet for HIV/AIDS. Spell casters have also advertised that they can cure HIV by having the client buy holy water or ritual gadgets upon which a healing spell has been placed. Other futile remedies include the Electromagnetism Complete Cure Device, chemicals like Virodene (from a toxic industrial solvent), and oxygen therapy primarily found in developing nations. As with other STDs from at least the fifteenth century, having sex with a virgin was supposed to cure HIV since her virgin blood was thought to cleanse the disease. This myth is thought to be a factor of HIV-positive men, particularly in Africa, raping female infants and children.

Skin Conditions

Hundreds of skin conditions exist. They often have signs and symptoms of redness, pustules, rashes, oozing, itching, and pain. Some are acute and some chronic. Others are associated with infectious diseases, which are discussed in other entries including smallpox, childhood diseases, typhus, the plague, and leprosy. Until the nineteenth century, many skin diseases were combined into a few conditions as clinicians could not accurately diagnose them as they appeared similar. They have many causes including microorganisms, as well as heredity and environmental factors. Skin diseases in the past were thought to be caused by an imbalance of the body humors or God's punishment. Although most were not deadly, they could cause psychological trauma and social exclusion due to the ugliness of the condition. Due to space limitations, only a few will be discussed in this entry.

Herpes Simplex and Herpes Zoster

For centuries, "cold sores" or fever blisters caused by the herpes simplex virus (HSV-1), genital herpes (discussed in the "Sexually Transmitted Infections" entry caused by HSV-2), and shingles caused by the varicella-zoster virus (VZV) virus have been considered a painful nuisance. When HSV-1 or HSV-2 infections appear, they generally start with a tingling sensation, then blistering, weeping from the blister, and finally crusting. They are triggered by stress, illness, fatigue, or sun and wind exposures. If the eruption is on the mouth, it is generally localized. However, shingles, which, in addition, has symptoms of pain and a red rash, is generally found along one side of the chest, side, or face and is a reactivation of VZV.

Until the early twentieth century, it was not known that childhood chickenpox caused shingles later in life as the herpes virus often migrated to the dorsal root ganglion near the spine. Up to 90 percent of adults worldwide have HSV, but not everyone will experience cold sores. About 30 percent of Americans will get shingles, which is usually found in people over 60 years or those who are immunocompromised. For some people, shingles pain continues long after the blisters have cleared. Vision and hearing loss, brain inflammation, and skin infections can be an aftermath of herpes zoster. Today vaccines prevent chickenpox and shingles, and various drugs can shorten the duration of both conditions.

The first use of the term "herpes" in English was during the fourteenth century and is derived from the ancient Greek word meaning "to creep" or "crawl." "Shingles" was also first coined in the fourteenth century and is derived from the Medieval Latin *cingulum*, "girdle." Herpes has been noted in the medical literature for centuries. The *Papyrus Ebers* (c. 1500 BCE), for example, mentions herpes on the genitalia of children, which begs the question: Was it caused by the virus or was it another skin condition as genital herpes is usually transmitted by sexual activity?

Greek physician Hippocrates (c. 460–370 BCE) describes the spreading of herpetic skin lesions and believed they were caused by a "disordered state of the abdomen." Cold sores and shingles were also noted by other ancient Greek and Roman healers. Greco-Roman physician Galen (c. 129–c. 216), for instance, whose teachings influenced medicine throughout the mid-nineteenth century, claimed that shingles was caused by yellow bile that had separated from the blood as did Byzantine physician Paulus Aegineta (c. 625–c. 690).

Boils and Carbuncles

Boils, or furuncle, and carbuncles (a cluster of boils) are skin abscesses that form in a hair follicle infected with a Staphylococcus or *Streptococcus* bacillus. They usually start as red painful lumps that are filled with fluid and pus and often leave scars. These sores tend to be found on the buttock, shoulders, and neck areas. Carbuncles can also have symptoms of fever and fatigue and are more common among men than women and among those with chronic illnesses or who are immunocompromised. Impetigo caused by staph infections form honey-colored crusts. The infectious nature of these skin diseases was discovered in the late nineteenth century. Today they are treated with hot compresses and antibiotics although some staph bacteria have become resistant to antibiotics.

The first-known use of the term "boils," in English, is from around the fourteenth century and is derived from the Latin *bullire*, meaning "to bubble." The term "carbuncle" was used before the twelfth century and is from the Latin *carbunculus*, meaning "small coal." Boils and carbuncles have been known since antiquity. For example, Galen in the first century believed the cause of these conditions was the overheating of "melancholic blood," and Paulus in the seventh century thought that boils arose from "gross and depraved humors." Boils have been mentioned in the ancient Hebrew and Christian scriptures. For example, the description of the sixth plague of Egypt appears to be of boils. Also, the account of the biblical King Hezekiah's malady was a classic description of a boil or carbuncle.

Atopic Dermatitis (Eczema) and Psoriasis

Both atopic dermatitis (eczema) and psoriasis are chronic relapsing autoimmune diseases caused by an overactive immune system. In eczema, patches of skin become inflamed, itchy, red, cracked, and rough. It is common on the hands, arms, behind the knees, and in babies on the face. Blisters may form, and the open wounds can develop staph infections and lead to impetigo. Sometimes it is difficult to distinguish eczema from contact dermatitis caused by allergens, and it was not until the 1930s when atopic dermatitis was first described as a separate condition. People with eczema often have hay fever, asthma, and food allergies, and it is likely genetic. The disorder is triggered by stress and allergens. Different stages and types of eczema affect about 32 percent of people.

Several types of psoriasis exist. The most common is plaque psoriasis, which causes itchy red skin with whitish scaly plaques. It is generally found on the scalp, elbows, knees, and other joints. The condition first appears between the ages of 15 and 35, and it tends to be more common among females. About 3 percent of the population worldwide is affected with the disorder. About 30 percent of people with psoriasis also develop psoriatic arthritis (PsA) discussed under the "Arthritis, Rheumatism, and Gout" entry. No cure exists for either atopic dermatitis or psoriasis. Medical management of these conditions includes topical steroids, antihistamines, and antiseptics, and some new drugs help to reduce the symptoms.

The first use of the term "eczema" was in the mid-eighteenth century. It is from New Latin, borrowed from the Greek *ékzema*, from *ekze-*, "to boil over, or break out." The term "psoriasis" was first used in the seventeenth century from the Greek *psōriasis*, meaning "skin itch." "Atopic dermatitis" was not coined until the early 1930s when it was recognized as a separate condition. Psoriasis has existed since antiquity and has been found in mummies. Physicians in antiquity thought these afflictions were caused by too much internal heat. During the European Middle Ages, eczema and psoriasis were sometimes confused with leprosy, and sufferers were required to carry a begging cup and ring a bell or clappers both to warn people they were coming and to attract people to give them alms. Around 1809, psoriasis was recognized as a specific disease. and in the 1960s, investigation of psoriasis as an autoimmune illness began.

Acne, Rosacea, and Erysipelas

Throughout the early nineteenth century, acne (acne vulgaris or pimples), rosacea (acne rosacea) and erysipelas were thought to be various stages of one skin disease. Although generally not life-threatening, these skin conditions—especially when on the face—can result in social isolation and psychological trauma. Acne vulgaris is a common skin disease in which hair follicles become clogged, resulting in inflammation. Acne is most common among adolescents and has genetics, hormonal, and other causes and may result in considerable scarring. It is the most common skin condition in North America affecting about 80 percent of the population at some point during their lives. Acne rosacea is a relapsing condition and may become worse with sun exposure, heat, alcohol, strong emotions, caffeine, and spicy foods. It is identified by the intense reddening of the skin and dilation of superficial blood vessels, which sometimes causes a bulbous nose and can be mistaken for erysipelas. Rosacea affects 5 percent of population and usually first appears when sufferers are in their 30s and 40s. In the past, it was thought to be caused by heavy alcohol consumption.

Unlike acne or rosacea, erysipelas—also known as St. Anthony's Fire, or rose—is a contagious skin disease caused by Group A hemolytic streptococci that affects the upper layers of skin. It causes fevers, shaking, chills, fatigue, headaches, general illness, and a burning heat over the eruption. Erysipelas is more common among the very young and old and among those who are immunocompromised or who have chronic diseases. Erysipelas affects the skin of the lower limbs and face, where it has a characteristic butterfly pattern on the cheeks and across the bridge of the nose. Today it is a rare disease with around 2 percent of the population being affected each year. In the past, it was more virulent. For instance, in 1089, a severe erysipelas epidemic, called St. Anthony's fire, is said to have killed all who did not pray to St. Anthony. In wars prior to the early twentieth century, erysipelas, in addition to tetanus, gangrene, and blood poisoning, killed more combatants than armaments. It is now medically managed by antibiotics.

These skin maladies have been described for years. The first-known use of the term "acne," for example, was in the early eighteenth century and was borrowed from the New Latin *acnē*, from the Late Greek *aknḗ*, meaning "eruption on the face." The term "rosacea," first used around 1813, is adapted from New Latin, meaning "rose-colored acne." The word "erysipelas" is from the fourteenth century from the Greek *erythros*, meaning "red," and Latin *pellis*, meaning "skin." Galen in the seventh century believed these eruptions were the result of swelling caused by hot blood that had mixed with bile. Earlier, Hippocrates had compiled the first written record of an erysipelas epidemic.

Skin Ulcers

An ulcer is an open sore on the skin or a mucous membrane (discussed in the "Gastrointestinal, Dysentery, and Diarrhea" entry), where the skin around it may become red, swollen, and tender. Ulcers that are caused by a lack of mobility, which leads to decreased blood circulation and a disintegration of tissue, are called bedsores or decubitus ulcers. These sores are common on the hips, buttock, and heels. Ulcers often have a crater surrounded by sharply defined edges, and the primary symptom is pain. Individuals with chronic diseases such as diabetes or poor blood circulation may develop ulcers on the skin of the lower extremities. They often become infected, have a foul order, and show few signs of healing, which can lead to amputations. In the United States, each year, more than 2.5 million people develop pressure ulcers. "Ulcer" comes from the Greek *elkos*, meaning "wound." They have been the bane of the bedridden for centuries. Throughout the eighteenth century, they were thought to arise from imbalanced humors and a lack of exercise. Medical management includes prevention measures such as movement, frequent change of position for the bedridden, and exercise to increase circulation in the extremities.

Scabies

Itchy skin has been known from antiquity, but in many cases, the cause was not known. Some causes of severe itchiness included insect bites, allergic reactions, and scabies. Scabies or "the itch," the "scab," or "seven-year itch" is caused by the small *Sarcoptes scabiei* mite that burrows and lays eggs under the skin of the wrists, finger webs, and other parts of the body. It causes a rash and severe itch that gets worse at night. Scabies is highly contagious and spreads quickly from person to person, and it has caused epidemics in residential facilities and in combat. It is now prevalent among homeless people. Globally it is estimated that from 130 to 300 million people are infected with the infestation. It has caused much annoyance and skin infections from scratching. The term "scabies" is from the Latin *scabere*, meaning "to scratch." It is treated today by lotions containing permethrin.

Skin diseases have been mentioned in scripture and the arts. For example, in the Christian *Old Testament* and Hebrew Bible (*Tanakh*) written from around 1200 and 165 BCE, Job was described as having sores, which may have been boils or some other skin disease. This story is depicted in English painter and poet William Blake's (1757–1827) illustration "Satan Smiting Job with Sore Boils" (1826). Rosacea, for example, has been illustrated in several paintings showing a person with a bulbous nose and red face such as the "The Old Man and His Grandson" (1490) by painter Italian Domenico Ghirlandaio (1448–1494). William Shakespeare (1564–1616) in *Romeo and Juliet* (c. 1595) writes that Queen Mab, the queen of nightmares and misfortune, in anger puts blisters on young ladies' lips because their breath smells like sweetmeats (candy).

UNUSUAL TREATMENTS

Skin problems for centuries have been treated with various folk remedies. In a folk treatment for warts in the Middle Ages, for example, a person would cut a mouse in half and apply it to the wart or place the dead hand of an executed prisoner on a wart as this was thought to eliminate it. Some folk remedies are still used today for skin issues and include placing cold cooked oatmeal; vinegar that has anti-inflammatory properties; and tars, iodine, and alcohol that have antimicrobial characteristics on eruptions. Other remedies include milk baths, mud, honey, and turpentine, and by the late twentieth century, fluoride toothpaste and diluted chlorine bleach, which have antimicrobial properties, were also used. Still other remedies have generally been confined to the past such as bathing in human blood in ancient Egypt. In the seventeenth century, a "cure-all" made of boiled-down tobacco juice, hog grease, and red wine was used. Also, "man's grease," from the rendered fat of executed prisoners, and mumia, from ground-up Egyptian mummies, were placed on skin diseases. Bloodletting and purging were also common for many skin eruptions to "rebalance the humors" throughout the mid-nineteenth century. In addition, specific cures were also carried out over the centuries for certain skin conditions.

Herpes Simplex and Herpes Zoster

The ancient Egyptian *Eber Papyrus* recommends that cold sores be treated by applying a dressing of the "inner-part-of-the-castor-oil-tree" and red lead as the lesion was red. For herpes on infants' and children's genitalia, healers boiled crushed almond and mixed it with flour and bull marrow and smeared it over the eruption. Byzantine physician Paulus Aegineta in the seventh century recommended lots of food with salt for herpes zoster, and if there was no fever, exercise and wine were also suggested. He also washed the vesicles with hot water or, if they spread, with hot wine. In addition, he opened them with a needle and dressed them with applications for "eating away putrid flesh." This treatment, however, could have introduced an infection. Clay mixed with the juice of the deadly strychnine plant was also placed on the lesion. Linseed oil boiled in wine, which is a drying agent, was applied and may have helped dry up the vesicles and quell the pain and itching. For severe pain, physicians gave opiates and belladonna to quell the pain.

Boils and Carbuncles

Throughout the ages, whole fat figs were split open and placed on the boil or carbuncle to draw out the infection. Paulus in the seventh century describes several treatments that were also used by the ancient Romans and Greeks. These included placing Egyptian mastic gum, salted seedless raisins, dried figs boiled in fermented honey, and linseed oil with honey on the eruptions. Some of these substances did have antimicrobial properties and may have helped to prevent further spread of the infections when the pustules broke. Paulus also writes that earlier healers also used natron (drying salt used in mummification) and rose oil. He suggested that carbuncles would fester and burst by placing the inner part of old walnuts and the leaves and shoots of cypress mixed with barley flour on the pustular. From at least Roman times into the mid-1950s in the United Kingdom and other nations, diachylon plasters that contained about 33 percent of lead oxide along with olive or linseed oil were used to treat boils and other skin infections, which may have helped as lead has antimicrobial properties but could have caused lead poisoning in long-term use.

In the nineteenth century, British physicians such as surgeon Edward John Waring (1819–1891) placed calcium sulfide and arsenic or the white precipitate of mercury (ammoniated mercury) ointment on boils and carbuncles. This was used into the 1950s when it was found to be too toxic. He also administered leeches, which had no effect. Others placed poultices sometimes with an onion to increase the substances' "stimulating properties" or hemlock or opium to allay severe pain. Saline purgatives were also given to eliminate toxic humors. Although today the folk remedy of "black drawing salve" is used, it has not been found to be effective.

Atopic Dermatitis (Eczema) and Psoriasis

Since these afflictions were often confused, their treatment was the same. By the eighteenth century, physicians used the newly invented Fowler's solution that contains arsenic for both conditions. This remedy was used into the early twentieth century, in particular, for psoriasis. Some physicians in the nineteenth century also applied lead water, limewater, olive oil, poultices of flaxseed meal, slippery elm bark, or breadcrumbs for eczema. Other substances that were applied included turpentine, kerosene, and sulfuric acid solutions, which likely harmed the patient. For eczema, a Turkish or dry hot-air bath was highly recommended in addition to wrapping the eruption with a rubber cloth, which had little effect. Sulfur as a topical treatment for eczema was fashionable in the Victorian and Edwardian eras and may have helped to quell the inflammation.

Hippocrates treated psoriasis with tar and topical arsenic, and Galen, along with arsenic, also boiled a snake and put the broth on the plaques. For psoriasis in the mid-twentieth century, Spectro-Chrome Light Therapy was touted as a cure with lemon light but was ineffective. However, by the late twentieth century ultraviolet B (UVB) light therapy was administered in a hospital or a special center, where a person stood in a light box several times a week. This often was effective in reducing the outbreak. A folk remedy, "Dr. Fish," with small fish has been used with some success. These small fish are members of the carp and minnow family and live in hot-spring pools in the Central Anatolia region of Turkey among other places. Bathers with psoriasis and atopic dermatitis claim that when the fish feed on their skin, their condition gets better as the fish only eat the scales.

Acne, Rosacea, and Erysipelas

Since acne, rosacea, and erysipelas in the past were often thought to be different phases of the same disease, the treatment was often the same. For example, the *Papyrus Ebers* directed healers to apply a poultice of a piece of lead or cat's and dog's dung to the eruption. Sometimes a hog's tooth was mixed with these substances. This concoction likely made the condition worse by introducing pathogens. Paulus in the seventh century mixed sulfur, opium, and acacia gum to place on the outbreak that may have reduced irritation. Applying leeches on the affected areas of the face was also common. During the eighteenth century, physicians used lotions made from lead acetate, and during the late 1880s, the standard treatment was to administer laxative purges to rebalance the humors. After the purging, they painted the patient's face with iodine or silver nitrate, which may have helped quell infections. An ammoniated mercury ointment—which could cause mercury poisoning—was also used on acne and rosacea as were sulfur ointments. Physicians also placed rhubarb on rosacea and other conditions if they did not believe it was infectious.

Skin Ulcers

The *Papyrus Ebers* gives several recipes for "stinking ulcers," likely those that had become infected or gangrenous. The healer would place "Egg-of-an-Ostrich Tortoise-Shell," or "Refuse-of-durra" (perhaps thyme), and thorns of an unknown tree mixed with hippopotamus or hog fat on the ulcer. Hippocratic recommended wine as a lotion for ulcers, which may have helped prevent infection. Galen preferred to apply oak leaves, willow, and cabbage. He claimed that cheese from acid milk cures even the large sores, and the wild pear repressed the discharge. On recent ulcers, Paulus placed wheat flour, or bread, or glue for books mixed with turpentine, frankincense, pitch, or pig, goose, or calf fat. For almost healed ulcers, he recommended prunes and gum resins from several plants. Buchan in the eighteenth century applied bread and milk, with boiled chamomile flowers. He mixed toxic mercuric chloride with brandy and placed this on the ulcer and had the patient drink a tablespoon of the concoction twice a day. Since these poultices often contained antibacterial materials, they may have sometimes been successful. However, some, such as mercury, were toxic and could have caused liver or kidney damage.

Scabies

The *Papyrus Ebers* describes many recipes for "the itch," although its cause was not known. For example, healers prescribed poultices composed of onion, crushed in honey, and added to beer. If the itch was confined to the neck, a chopped-up bat was applied, which was likely harmless but not effective. If they were on every limb, healers applied a poultice of clay from a wall, wheat flour, and animal fat mixed with yeast for making sweet beer. By the seventeenth century, many clinicians prescribed bleeding and purging to balance the humors. In the nineteenth century, they applied an ointment of sulfur mixed with lard or butter, which may have killed mites, and sulfuric acid, which is highly corrosive.

Patent medicines were also touted into the early twentieth century for scabies such as Dr. Mason's Indian vegetable panacea, which likely was not effective. Some folk remedies used today may be effective in destroying the mites and eggs and include clove, anise seeds, rosemary, and neem oil. Cayenne pepper can relieve itching, and a mixture of turmeric and neem oil has been found to have insecticidal activity.

Smallpox

Smallpox, or variola, a major scourge for centuries, was caused the *variola* virus. Symptoms of the disease included fever, exhaustion, headache, and backache. A macular (flat) red rash developed, first on the face and then spread over the rest of the body to the palms and the bottom of the feet. Out of the rash, small pink bumps (papules) simultaneously arose, which progressed to hard pustular (pus-filled) sores. They crusted over, and in about three weeks, the scabs fell off. The virus was transmitted through personal contact, clothing, and contaminated objects.

Smallpox was a lethal disease and killed, scared, or blinded untold millions until the late 1970s when the World Health Organization sponsored a successful global campaign to eliminate the disease. There were two forms of smallpox: *variola minor* (discovered in the late 1800s in Florida, USA) killed about 1 percent while *variola major* killed about 30 percent of patients in endemic (established in) populations. Among peoples who had not previously been exposed to the disease, the case fatality rate could be up to 90 percent. The disease affected all levels of society, and in Europe, it was a leading cause of death in the eighteenth century, killing an estimated 400,000 Europeans each year. When it became endemic in the 1400s, it primarily affected children as adults had immunity to it from previous exposures.

The word "variola" for smallpox was first used in 571 by a Swiss bishop when the disease initially entered Western Europe. The term "small pockes" was first used in England at the end of the fifteenth century to distinguish it from the new disease, the *great pockes*, syphilis, discussed in another entry. There is some disagreement among scholars as to the earliest evidence of smallpox. Recent genomic research suggests it likely emerged in human societies between 3000 and 4000 years ago in eastern Africa from a similar disease found in rodents, camels, and cattle. It spread into the Middle East, South Asia, and East Asia through trade and military incursions. The earliest description of smallpox is found in an ancient Indian Sanskrit text from around 1500 BCE. In ancient Egypt, smallpox-like lesions have been found on the face of the mummy of Ramses V (died 1156 BCE). The virus has been identified in Vikings remains from around 1000.

Several theories were advanced as to the cause of the disease. Egyptian, classical Greek and Roman, and other Western physicians into the early 1800s thought smallpox was due to an imbalance of humors, resulting in hot infected blood emerging up through the skin. Until around 900, measles (described in the "Childhood Diseases" entry) and smallpox were thought to be different forms of the same disease. However, Persian physician Rhazes (854–925?) described measles and smallpox as separate diseases. By the late 1800s, physicians realized that smallpox was caused by an unknown microorganism, and in the twentieth century, the variola virus was identified.

Societies had noted for centuries that once someone had survived smallpox they did not get it again. To prevent the disease, some African, Asian, and Middle Eastern cultures developed a technique called "inoculation" or "variolation." Material from a pustule or scab of a recovered smallpox patient was placed in a scratch on the skin of a child or adult. The technique was introduced to England by the aristocrat Mary Montagu (1689–1712) in the early eighteenth century, and African slaves taught the technique to the American colonists.

However, a better method for prevention was found. Many had noted that dairy maids who developed cowpox did not get smallpox. Therefore, British physician Edward Jenner (1749–1823) tried an experiment and used cowpox matter to vaccinate a child, who was then exposed to smallpox and was found to be immune to the disease. Jenner's successful experiment was slowly accepted, and by the late 1800s and early 1900s, vaccination was made mandatory for children prior to attending school in many developed nations. In Germany, for example, during 1906 only 26 cases were found, which had been imported from other countries. By 1981 due to the World Health Organization's (WHO) worldwide vaccination program, smallpox had been eradicated globally.

Smallpox has affected the course of history and found in the arts. The devastating Antonine Plague (165–180) in the Roman Empire is estimated to have killed from 3.5 million to 7 million people, or about 25 percent of the population. In its aftermath, the epidemic led to a decline in literature, the arts, and trade along with social and political upheaval and was a factor in the demise of the Roman Empire. Up to 90 percent of the indigenous people in the Americas were wiped out by smallpox and other diseases brought by European explorers, colonists, and African slaves in the 1500s. Smallpox has also been depicted in the arts. For example, a classic panting by Eugène-Ernest Hillemacher (1818–1887), *Edward Jenner Vaccinating a Boy* (1884), shows a reluctant boy being vaccinated. Likewise, Italian sculptor Giulio Monteverde (1837–1917) rendered a sculpture of Jenner vaccinating his own son against smallpox (1873).

UNUSUAL TREATMENTS

Prevention and treatments methods for smallpox have been attempted since at least Egyptian antiquity. In ancient Egypt, to force all the smallpox lesions to emerge, healers placed a poultice of cat and dog dung mixed with a herb on the skin. Once the pustules had emerged, they applied crushed grains mixed with semen to help form scabs. The dung could have caused an infection. In the ninth century, Persian physician Rhazes suggested that once the pustules had started to break out, patients should be bled until they fainted as this would help eliminate the bad blood. Another physician of the time recommended that when the pustules had broken open, they should be anointed with oil and salt after which the patient should stand in the sun for an hour. After this ordeal, they were washed with a mixture of figs and myrtles and finally fumigated with smoke from the tamarisk shrub to help draw out more poison and cause scabs to form. Needless to say, these methods were ineffective and may have caused infections.

Medieval physicians covered patients with red material, placed red balls in patients' beds, and gave children red toys. Doctors had observed that covering the patient with red material and preventing light from shining on the patient's skin reduced pustules. In the late 1800s, health workers placed patients in rooms with sunlight filtered through dark red glass or curtains. Some physicians claimed this prevented "chemical rays" (ultraviolet) in sunlight from hitting the skin, which caused more pustules or scarring and may have been somewhat effective.

In the seventeenth century, based upon earlier traditions, the noted English physician Sydenham (1624–1689) recommended that patients be heavily bled to reduce the "hot infected blood." Bleeding was done from a vein in the arm, under the tongue, or by cupping (heated glass cups placed on the skin to draw blood). This massive bleeding resulted in more rich people dying compared to poor people as the poor could not afford a physician. Physicians carried out other treatments to cool the patient. They forbade hot drinks and food and did not allow fire or heat in the patient's room. The patient's windows were kept open even in the winter, and bedclothes were not allowed any higher than the patient's waist. Enemas, emetics, and purges were also done to "cool the blood." Another cooling treatment was to "dilute the blood." To accomplish this, Sydenham had the patient drink 12 bottles of weak beer over a 24-hour period that contained several drops of Spirit of Vitriol (sulfuric acid). These had little effect on the disease but did hydrate the patient.

By the mid-nineteenth century, physicians recognized that no treatment existed for smallpox—just prevention and supportive care. However, to prevent scarring, several remedies were followed. Many had observed that patients kept in cool dark rooms were less likely to scar. Therefore, caretakers would place a poultice made of bread and milk, flaxseed meal, and slippery elm bark (which has anti-inflammatory qualities) over the face to seal out light and air and to soothe the inflammation. Physicians would touch each pimple on the face with the point of a "lunar caustic" (silver nitrate stick) to seal it. To relieve itching, a white paint made of lead oxide and linseed oil was applied to the eruption, which would have also kept out the light. As previously discussed, blocking ultraviolet light may have been a factor in preventing disfiguring scars.

Sore Throats

Sore throat or pharyngitis is an inflammation of the pharynx (throat). If caused by a virus such as the rhinovirus or Epstein-Barr virus, it develops over the course of a few days and is usually accompanied by a runny nose, postnasal drip, and a cough with clear or greenish-gray sputum and sometimes fever. A "strep sore throat" is caused by *streptococcus pyogenes*, a Group A *Streptococcus* bacterium, and develops within hours. The patient has difficulty in swallowing, severe pain, tender lymph nodes under the jaw, foul-smelling breath, and fever. It is primarily found in younger individuals and children and can lead to rheumatic and scarlet fever discussed in another entry. Another painful condition is tonsillitis. The tonsils in the throat become inflamed and swollen. If caused by bacteria, pus is often found on the tonsils. Another condition, herpangina, is caused by a coxsackie virus. Small painful blisters develop on the tonsils and the roof of the mouth. The deadliest of throat conditions is peritonsillar abscess (also called "quinsy"). An extremely painful pus-filled abscess forms near a tonsil, and the throat sometimes becomes so swollen that breathing is difficult. Chronic swollen tonsils with extreme fatigue is a sign of infectious mononucleosis (older terms glandular fever, glanders, and the "kissing disease") caused by the Epstein-Barr virus that primarily affects young adults and may take several months to resolve.

Although streptococcus throat infections can lead to serious outcomes, they only account for about 5–15 percent of all sore throats in adults and 15–30 percent in children. In developed nations, about 10 percent of people seek medical care for a sore throat each year. Without treatment, symptoms of pharyngitis disappear in 3 days in about 40 percent of people. By one week, 85 percent of people are symptom free without treatment. However, for untreated peritonsillar abscess, the mortality rate is around 30 percent.

Sore throats have been described since antiquity, and depending upon the culture, their cause has been thought to be an imbalance of humors, evil forces entering the body, change in wind direction, or a contagion. Until around the sixteenth century, tonsillitis and quinsy were often confused with diphtheria discussed in another entry. In 1874, Austrian surgeon Theodor Billroth (1829–1894) first described the streptococcal bacteria, and in the 1920s, American physicians George (1881–1967) and Gladys Dick (1881–1963) showed that streptococci were the cause of a strep sore throat and scarlet fever. The term "pharyngitis" (inflammation of the pharynx) was first used by Greek physician Hippocrates (460–c. 375 BCE), "quinsy" from the Old

French *quinancie* in the fourteenth century, and "sore throat" at the end of the eighteenth century. "Tonsillitis" was first used in the early nineteenth century and is based upon Latin.

Some ancient remedies for inflamed throats are still endorsed by health professionals today to manage sore throats. These include gargling with warm salt water and ingesting honey, hot teas, and soups, which often provide symptomatic relief for sore throats. The ancients also cut into the tonsils and abscesses to drain the pus while today a needle and syringe is used. For pharyngitis, today physicians recommend anti-fever and pain medications such as acetaminophen and ibuprofen. Antibiotics began to be prescribed in the 1950s but had little effect on the duration or the severity of symptoms of pharyngitis. Antibiotics also caused some bacteria to develop resistance. Therefore, they are generally only administered to patients with severe cases of streptococci tonsillitis, quinsy, a history of rheumatic fever, or those who are immunocompromised.

Sore throats may have affected the course of history. For example, it is unknown what the implications of the early death of the first American president, George Washington (1737–1799), might have had on the new republic. He died from too much bloodletting as a treatment for a serious throat inflammation, which may have been quinsy.

UNUSUAL TREATMENTS

Treatments for inflamed throats were first recorded in Egyptian papyri (c. 1500 BCE). Some of these ancient remedies are still part of folk medicine; however, research is mixed as to their effectiveness. Egyptian healers, for example, crushed raw garlic in a mixture of vinegar and water to use as a gargle or sip along with performing certain incantations to the gods. Garlic was also added to oil and used as a liniment on the throat. Various aromatic herbs were combined, put into a cloth, and used as a plaster that was placed on the throat. Healers also steeped spicy cubeb pepper and other herbs in wine for patients to drink. They also recommended frankincense for throat issues. Some of these remedies may have helped soothe soreness. However, another treatment used by ancient Egyptians was to have the patient eat dried dog manure, which likely led to parasite infections and other illnesses.

Hippocrates, in ancient Greece, for the treatment of tonsillitis and peritonsillar abscesses recommended wet cupping (skin is cut and a warm cup placed over it to draw out blood) on the throat and neck as it was thought to draw out poisons in the blood. He also prescribed a combination of apple cider vinegar and honey—known as oxymel—to treat sore throats, which is still used as a folk remedy. For herpangina, he practiced bleeding and prescribed ointments for the throat that included rosemary, wild raisin, absinthe, and honey. He also recommended that the patient inhale a mixture of vinegar, herbs, sulfur, and tar. Some of these concoctions may have temporarily soothed soreness and were for the most part harmless to the patient.

By the Middle Ages, many plants were used to treat sore throat. For example, licorice root, a plant native to Europe and South Asia has been part of traditional medicine. As a tea, it reduced throat pain as it has analgesic properties. Cayenne pepper takes away soreness as it contains capsaicin, which has some anti-inflammatory and analgesic properties. Some thought wearing onions around the neck "drew out poisons."

Another popular treatment was a poultice to "draw out the infection." In the mid-nineteenth century, for example, a poultice made of flaxseed meal, lard, and laudanum (an opiate) was placed over the swelling on the neck. For serious inflammation, a small blister, with a blistering agent such as ground cantharides beetles, was applied to the swollen tonsils until a blister formed. Although the opiate may have reduced some pain, these remedies were not effective.

Various folk treatments that involved wrapping the neck were used to treat sore throats. An old English custom was rubbing lard onto the neck and then wrapping it with a dirty sock. At least by the Middle Ages, cloths were soaked in hot water, wrung out, and applied to the neck. In the nineteenth century, particularly in Germanic nations, cold water on a cloth was applied to the neck and a towel wrapped around it at bedtime to chase the infection away from the throat. Even scarves were worn at night to keep the throat warm. Another treatment consisted of a cloth or paper saturated with kerosene and placed around the neck. Some reduction in inflammation or soreness may have been due to the placebo effect or by extra immune cells drawn to a warm neck.

Although leeches had been used in medicine since antiquity, the nineteenth century was a high point for leech therapy for several diseases. Physicians, for example, used leeches as a method of bloodletting to eliminate supposedly excessive amounts of blood from inflamed tonsils and abscesses. Healers would place 20–30 leeches on the swollen neck. In 1822, Irish physician Philip Crampton (1777–1858) suggested another method. A silk thread was passed through a leech, which was then placed on an inflamed tonsil. When the leech was engorged, it dropped off and was pulled out of the patient's mouth. These treatments were not effective in curing throat issues and often caused death from blood loss, especially if bloodletting was also done from a vein.

Physicians used leeches as a form of bloodletting for many diseases. For a bad sore throat, a leach was attached to a silk thread and placed on the tonsils so it could be pulled out after it had fed. (© Alexan24/Dreamstime.com)

By the mid-nineteenth through the mid-twentieth century, health professionals "painted" sore throats and tonsils with a brush or cotton swab that had been dipped in iodine, silver nitrate, mercurochrome, and other toxic chemicals. This treatment is still used in folk medicine or in alternative therapies. It may have relieved some soreness as the solutions acted as an antiseptic and analgesic but could cause other health problems.

Sterility and Male Reproductive Issues

Male sexuality and reproductive issues include, but are not limited to, infertility (sterility), erectile dysfunction (impotency, ED), and enlarged prostate (benign prostatic hyperplasia, BPH). The inability to impregnate a woman due to abnormal sperm, no sperm, or a low sperm count can be traumatic for a man and cause self-esteem and other psychological issues. Since the 1970s, in Australia, Europe, and North America, sperm counts have declined by almost 60 percent. Researchers believe this might be caused by health problems including diabetes, obesity, lack of exercise, and sexuality transmitted diseases or even hot baths and tight clothing. Other risk factors are anabolic steroids, heredity, medications, and estrogen mimicking chemicals in drinking water. However, varicose veins (varicocele) in the testes cause about 40 percent of male infertility, which surgery today can reverse in about 50 percent of the cases.

The word "infertile" was first used in medical texts in the late seventeenth century and is from Middle French, from the Late Latin *infertilis*, meaning "not fertile or productive." The earlier term "sterility" was first used in fifteenth century, from the Latin *sterilis*, "inability to conceive children." In a twelfth-century gynecological treatise, the *Trotula*, the anonymous likely female author was one of the first to suggest barrenness might also be the fault of the man and not just the woman as had been largely believed for centuries. Also, based upon humoral theory—accepted from ancient Greece into the late eighteenth century—men with excessively cold and dry testicles were thought to have seeds that were "useless for generation." In addition, few successful treatments existed until the mid-twentieth century. They now include surgery, hormone treatments, antibiotics for infections, or assisted reproductive technology (ART) such as artificial insemination, in which the infertile man's sperm is mixed with another man's fertile sperm.

"Sterility" into the twentieth century was often used interchangeably with "impotence," now called "erectile dysfunction," which was coined in 1970. ED is defined as the chronic inability to have or keep an erection firm enough for sexual intercourse. For many men, ED leads to severe psychological distress and humiliation. Although often a taboo subject, about 5 percent of 40-year-old men have complete erectile dysfunction, and it increases to 15 percent by age 70. The major causes of ED include diabetes, heart disease, hypertension medication, and high cholesterol. About 10–20 percent of cases of impotency are attributed to psychological and emotional factors.

The word "impotence" is derived from the Latin word *impotencia*, which means "lack of power." The term was first used in the mid-seventeenth century, meaning sexual dysfunction. Impotence has been noted for centuries. Ancient Egyptian and Abrahamic religious writings recorded that it could be caused by a curse. In the Old Testament, for example, a king was made impotent as punishment for having sex with Abraham's wife, Sarah. In the Middle Ages, and for many years thereafter, impotence was believed to be caused by witches. From the thirteenth to the end of the seventeenth century, particularly among the upper classes, impotence was the only grounds for divorce, and the Church considered it was a deadly sin for an impotent man to marry. Although there was little effective treatment, today ED can be treated with psychotherapy, hormone replacement therapy, vacuum erection devices, and penile implants. In 1998, oral ED drugs were approved, and within five years, erectile dysfunction diagnoses had increased by 250 percent.

An aspect of ED is Peyronie's disease. If the penis is not as firm as it needs to be during intercourse, it can be physically damaged with the formation of scar tissue. This results in a sharp curve or even a shortened penis. In addition, the man may experience a painful erection or develop ED. The malady affects around 9 percent of men between the ages of 40 and 75. It was named after a French physician who first described it in 1743, although it was largely ignored as a health issue until the early twenty-first century. It is medically managed by various medications, surgery to remove the scar tissue, and even penile implantations.

A common issue of older men is benign prostatic hyperplasia (BPH). In BPH, cells in the gland multiply, enlarge the prostate, and squeeze the urethra, which causes frequent urination. In addition, difficulty in emptying the bladder and incontinence may occur. About 50 percent of men by age 50 have BPH, and as the man ages, the probability of the condition increases, which can be very frustrating and cause embarrassment. It is managed by self-catheterizing or several surgical techniques including cutting, heat, cold, lasers, and ultrasound to remove tissue.

Another condition—in opposition to a flaccid penis—is a prolonged erection or priapism. It occurs with or without sexual stimulation. The malady can be extremely painful and cause damage to the erectile tissue from lack of oxygen. It usually results from blood becoming trapped in the erection chambers. Risk factors are leukemia, malaria, or sickle-cell disease, and it can result from injury to arteries going into the penis. Priapism also occurs from spinal cord neck injury. The Edwin Smith *Surgical Papyrus* (c. 1600) described such a fracture, which resulted in priapism, and it has been observed during executions by hanging. However, ED medication rarely causes priapism. The term "priapism" was introduced by Greek physician Galen (c. 129–c. 216). It was inspired by the name of the god Priapus who was a small man with a large erect phallus. Some medieval priests believed it was a consequence of a curse or witchcraft. In managing the disease today, physicians drain excess blood from penis and sometimes use medications or surgery.

Masturbation, or the older terms of onanism and self-pollution, is the sexual stimulation of one's own genitals usually to orgasm. It has likely been commonly practiced from prehuman times in both men and women. Societal and religious views regarding self-stimulation have varied over time and in different cultures. The Sumerians (c. 4000 BCE), for example, believed that masturbation enhanced sexual potency. The ancient Greeks regarded masturbation as a normal and healthy substitute for other forms of sexual pleasure, and it was depicted on pottery and other art forms. The term "masturbation" was first used in the early seventeenth century. It is likely from the Latin *masturbor*, from *manus*, meaning "hand," and *turbō*, meaning "to disturb." However, the Christian Church deemed that any sex without procreation was sinful, and a man's duty was to replenish the earth. The term "onanism" first used in London in 1716 in a pamphlet entitled "Heinous Sin of self-Pollution." It is from the Latin *onanismus*, from the name *Onan*, the biblical man who "spilled his seed on the ground." However, although the term has been used for masturbation, Onan was likely using *coitus interruptus*, or withdrawal to prevent conception. By the nineteenth century's Victorian age, masturbation and nocturnal emissions were thought to have devastating effects on the body, including insanity, blindness and retardation, erectile dysfunction, and a decrease of a finite amount of semen. This inaccurate view mostly waned over the twentieth century into the twenty-first century as it began to be viewed as a normal part of human sexuality.

Numerous male reproductive and sexual issues have been found in art works and literature from early times. The Roman fertility god, Priapus, was portrayed in sculptures, frescos, and paintings. Injuries to the penis or psychological erectile dysfunction due to war trauma have been presented in fiction and memoirs. A few examples include British novelist D. H. Lawrence's (1885–1930) *Lady Chatterley's Lover* (1928) and American novelists Ernest Hemingway's (1899–1961) *The Sun Also Rises* (1926) and Ron Kovic's (1946–) memoir (1976) and later film *Born on the Fourth of July* (1989). Besides ED, masturbation has been described. In the Roman Empire, for instance, the poet Martial (c. 41–c. 104), when he could not afford a beautiful slave boy for sex, admits he masturbated instead. American novelist Philip Roth's (1933–2018) *Portnoy's Complaint* (1969) focuses on sexual dysfunction and compulsive masturbation. Even the extremely rare four-hour erection from taking Viagra is found in the film *I Think I Love My Wife* (2007).

UNUSUAL TREATMENTS

Some remedies have been recommended for male infertility and other issues over time. In ancient Greece, the Hippocratic works suggests that incantations and ritual were important for fertility problem no matter what its cause. Archaeological evidence suggests that votive offerings in the shape of male genitalia and inscribed enquiries on tablets about having children were placed by men likely wanting children. Since having a male heir was so important, men who wanted to bear a male child bound their left testicles during intercourse as it was believed that male semen was formed in the right testicle. These remedies were not likely effective.

Physicians by the end of the Middle Ages had described recipes for treating male infertility, but little information was given why they thought a remedy would work. Healers believed that ingesting animal testicles may have given the man extra virility. The Swiss physician Jakob Rueff (1500–1558) suggested that the patient drink a concoction containing fox testicles, mint, satyrion, a supposed aphrodisiac, wild rocket (arugula), and other herbs mixed in sheep's milk. Another remedy to assure impregnation was for the man to boil catmint with wine and drink it on an empty stomach for three days or drink ground-up dry pig's testicle added to wine. Castoreum from the beaver scent glands and ambergris from sperm wales—both used for making perfume—were also ingested to encourage fertility. Even in the twenty-first century, castoreum was advertised for fertility and ED issues, but research is scanty as to the effectiveness of these nostrums.

In the late eighteenth century, a London entrepreneur rented out a "celestial bed" with magnetic lodestones underneath to provide "celestial fire." He also played ethereal music and shone soft lights; the experience was "guaranteed" to lead to conception. Needless to say, it was not effective other than placebo effect. By the early twenty-first century, acupuncture, herbal and testosterone supplements, lifestyle changes, and various diet regimes had been tried, and sometimes they were marginally effective. In the future, some researchers speculated that sperm cells may be able to be produced out of a man's other cells and used with IVF procedures.

Erectile dysfunction has also been treated since antiquity. The Egyptian *Papyrus Ebers* (c. 1500 BCE) gave numerous recommendations. Incantations to the gods were a common remedy for both an evil curse and natural causes. Many recipes were suggested, but healers gave no reason as to why they thought they would work. One remedy consisted of smearing the penis with a mixture of the deadly nightshade plant, beans, sawdust from several trees, pig's dung, salt used for embalming, and other substances. Healers also smeared the penis with baby crocodiles' hearts mixed with wood gums and oils. They may have caused skin irritation and could have had a placebo effect.

In biblical times when an elderly King David could not have an erection, a beautiful young virgin was told to "lay with him." However, it was not successful although healers may have recommended this activity to cure impotency. Penis rings were developed around 1200 in China for keeping the penis erect and are still used by some men today. Ancient Romans claimed that eating fish and fish organs were sexually stimulating, and folk medicine today suggests oysters, which contains a high level of zinc and helps testosterone formation, might be useful. Spanish fly (cantharis beetle) was, and occasionally is still, used as an aphrodisiac, but it is very toxic and can lead to death although it may cause an erection.

In the Middle Ages, the Church urged men with ED to track down the women who had "bewitched" their penises and force them to restore their erections. In the early nineteenth century, since electricity was seen to stimulate nerves, electric belts were advertised to shock the penis into an erection. Any erection was likely due to placebo effect. This was also the case for several other devices advertised in the early twentieth century. These included the "Stringer self-treating device," which combined a vacuum, moist heat, vibration, and electricity. Another device, the "rectal prostate gland warmer," advertised it stimulated blood vessels. In the 1920s and 1930s, slivers of goat testicles and "monkey glands" were transplanted onto testicles, which would likely have led to a rejection reaction or infections and did not increase virility.

One of the most damaging treatments for ED in the 1920s was the use of radium, which was considered a "cure-all" until its dangers were realized. Radium was soaked in material that was placed under the scrotum at night and added to suppositories and a drink called "Radithor." These caused cancer, generally leading to death. By the 1970s, a penal implant that could be pumped full of air was invented, which worked but involved the risk of surgery. In the second decade of the twenty-first century, human "fetal pills" were smuggled from China to Korea, and perhaps North America, that folk medicine claimed restored sexual vitality, but they were ineffective other than placebo effect.

Prostate issues have been a problem for older men, and in the nineteenth century, physicians prescribed cream of tartar, a laxative; the toxic sweet spirit of nitre; and squill bulb to decrease swelling, which was not likely effective. In addition, leeches were placed on the prostate to reduce its size and eliminate "poisons." If there was pain, opiates, hemlock, and other narcotics were injected into the anus, which could be toxic but did relieve pain. Physicians also recommended bloodletting, enemas, and a catheter to relieve retained urine. In the late twentieth century, folk and alternative herbal concoctions were tried including curcumin, nettles, and saw palmetto. However, research has shown they were no more effective than a placebo for relieving BPH symptoms.

Physicians over the centuries have suggested remedies for other conditions of the male reproductive system. Ancient Egyptians realized that priapism was incurable if it was from a spinal cord neck injury. However, for other causes the penis was bandaged with a mixture of myrtle, grease, ox spleen, frankincense, beans, and other ingredients to soften the skin. In the Middle Ages, a short penis caused by priapism was beaten gently with rods and then covered with pitch. This did not lengthen the penis but may have reduced the erection.

Since Victorians feared that masturbation caused serious problems, they attempted to prevent it among young men. Treatments included quinine, opium, and digitalis, and sponging the male with cold salt water early in the morning. For those who did not stop the practice, physicians scarified the perineum and placed suction cups to draw out several ounces of blood. They also placed rings on the penis that would hurt if the male had an erection or pull on pubic hairs. American Surgeon and health reformer John Harvey Kellogg (1852–1943), to cure masturbation, circumcised the male without anesthesia. None of these prevented the practice.

Stomach Ailments

Common stomach or gastric issues range from indigestion (dyspepsia), gastritis, and ulcers to gastroesophageal reflux disease (GERD). Heartburn, nausea, and vomiting are often manifestations of these maladies. In past centuries, these symptoms were seen as different aspects of one illness. The most common malady is indigestion—also called dyspepsia, or upset stomach—which is a group of symptoms that cause discomfort in the stomach area. It is usually associated with difficulty in digesting food. Symptoms include a feeling of bloating, abdominal pain above the navel, burping, nausea, sometimes vomiting, and heartburn (acid indigestion), which feels like burning in the esophagus or chest. These symptoms are now classified as functional dyspepsia. Almost everyone will experience indigestion at some point; it tends to increase with age and is more common among women. Causes of indigestion include overeating, consuming fatty or spicy foods, eating too quickly, or consuming too much alcohol or caffeine. However, frequent indigestion is sometimes a sign of an underlying problem, such as GERD or ulcers.

Ulcers are open sores in the lining of the stomach, duodenum, and sometimes the esophagus. Symptoms include gnawing pain below the breastbone, which often radiates to the back, especially on an empty stomach. Food or antiacids usually relieve the symptoms. Ulcers are generally nonfatal and until the twentieth century were difficult to diagnose. In severe cases, they can cause vomiting of blood and fatal bleeding. Both gastric and duodenal ulcer became more frequently diagnosed in Western nations in the nineteenth century. In the past, physicians believed that ulcers were largely due to stress, spicy foods, and heavy alcohol consumption. However, in 1982, two Australian physicians identified the *Helicobacter pylori* bacterium in the stomachs of patients with ulcers and gastritis—inflammation of the stomach lining that can lead to ulcers. This bacterium causes about 80 percent of ulcers, but the overuse of over-the-counter painkillers and long-term corticosteroids use can also enable ulcer formation. About 10 percent of adults in industrial nations develop the condition.

Another common stomach malady is GERD, which has frequent symptoms of heartburn. GERD is a result of a weak lower esophageal sphincter that allows acidic stomach contents to flow back up into the esophagus. The condition was first identified in 1935. About 5–10 percent of people with GERD develop Barrett's esophagus, in which stomach acid causes changes in cells that can sometimes leads to esophageal and stomach cancer. GERD can also cause dental, throat, and lung issues and is mostly found in adults. It affects about 20 percent of the population and is more common in women. Risk factors for this annoying syndrome include obesity; pregnancy; drugs including antihistamines, anti-inflammatory, and painkilling medications; antidepressants; and smoking.

Stomach maladies have been seen from antiquity and have been blamed on heavy eating and drinking. Ulcers, for example, have been found in a 2,000-year-old Chinese mummy. In the eighteenth and nineteenth centuries, dyspepsia was thought to be one of the "nervous disorders" along with hypochondria and hysteria. In the nineteenth century, some physicians such as American Elijah Hammack (1826–1888), believed that "literary men" were more likely to have dyspepsia. The term "dyspepsia" originates via Latin from the Greek *dys-* and *pepse*, "difficult to digest," and was first recorded in the early eighteenth century. "Indigestion" is from late Middle English, from the Latin *in-* and *digestio*, "not digesting."

Today many treatments for stomach complaints are found. For example, antacids that have been used for centuries include bicarbonate of soda or antacid tablets made with salts of magnesium or calcium. In the nineteenth century, bismuth subsalicylates—a thick pink liquid—was used to treat stomach upsets. In 1901, it was marketed as Pepto-Bismol and is still considered a first line of defense against many stomach maladies. Since the late twentieth century, histamine receptor blockers and antibiotics have been used to treat ulcers caused by *H. pylori*.

Stomach ailments have been mentioned in fiction. For instance, British author Geoffrey Chaucer (1343–1400) in his *The Canterbury Tales* (1392) remarks that nightmares were a symptom of indigestion. Charles Dickens (1812–1870) also alludes to this in his *A Christmas Carol* (1843) when Ebenezer Scrooge wonders if he is having dreams due to the food he ate. During the nineteenth and early twentieth century, novels and short stories included characters with a "delicate stomach." Early-twentieth-century American author Upton Sinclair (1878–1968), for example, not only described his own dyspepsia but also several of his fictional characters suffered from stomach ailments.

UNUSUAL TREATMENTS

Numerous folk treatments going back centuries have existed for stomach ailments and are sometimes used today as home or alternative medicine remedies. For example, a folk cure for indigestion is to chew and swallow a teaspoon of fennel or caraway seeds. The root of the ginger plant has been a remedy for heartburn and nausea for centuries as has licorice, which increases the mucous coating of the esophageal lining and may help protect it from stomach acids. In the late twentieth century aroma therapy and cannabidiol (CBD, an extraction of hemp plant) in the early twenty-first century became popular as home remedies for nausea and other conditions. These remedies may be effective for some stomach issues.

In Egypt around 1500 BCE, the *Papyrus Ebers* gave recipes for treating indigestion based upon tradition. One included mixing figs, caraway seeds, "resin-of-Acanthus plant," ink, peppermint, and beans with sweet beer, which the patient consumed. The peppermint may have been helpful. Another concoction consisted of crushing a hog's tooth into a powder, which was put inside four sugar cakes and eaten for four days. These likely caused further stomach issues. If the patient had a hard abdomen, cat's dung, red lead, watermelon, and the mealy fruit of the Zizyphus-Lotus plant was mixed with sweet beer and wine and applied to the stomach as a plaster, which was not helpful.

In ancient Rome and Greece, physicians gave patients rennet—an enzyme from the stomach of animals used for making cheese and the dessert Junket—as a cure for indigestion. They also administered mastic (aromatic resin used in chewing gum). These remedies likely helped in some cases.

In the fourteenth century, British physician John Arderne (1307–1392) made a plaster of hyssop, absinthe, and dill soaked in wine that was placed on the stomach for indigestion. He also recommended anointing the patient's back and front from the stomach to the flanks with honey mixed with mint, mastic, and mustard. To stop vomiting, he suggested bandaging the limbs and rubbing them. In the fifteenth and eighteenth centuries, physicians gave mumia (from ground-up mummies taken from Egypt), considered a cure-all, to patients. These methods were ineffective.

During the eighteenth century, Scottish physician William Buchan (1729–1805) claimed that vomiting was quelled by having the patient drink cinnamon and mint teas, which may have soothed the stomach. However, since he believed that vomiting was caused by the movement of gout (discussed in another entry) or other diseases to the stomach, he required the patient to undergo bleeding, purging, setons (an open wound kept open with thread and usually a pea), and perpetual blisters (caused by an irritant) to draw the disease out of the stomach, which could have caused infections.

American physician Elijah Hammack (1826–1888) administered Seidlitz powder (a laxative preparation that contains tartaric acid, sodium potassium tartrate, and sodium bicarbonate) that effervesces when mixed with water, which may have helped indigestion. He also recommended purging with "blue mass" (a toxic mercury-based pill) and rhubarb at night and castor oil in the morning as a treatment for most stomach complaints, which could have led to dehydration from copious diarrhea and mercury poisoning. For persistent cases, he suggested "a change of air" along with giving up alcohol, coffee, and tea, which might have helped. For vomiting, Hammack suggested rubbing the patient's legs with chloroform, which could have put the patient to sleep and caused liver damage in long-term use.

In the 1840s, pepsin, a stomach enzyme that digests protein, was discovered, leading physicians to treat stomach issues with pig pepsin, which was effective in some cases. American physician Henry Hartshorne (1823–1897) listed a litany of remedies for nausea and vomiting in his medical text. They included cinnamon water, limewater, peppermint, an infusion of cloves, ice bags to the spine, and morphine enemas. Some of these are still used as folk and alternative medicine remedies and can be effective. Physicians also suggested milk and cream from the mid-nineteenth to mid-twentieth centuries for ulcers. However, later research showed that this increases the secretion of stomach acid.

In the early twentieth century, Sinclair described various diets that were "cure-alls" such as the milk, anti-protein, cereal, and raw vegetable/fruit diets along with fasting and extreme exercise for dyspepsia. These regimes may have worked in some cases due to placebo effect. In the late twentieth and early twenty-first centuries, alternative cures became popular. Daily consumption of fresh cabbage juice appeared to help heal stomach ulcers. However, conflicting research evidence as to the effectiveness of many alternative regimes and folk medicines has been found.

Stress and Anxiety

Stress is the body's reaction to a threat when stress hormones including cortisol and epinephrine are released triggering the "flight or fight" response to react toward the immediate danger. This response is generally short-lived. But if a stressor is consistent or the stress response remains once the threat is gone, these hormonal changes over time can lead to cardiovascular diseases, high blood pressure, gastrointestinal issues, arthritis pain, kidney damage, cancers, and allergic reactions. In addition, these hormones can also trigger anxiety, which is a reaction to stress with fear, worry, apprehension, unease, or nervousness.

Anxiety can persist even if the stressor has passed and when someone cannot identify any significant stressors in their life. Stress and anxiety are normal reactions, and everyone experiences them such as the stress of getting married or anxiety about taking an exam or finding a job. Although clinicians generally define stress and anxiety as separate entities, popular culture by the late twentieth century often used the term "stress" for both. This led to "stress reduction techniques" to bring calmness and relaxation through regular, alternative, and folk medicine.

Even though the first use of the term "stress" was in Middle English based on the Latin *strictus*, meaning "drawn tight," Hungarian endocrinologist Hans Selye (1907–1982) was the first to give a scientific explanation for biological stress in the 1920s. Selye explained his stress model, called the general adaptation syndrome, based on physiology and psychobiology. In this syndrome, the body goes through the stages of alarm, resistance to the stress, and exhaustion or long-term reaction to the stressors that lead to health issues and anxiety.

If anxiety remains for at least six months, it is considered generalized anxiety disorder. Persistent anxiety is often a sign of serious mental health issues such as posttraumatic stress disorder (PTSD), obsessive-compulsive disorder (OCD), panic disorder, or phobias discussed in the "Mental Disorders ('Neurotic'): Anxiety and Functional Disorders" entry. In addition, American psychologist Charles Spielberger (1927–2013) in the 1970s defined two types of anxieties: "state" anxiety is acute or temporary unease at a specific moment while "trait" anxiety is a personality characteristic that is often chronic. He also developed a scale to measure them. Trait or persistent anxiety affects from 4 to 7 percent of the population and is more common among women.

The first use of the term "anxiety" in English was in the early sixteenth century and is derived from the French *anxiété*, from the Latin *anxietas*, from *anxius*, meaning "uneasy or disturbed." In antiquity, Greek and Latin healers and philosophers identified anxiety as a medical disorder. Roman philosopher and statesman Marcus Tullius Cicero (106 BCE–43 BCE) distinguished between *anxietas* that designates trait anxiety and *angor* that means state anxiety "reidentified" by Spielberger. However, from classical times into the early twentieth century, the concept of anxiety as being a mental disorder was rarely seen in medical texts. In the early nineteenth century, for example, Scottish physician Alexander Macaulay (1783–1868) defined it as "restless uneasiness attendant on several diseases" that was best relieved by the elimination of the cause. Generalized anxiety disorder (GAD) that came in sporadic attacks was sometimes described as "panophobia," "anxiety neurosis," or "neurasthenia" from the late eighteenth into the twentieth century.

In 1952, anxiety was finally classified as a separate condition in the first edition of the *Diagnostic and Statistical Manual of Mental Disorders* (*DSM*). This publication was sponsored by several organizations including the American Psychiatric Association. The DSM was a glossary of descriptions of the diagnostic categories, and anxiety was considered similar to a "psychoneurotic disorder." In 1949, the World Health Organization (WHO) published the sixth revision of the *International Classification of Diseases* (*ICD*), which now included a section on mental disorders for the first time and similarly defined anxiety. GAD appeared as a diagnostic category in *DSM-III* in 1980. Management of anxiety and chronic stress now includes stress reduction techniques, cognitive behavior therapy, counseling, antianxiety medication, and a focus on the present in techniques such as mindfulness meditation.

Characters exhibiting stress and anxiety have been featured in the arts. In the painting *The Scream* (1893) by Norwegian artist Edvard Munch (1863–1944), Munch reveals his inner feelings of anxiety. Also, the painting symbolizes the angst of humans living in the modern age of stress and anxiety. In addition, most novels and films foster stressful situations that induce anxiety in their characters to move the plot along. As an example, a post–WWII novel by American Sloan Wilson (1920–2003), *The Man in the Gray Flannel Suit* (1955) and made into a film the following year, portrays a married veteran in a stressful job while secrets from the war including a child he has fathered and accidentally killing his best friend haunt him, resulting in extreme anxiety. In another film, *The Devil Wears Prada* (2006), the young female protagonist is put through much stress by her boss in the fashion industry, and she finally leaves the intolerable situation for the journalist job she had really wanted.

UNUSUAL TREATMENTS

Numerous folk and alternative treatments abound for stress and anxiety, including several herbs used at least from the Middle Ages. Lemon balm mint, for example, has been a popular tea for nervousness for centuries. Some research suggests that subjects who consumed lemon balm tea were calmer compared to those who ingested a placebo tea. Likewise, other herbal teas have been found to have a calming and antianxiety effect including chamomile, valerian—which contains a chemical like benzodiazepine tranquilizers—and passionflower. L-theanine, an amino acid found naturally in green and black tea, as well as in mushrooms also has been reported to have a calming effect. In the early twenty-first century, proponents of CBD (cannabidiol), derived from the hemp plant, have claimed it helps alleviate symptoms of anxiety and insomnia, and some research shows it might be effective.

Another substance that was used from at least the time of the Mesopotamian civilization (c. 3100 BCE) for relaxation and nutrition was beer. Not only was it drunk but also some propose that ancient Egyptians bathed in it to reduce stress and anxiety. Folklore suggests that people in the Czech Republic have been bathing in oak tubs full of warm beer for centuries to reduce stress. Moreover, by the 2020s beer spas or beer baths became a new fad, particularly in Europe. The person sits in a bath of warm beer, which also has a spigot for drinking a cooler version. Proponents of the spas claim that soaking in beer helps reduce anxiety, stress, tension, and muscle aches and pains. Mud baths have also been popular in recent years for the same reasons. Another bath is a "forest bath" or a walk in the woods, which is popular in Japan. Some research has suggested that people who walk in the woods with the sounds of nature had lower levels of stress hormones compared to a similar walk in an urban area.

Another alcoholic beverage to reduce stress has been wine. Scottish physician William Buchan (1729–1805), for instance, suggested that nervous people should drink wine and water with meals. However, if it caused stomach problems, he recommended brandy and water. Alcohol in small amounts does have a relaxation effect, but Buchan also realized that too much could cause alcohol dependency. For nervous disorders, Buchan also recommended exercise, in particular riding on horseback, as it gave "motion to the whole body, without fatiguing it." He primarily favored long sea voyages or journeys for those who could afford them as the patient could focus on things other than what was causing stress and anxiety.

Some studies have found that vacations are helpful in reducing stress and anxiety as it gets the person out of the environment, which is causing these issues.

Pharmaceutical drugs have also been used to manage stress and anxiety, but they tend to cause dependency. American physician Henry Hartshorne (1823–1897), for example, prescribed opium, chloroform, chloral hydrate, and morphine as calming agents for nervousness. He also recommended some herbal teas such as valerian mixed with morphine. These would have had a relaxing effect but also likely produced dependency. In the 1950s, physicians prescribed barbiturates to frazzled housewives, who then developed addictions and withdrawal seizures when they ran out of their medicines. In the 1960s, clinicians prescribed the newly developed benzodiazepine tranquilizers such as Valium, Librium, and Xanax. But they also could cause dependency and were misused leading to addiction and social problems. Consequently, many physicians were hesitant to prescribe them in the twenty-first century.

A popular alternative treatment by the early twenty-first century was "sound healing." Proponents of this remedy claim that listening to certain sound vibrations will relax the mind and body, lower blood pressure, and relieve symptoms of anxiety and insomnia. They claim that certain sounds such as Tibetan singing bowls, tuning forks, gongs, or vocal resonances supposedly realign the "natural vibrations in the body," which results in healing. Practitioners consider vocalization of the sound "AUM" or "OM," the perfect sound for this healing practice. Calming sounds also include recordings of running water, ocean waves, and rain. However, the research is mixed concerning the effectiveness of sound therapy. Another popular therapy in the early twenty-first century was a heated stone massage for relaxation and stress reduction. During a stone massage, flat, smooth heated stones are placed along the spine and other parts of the body. This may have been effective as massage has been shown to temporarily reduce stress.

Some folk remedies for stress and anxiety recommend that people look at fractals, or repeated patterns, found in nature and art. For example, the mandala is a geometric figure representing the universe in Hindu and Buddhist symbolism. During mediation, its purpose is to help focus on the mediation to reduce stress reduction and tension. Another folk tradition is rubbing both ears when feeling stressed. Proponents claim that this acupressure can relax muscles and reduce anxiety and stress. A popular stress reduction technique by the early twenty-first century was aroma therapy. Advocates claim that inhaling the scent of lavender, lemon, or mango will reduce stress, and some research supports this. Some studies have also suggested that an orgasm can reduce stress as it increases dopamine and serotonin that increase feelings of well-being. Some research has even found that chewing gum can reduce stress and anxiety.

By the late twentieth century, crystal therapy became popular for stress reduction. Proponents claim that black tourmaline due to its "balancing nature" is helpful. It is held in one hand during medication to release tension from the muscles and to help calm the mind. Another crystal is blue lace agate, which advocates claim is one of the best crystals for anxiety and stress relief as it supposedly releases soothing and calming vibrations. No evidence as to the effectiveness of crystal therapy exists other than the placebo effect.

In the 1920s, nurses gave women mud baths at a Russian sanatorium to restore health. Mud baths had a revival in North America by the early twenty-first century for maintaining health and reducing stress. (National Library of Medicine)

On the other hand, some studies have shown that light therapy can affect anxiety and stress. Light therapy has been attempted since the mid-nineteenth century. Research has found that blue light, found in midday sunlight, was more likely to cause relaxation after a stressful situation compared to white light that contains all the colors. But bright light can help with seasonal depression discussed in the "Depression and Melancholia" entry. Blue light in the evening, such as from a computer, on the other hand, may disrupt the circadian rhythm, which in turn causes sleep disturbances leading to stress and anxiety. Exposure to sunlight, or sun lamps, produces endorphins that activate opioid receptors and make people feel better. However, a side effect is that some people become addicted to sunlight, and much exposure to the sun is a risk factor for skin cancer. It has also been found that people exposed to morning light are better at coping with anxiety-provoking experiences.

Substance Use Disorders

Drugs such as alcohol, opiates, cannabis, coca, tobacco, mushrooms, and other substances have been used since prehistoric times for medicinal, religious, and recreational purposes, and most have been subject to abuse. The depressant drugs such as alcohol and opiates have been used to relieve pain, promote feelings of well-being, induce sleep, and cure mental illness. However, all of them can cause severe physical dependency, and abrupt withdrawal can sometimes produce life-threatening symptoms and psychotic behavior. Stimulants such as coca, tobacco, and coffee have also been used, but until the twentieth century—other than coffee—they were not seen as a problem to the individual or society as they made difficult work easier for workers.

Alcohol Use Disorder and Alcohol

Alcohol use disorder, alcoholism or alcohol dependency, is a chronic relapsing brain disorder in which the person loses the ability to stop or control their consumption of alcohol despite adverse social, physical, occupational, financial, legal, or health consequences. Around 10 percent of people who drink are thought to have the disorder. It is more often found among older individuals and is twice as common among males as females. Alcohol abuse includes drinking to intoxication that is sporadic and results from heavy drinking such as found in young male drinkers under the age of 25. However, it can also result in social, school, and legal problems, but the individual may not be physically dependent upon the drug. When people are in a drunken state, they can harm family members, drive irresponsibly, get into fights, and are at risk for suicide.

After a bout of heavy drinking, hangovers often occur the next day and can last up to 24 hours. This side effect of abusing alcohol may cause a dry mouth, fatigue, excessive thirst, headaches, nausea and vomiting, shakiness, dizziness, and muscle aches. One of the best treatments is eating a good breakfast to increase blood sugar, taking an over-the-counter painkiller, and rest. A dangerous effect of alcohol withdrawal for those with alcohol dependence is delirium tremens (DTs). This is characterized by fluctuations in blood pressure, tremors, hallucinations, and convulsions. Mortality without treatment is around 25 percent. Physicians prescribe minor tranquilizers to prevent these symptoms.

The term "alcoholism" was first used in English in the mid-nineteenth century. It is from the Latin *alcoholismus*, "disease of alcohol addiction." Alcohol dependence was also called "habitual drunkenness" into the early twentieth century. The alcoholic or "inebriate" was thought to have a lack of "will power" and was stigmatized.

Drunkenness or inebriation has been described throughout history. Several Roman authors, including the noted historian Tacitus (56–c. 120), reported that the northern Germanic tribes always drank to intoxication, which was against the Mediterranean norm. Roman conquerors found if they could get these barbarians drunk, they could easily slaughter and subdue them. Greece in antiquity had the religious cult of Dionysus, the god of wine, which was adopted by the Romans as the god Bacchus. In these cults, drunkenness was condoned in ritual gatherings. As the Western Roman Empire declined in the fourth and fifth centuries, the Christian Church with a philosophy of moderation influenced patterns of drinking and other behaviors. Wine was consumed as part of the eucharistic rite.

However, some Protestant sects, during the Reformation (1517–1648), began to condemn the abuse of alcohol and viewed drunkenness as sinful and a vice. In the United States, a temperance movement arose in the late nineteenth century leading to National Prohibition in 1920. Little treatment was given to alcoholics other than in mental institutions, and alcoholics were considered incurable until 1935 when Alcoholics Anonymous (AA) was founded in Akron, OH, as a self-help group. With the rise of AA's popularity, and success, the American Medical Association in 1956 declared that alcoholism was a treatable illness. By the late twentieth century, many treatment modalities including individual and group talking therapy, detoxification to address withdrawal symptoms, cognitive-behavioral therapies to avoid relapses, and certain drugs to reduce craving had been instituted. To reduce stigma surrounding the condition, the term "alcohol use disorders" (AUD) was first used in 2013 in the DSM-V (*Diagnosis and Statistical Manual*) that is used by mental health professionals. The disorder combined alcohol dependence and alcohol abuse.

Other Drug Use Disorders

Opium has been widely used for centuries, and addiction to its resin and derivatives has likely been common since humans first discovered it. In 2020, it was estimated that roughly 20 percent of people who use opiates as a prescription or for recreational use have opiate use disorder. The term "opium" is borrowed from Latin, meaning "drug made from the latex of the opium poppy," borrowed from the Greek *ópion*, and the term was first used in English in the fourteenth century. Evidence for opium use is found in the Egyptian *Papyrus Ebers* (c. 1500 BCE), which gives numerous recipes for opium to treat many maladies. In Greece, opium was used for sleep, pain, recreation, and as an aid for suicide. The active opioid alkaloids morphine and codeine are derived from the pod of the poppy flower, *Papaver somniferum*, which in Latin means "sleep-inducing poppy." It was first cultivated by humans in Greece and the Middle East from at least 10,000 BCE and soon spread to the rest of the Old World. This sleep-inducing poppy was also mixed with alcohol, later called laudanum, and is mentioned in late eighth century BCE epic Greek poem the *Odyssey*. Opium and its mixtures were primarily used as medication and in antiquity likely in religious rituals. Throughout the Middle Ages, opium was also a common ingredient in remedies.

By the early nineteenth century among the upper classes and art and literary circles, "opium eating" became popular. Opium was taken in many forms including smoking and as laudanum. However, reports of dependency became more common. Detailing this addiction process was British writer Thomas De Quincey's (1785–1859) essay "Confessions of an English Opium-Eater" (1822) about his laudanum dependency and its effect on his life.

In the nineteenth century, morphine was purified from opium. Once the hypodermic needle was invented, many soldiers became addicted to morphine in the post–Civil War era for war wounds, and it was known as the "soldiers' disease." By the turn of the twentieth century, heroin was synthesized, and it was first thought to be nonaddictive. Until WWI in the West, many middle-class women became dependent on heroin and morphine in patent medicines; they were looked at with sympathy. But this view changed when heroin began to be used recreationally by "night people" and supposed "criminals" when morphine was not readily available. The Harrison Tax Act was passed in the United States in 1914 to maintain control over opiates and coca, making it illegal for recreational use. Addicts were stigmatized as criminal, arrested, and often imprisoned. A heroin epidemic occurred in the post–WWII years in urban areas and again in the 2010s among young urban and rural youth, leading to an epidemic in overdose deaths. By the early twenty-first century, more nations realized that opiate addiction was a medical issue, and treatment modalities were established to treat substance use disorders.

The use of cannabis was not considered a substance abuse issue until the early twentieth century in Western cultures. It is estimated that 30 percent of those who use marijuana in the 2020s may have some degree of marijuana use disorder in which people find it difficult to stop using the drug even though it interferes with relationships, jobs, or social activities. The term "marijuana" is from Mexican Spanish *mariguana, marihuana.* The term "cannabis" was first used in the late eighteenth century. It is derived from Latin, meaning "hemp," from the Greek *kannabis.* The ancient Greeks used cannabis as a remedy to treat inflammation, earache, and edema. Marijuana may have been the incense that was used in the cult of Asclepius, the god of healing, and also by the oracle of Delphi, but since these cults had secret rituals, it is not clear as to what psychedelics they may have used. However, for much of human history, the *Cannabis sativa* plant was primarily cultivated for rope and clothes in many regions of the world. Around the eleventh century when alcohol was first distilled, cannabis was added to it to form a tincture, called tincture of cannabis. Clinicians prescribed it for sleep and pain into the mid-twentieth century. In the early twentieth century, seasonal immigrant workers from Mexico brought marijuana with them, and its use became associated with crime. Under the Marijuana Tax Act of 1939, it was made illegal for any use.

The coca leaf, native to Peru, was chewed for centuries as a stimulant to ease off cold and hunger and to increase work production. Cocaine was extracted from coca leaves in the mid-nineteenth century and was used in medications. The term "cocaine" was first used in 1860 for this extracted drug. It did not become a popular recreational drug until the late nineteenth century and was highly psychologically addicting. When it began to be associated with crime, its use was also criminalized. In the 2020s, the age group most likely to abuse cocaine are males between ages 18 and 25.

By the end of the nineteenth century, in Europe, North America, and some former British colonies, a public outcry against the recreational use of these drugs arose as they were thought harmful to society. International and national laws were passed to control the manufacturing, sale, and distribution of many of these substances to prevent recreational use. However, a lively international black market arose, and they continue to be used illegally.

Numerous depictions of alcohol and other drugs have been presented in literature, scripture, and the arts over the centuries. For example, the Christian Bible contains hundreds of references to wine and vineyards. These references laud the virtues of drinking in moderation and condemn drunkenness. Numerous works of art on pottery and frescos from Greek and Roman antiquity picture drinking scenes and wine. Several film and theatrical performances have drugs as part of the plot line. For example, in American writer Frank Baum's (1856–1919) children's book *The Wonderful Wizard of Oz* (1900), the main characters walk through a field of poppies poisoned by a wicked witch, where they fall asleep, but they are saved by a good witch. In the film *A Star Is Born*—which has three versions—the male lead descends into alcoholism and drug use and commits suicide.

UNUSUAL TREATMENTS

Since the use of most drugs now considered to be problematic and leading to substance use disorder was acceptable for most of human history, treatment was not considered. When a problem was perceived, such as drunkenness, the treatment was usually punishment.

Alcohol Use Disorder and Alcohol

On the other hand, healers down through the centuries have recommended remedies for hangovers. The Egyptian *Oxyrhynchus Papyri* (c. third–seventh century BCE), for example, reports several cures. One involved wearing strung leaves of the Alexandrian laurel shrub around the head. It likely had no effect on the hangover. Roman philosopher Pliny the Elder (c. 24–79) recommended eating two raw owl's eggs while Greco-Roman physician Galen of Pergamum (129–200) suggested that people should wrap their head in cabbage leaves. Cabbage leaves do have a mild analgesic effect.

In the Middle Ages, a folk remedy was the ingestion of eels as they were thought to suck up the remaining alcohol causing a hangover. It was not effective. Since the mid-sixteenth century, the term "hair of the dog" has meant drinking more alcohol to ease a hangover. It is effective but can lead to alcohol dependency. In the seventeenth century, Goddard's drops made of the distillate of a mixture of dried snakes, pieces of skull from a recently hanged man, and ammonia boiled in a glass container were used as a hangover cure. It was not effective but largely harmless.

In the nineteenth century, one folk tradition advocated eating rabbit feces. They did contain salts such as potassium that is depleted in heavy drinking but likely could lead to parasites and other infections. By the late twentieth century, other folk remedies were suggested to cure hangovers. They included cold showers, or conversely hot showers, drinking a large amount of green tea, various juice cleanses, or inhaling oxygen. In the eastern United States, a folk treatment was ingesting a raw egg yolk mixed with Worcestershire sauce, celery salt, and pepper. None of these had much effect as a hangover cure but were generally harmless.

For alcohol withdrawal, or DTs, in the mid-nineteenth century, physicians administered liquid ether; although it calmed the DTs, it could explode. British surgeon John Waring (1819–1891), from his experiences in India, noted that some physicians and folk practitioners recommended cannabis, which helped to prevent DTs.

As a punishment for drunkenness throughout the sixteenth and seventeenth centuries, "stocks and pillory" were used in England and New England. This was a contraption in which legs and feet were locked in a frame, and the public was encouraged to throw rotten food at them. This punishment was ineffective. By the late eighteenth century, inebriates who caused repeated problems were put into asylums with the mentally ill, where they often were abused. By the late nineteenth century, physicians also treated alcoholism with cocaine and opiates, which often resulted in dependence on these other drugs.

Other Drug Use Disorders

Many treatments beginning in the late nineteenth century were used for all substance use disorders. In the early 1900s, middle-class patients in Western nations with conditions including mental breakdowns, depression, alcoholism, and drug abuse would go to sanatoriums or spas to "take the waters" for a cure, away from the area in which they lived.

American physician Leslie Keeley (1836–1900) developed the "Leslie Keeley's Double Chloride of Gold Cure" for the treatment of "alcoholism, drug addiction, and the 'tobacco habit.'" He believed that they were diseases and curable. It was a commercial medical venture that had branches around North America and was operatable from 1879 to 1965. A patient spent a month at the institute and received four injections a day of a secret formula. Some of the ingredients in the injections may have been strychnine and opiates. The treatment may have been effective based upon placebo effect, informal peer counseling, and group dynamics. A few patients did have signs of allergic reactions. However, many physicians considered his treatment quackery.

In the 1930s, lobotomy—destruction of brain tissue—was developed for the management of depression, mental illness, and alcohol and drug use disorders. It was popularized by American physician Walter Freeman (1895–1972). Freeman would push a sterile icepick-like instrument through the eye orbit into the prefrontal area of the brain, wiggle the pick around, and destroy the tissue. Lobotomy was extensively used in the 1940s and 1950s but fell into disrepute by the mid-1960s as it could cause the patient to be dysfunctional in activities of daily living.

By the late twentieth century, a crystal therapy folk remedy became popular for treating many illnesses including substance use dependencies. Proponents believed that "vibrations" from crystals correct "imbalances" in the body. Amethyst was suggested for recovering alcohol dependency, carnelian for marijuana, selenite for cocaine, and lepidolite for heroin addiction. The person carries the stone with them or near them to obtain their power. However, there is no evidence that crystals or stones are effective for any condition, including substance abuse disorders.

T

Tetanus

Tetanus, or lockjaw, is found in humans and animals. The first sign of the disease is trismus—spasms in the jaw muscles. This is followed by neck stiffness, difficulty in swallowing and breathing, and generalized rigidity and convulsive spasms of skeletal muscles. In extreme cases, the whole body may arch backward, called opisthotonus, which can break spinal bones. Spasms can continue for three to four weeks; the condition is often fatal and has no cure. It is caused by the bacterium *Clostridium tetani*, whose spores are found in soil, dust, and animal feces. When the spores enter a wound—in particular, puncture wounds—they develop into bacteria that can produce a powerful neurotoxin that damages the nerves that control muscles. Unlike many infectious diseases, recovery does not usually result in immunity.

In 1891, Japanese bacteriologist Kitasato Shibasaburō (1853–1931) first isolated *C. tetani* from a human victim. In 1897, French veterinarian Edmond Nocard (1850–1903) demonstrated that tetanus antitoxin from horse serum could be used for prevention and treatment in domestic animals and humans, and it was first used during WWI. Tetanus toxoid (TT) vaccine was developed in 1924 from killed bacteria and was first widely used to prevent tetanus from battle wounds during WWII, after which it was introduced into routine childhood immunization, particularly in developed nations.

The use of these vaccines and better cleaning of wounds resulted in a decreased incidence of lockjaw worldwide beginning in 1900. Before this time, the death rate from tetanus was around 30 percent. Now it causes death in about 11 percent of cases even with supportive treatment. In developed nations, it is higher in the elderly as many have not received a recent booster immunization. The European Centre for Disease Prevention and Control estimates that tetanus causes 213,000–293,000 deaths worldwide each year and is responsible for 5–7 percent of all neonatal and 5 percent of maternal deaths. Tetanus in infants is due to the umbilical cord not being cut in a sterilized manner in unvaccinated mothers. It is more prevalent in developing areas, in particular, some African and South Asian nations.

The term "tetanus" was first used in the fourteenth century from the Latin based upon the Greek *tetanus*, meaning "ridged" or "muscular spasm." The term "lockjaw" was first recorded in the late eighteenth century as another name for the condition. However, the disease has been described since antiquity. It was first documented around 1550 BCE in Egypt in Edwin Smith's *Surgical Papyri* as a frequent aftermath of battles along with gangrene from wounds (discussed in another entry). The papyrus mentions that tetanus from a head wound was always fatal. It was also noted that even a small wound or cut could cause the deadly muscle spasms. Greek physician Hippocrates (c. 460–370 BCE) in *Genuine Works* also detailed tetanus, and throughout history, it was a dreaded aftermath of wounds, in particular war wounds. In the nineteenth century, military surgeons and many physicians, such as Henry Hartshorne (1823–1897), felt that tetanus was due to exposure to cold, dampness, excessive heat, injury, bullets, and pressure on nerves by bandages. On the other hand, American Physician Elijah Hammack (1826–1888), with some foresight, believed it was caused by a poison generated in lacerations and puncture wounds. Since there is no cure for lockjaw, treatment today is focused on supportive care with oxygen, muscle relaxants, and pain medication until the body can grow new nerve endings. Human tetanus immune globulin and TT booster are also administered. Prevention by immunizations starting in infancy and booster shots every 10 years is the only effective measure against this ancient malady.

Tetanus was a major causality of war wounds and has been depicted in the arts. During the Napoleonic Wars in the early nineteenth century, tetanus was one of the deadlier complications from a bullet. The agony of lockjaw in this war is depicted by British surgeon and artist Charles Bell (1774–1842) in an 1809 painting *Tetanus Following Gunshot Wounds*. It shows a man with an extreme arched back suffering from the condition. In the American Civil War, about 89 percent of soldiers who were infected with tetanus died, which greatly reduced troop strength and led to high casualty rates in this war.

UNUSUAL TREATMENTS

Throughout history, many treatments have been tried to relieve the spasms. The Smith *Surgical Papyri* (c. 1550 BCE) notes that healers would place hot cloths on the mouth until it opened and then bind it with "grease, honey, and lint" to keep it open so that liquid food could be given. Ancient Greek physicians, such as Hippocrates, treated the patient with liniment rubbed on the limbs, hot moist herbal packs to ease pain, and strong purgatives to eliminate imbalance in the humors. During the eighteenth century, some physicians recommended mixing cantharides—a blistering agent—with turpentine and placing it along the patient's spine to draw out poisons so as to stop the spasms. This treatment did not work.

In the late nineteenth century, American physician Alphonso Rockwell (1840–1933) suggested passing an electric current through a gold leaf placed on the skin and connected to a battery such as the Leyden jar to treat paralysis, cramps, and tetanus, which was ineffective. In the late century, many physicians, including Hartshorne, tried chloroform by inhalation, with variable effect. In 1942, the noted Canadian physician William Osler's (1849–1919) medical text still suggested this dangerous gas that caused liver damage to reduce tetanus spasms. Other methods to help in preventing spasms included potassium bromide (used to control seizures) along with opium or morphine, but the results were varied.

Many observed that if patients with lockjaw were kept in a quiet room it prevented contractions. Some physicians, such as Hammock in the nineteenth century, suggested cauterizing a wound with a stick of silver nitrate to prevent the disease. Physicians recommended other remedies including cold baths, ice to the spine, mercury, cannabis, tobacco, and ether. The ether may have temporarily stopped the spasms. Some physicians thought opium and brandy to be helpful. However, few patients survived any treatment regime.

Physicians prescribed other substances including potentially toxic plants such as belladonna and monkshood—used for heart conditions—to treat lockjaw. They administered deadly hydrocyanic acid (the gas hydrogen cyanide in water) in addition to quinine, which is only effective for malarial-type fevers. Since patients found it difficult to eat, wine was recommended for nourishment in the late nineteenth century. In addition, "brandy and good beef tea" were also administered as enemas as they were thought to stimulate the system and breathing. These treatments generally did not save lives. In the early twentieth century, some physicians claimed that intra-spinal injections of magnesium sulfate relieved spasms for hours.

Amputations of limbs in many wars throughout history may have prevented lockjaw in some cases. Physicians in the American Civil War, for example, often hastily performed amputations rather than attempting to save the limb. If this was accomplished before the toxin was produced, it may have prevented lockjaw. Some physicians, including Hartshorne, in the late nineteenth century claimed that amputation of infected fingers and limbs sometimes saved the patient.

By the early twentieth century, patent medicine salesmen touted opium added to alcohol as a cure for "nervous irritability" as well as rabies and tetanus. Unscrupulous drug manufacturers advertised ineffective patent medicines to physicians for distribution, which mostly contained alcohol. In the late twentieth century, various treatments and home remedies were touted. For example, in the 1960s, hyperbolic oxygen was administered as a remedy for tetanus and may have lessened the progression of the disease. Likewise, botulinum toxin (used in facial cosmetic procedures) was injected into the jaw muscles, which sometimes reduced the spasm. American home remedy herbalist John Heinerman (1946–) reported that Aztec healers cleaned wounds with marigolds that had been steeped in hot water as they thought it was a cure for tetanus. Curare was also used to relax muscle contractions. High doses of ascorbic acid (vitamin C) have also been used along with regular treatment, which some claim has reduced mortality. In the early twenty-first century, tiny nano-robots that could heat and destroy *C. tetani* and its spores have been suggested as a method to reduce the effect of the disease, but they are still under investigation.

Tooth and Mouth Issues

Poor diet, trauma, and lack of dental and oral hygiene leads to tooth decay and infections, gum disease, abscesses, and halitosis, which sometimes results in serious health consequences. Tongue and mouth issues such as stuttering can cause psychosocial problems. In addition, decayed, crooked, and yellowed teeth have led to social problems, due to which people, especially women, were considered less desirable as a mate.

Dental Caries and Gum Disease

Dental caries (cavities) are caused by the decay-causing bacteria, *Streptococcus mutans*, which metabolize sugars to produce acid. Over time, the acid attacks the tooth enamel, which results in a cavity. An explanation of tooth decay for centuries has been the "tooth worm." This concept originated in Samaria around 5,000 BCE. The worm was thought to bore into, and then decay, the tooth. In 1881, American dentist Willoughby D. Miller (1853–1907) presented the "chemo-parasitic theory" of tooth decay, which became universally accepted. With some modifications, today it includes a combination of poor diet, dental plaque (microbial load), characteristics of the host and time.

If tooth decay is not treated, it can cause pain, infection in the tooth or gum, and even death. For example, London statistician John Graunt (1620–1674) reported that in 1632, out of 7,535 deaths, 470 people had died of "teeth problems." Almost all people today have had a cavity by age 60. Those between 12 and 19 years of age have the highest number followed by children and then adults. Infants are prone to "nursing bottle caries." The first use of the term "carries" in English was in 1634. It is from the Latin *caries*, meaning "decay" or "rottenness." The term was initially used to describe holes in the teeth. Dental caries are one of the oldest and most common diseases found in humans. Cavities are thought to have increased with the beginning of farming about 12,000 years ago when carbohydrates such as oats and barley became domesticated. Archaeological evidence suggests that among hunter-gatherers roughly 1–5 percent had caries compared to 80–85 percent among early farmers.

From antiquity into the twentieth century, the most common procedure for an infected and painful tooth was to extract it. However, the Etruscans (c. 750 BCE) developed gold inlays and crowns, but they have always been expensive. It was not until the mid-nineteenth century that tooth restoration became more common by filling cavities with amalgam, but it was expected that most people would lose their teeth as they aged. In 1938, nylon-bristled toothbrushes, and in 1945 the beginning of water fluoridation and fluoride-fortified toothpaste emerged to help prevent tooth decay. By the mid-twentieth century, dentistry focused on preventing tooth loss by promoting flossing in addition to brushing. Root canal surgery—the pulp and nerves are removed—was also routinely practiced, saving a tooth rather than pulling it. By 2000, dental implants for an extracted tooth and cosmetic procedures to replace dentures became less expensive and popular.

Besides cavities, another dental issue is periodontal or gum disease that increases as people age. Periodontal disease has been recognized since ancient Egyptian times and includes gingivitis, a nondestructive infection, but left untreated, it can progress to periodontitis, leading to loss of teeth. Risk factors include infection under the gum, scurvy from months at sea without fresh food, tobacco use, and eating inexpensive sugary foods. From the sixteenth through the late nineteenth century, a risk factor for gum disease was the increased use of mercury as a purgative and as a cure for diseases, especially syphilis. Sometimes a tooth abscess erupts through the gum called a "gum boil." Today good oral hygiene and antibiotics are used to treat these conditions.

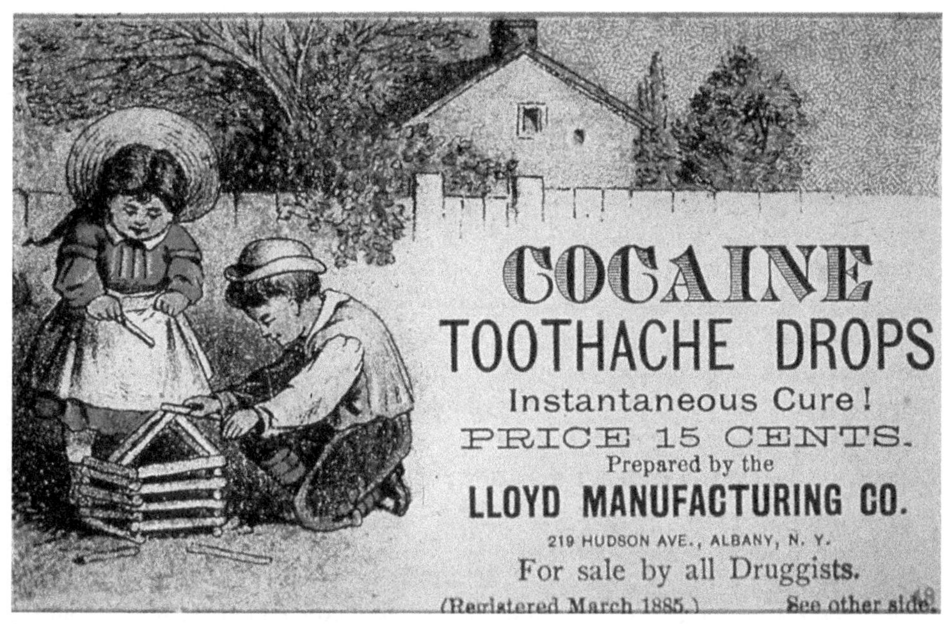

Patent medicines containing cocaine were used by adults and children to numb toothache pain in the late nineteenth century. (National Library of Medicine)

Bad Breath (Halitosis)

Halitosis that produces repulsive mouth odors is a crippling social problem. It is caused by sulfur-producing bacteria in the mouth. They feed on microscopic bits of food stuck in teeth, gums, dentures, cavities, or on the tongue. Risk factors include a dry mouth, periodontal disease, and infected tonsils. Today about a third of the population has halitosis. The term "halitosis" was first used in English in 1874 and is derived from the Latin *halitus*, meaning "breath," along with the Greek suffix *osis*, meaning "pathological process." Bad breath has been mentioned in ancient writings. To eliminate foul breath, for example, the Chinese around the eighth century developed a hog-bristle toothbrush, which was adopted in Europe during the seventeenth century. Today treatment for bad breath is brushing twice daily with toothpaste along with scraping or brushing the tongue and using oral rinses containing antimicrobial agents.

Stuttering

A major mouth and tongue issue is stuttering. Stuttering is a communication disorder involving an involuntary disruption in the flow of a person's speech. It is about three or four times more common in males than females and is found in about 1 percent of the population. Risk factors include delayed childhood development, genetics, and stress. For centuries, people who stutter have been viewed as weak, timid, unintelligent, or even amusing. Stuttering can cause self-esteem issues and affect interaction with others and career opportunities. People who stutter may experience repetitions (B-b-b-b-ed), prolongations (Nnnnot), or blocks (no sound).

The term "stutter" in English was first used in the late sixteenth century and is from the Middle English *stutten,* from the Germanic *stossen,* "to strike against." One of the earliest historical references to stuttering was to Greek orator and statesman Demosthenes (384–322 BCE). Over the centuries, many theories as to its cause have been proposed. Greek philosopher Celsus (43–37 BCE) believed that stuttering was a disease of the tongue. In the Middle Ages, it was thought to be a curse from God or from demon possession. In 1817, French physician Jean Mark Gaspard Itard (1775–1838) suggested that stuttering was caused by a problem with nerves associated with the movements of the larynx and tongue. German surgeon Johann Frederick Dieffenbach (1795–1847) claimed that stuttering was caused by spasms in the voice box that resonated up the length of the tongue. No therapy, device, or drug exists for stuttering that are effective all the time or for everyone who stutters. Changing the way in which a person speaks such as talking slower or softer or even singing may induce temporary fluency. Today some devices for stuttering attempt to alter speech patterns, provide feedback about the patterns, help to pace speech, and provide auditory feedback.

Tooth and mouth issues have been illustrated in the arts and may have influenced history. For example, George Washington (1732–1799) wore uncomfortable dentures made of human teeth, ivory, and possibly cow and horse teeth—but not wood. A letter from Washington to his dentist was intercepted by the British, misleading them to inadequately defend Yorktown, VA. This action allowed Washington to defeat the British on October 19, 1781, and the Battle of Yorktown became the last major battle for American independence. Italian painter Cosola Demetrio's (1851–1895) *Mal di denti* (1875) shows a young girl with a bandage around her chin and head to ease the pain of a toothache. In British author J.K. Rowling's (1965–) *Harry Potter and the Sorcerer's Stone* (1997) novel and later film (2001), the wicked Professor Quirrell pretends to stutter and be harmless in order not to be suspected of evildoing. He is taken over by the evil wizard Voldemort, and when he stops pretending to be timid and meek, he also stops stuttering and is shown as an evil villain.

UNUSUAL TREATMENTS

Numerous remedies for dental and mouth issues have been tried over the centuries, particularly among the upper classes. To clean teeth, for example, "tooth powder" has been used for most of history. The Babylonians, along with the ancient Egyptians, used a ground mixture of hooves, ashes, burnt eggshells, and pumice and most likely used their fingers to scrub their teeth with this powder. The Romans made a tooth powder by mixing charcoal with the above substances and cleaned their teeth with rags. Some of the abrasives in these powders could damage the teeth. By the nineteenth century, tooth powder contained chalk and soap and was less abrasive.

Dental Caries and Gum Disease

Dental caries throughout history were sometimes treated. For example, in Italy, a cavity from the tooth of a male skeleton from around 14,000 was scraped likely to remove the decayed tissue. A 6,500-year-old tooth in Slovenia with a deep cavity was coated with beeswax, which may have helped reduce pain. The Egyptian *Papyrus Ebers* (c. 1500 BCE) for a "lost tooth" recommends a mixture of ground barley, honey, and yellow ochre be used as a filling or splint to keep the tooth in place. The ochre and honey acted as an antiseptic and the barley as a binding agent. Ancient Greek healers attempted to treat toothaches and cavities by soaking linin in herbs and packing it into cavities. The cloth prevented food from entering and causing more decay and likely had some effect. In ancient Rome, to stop toothaches the patient inhaled wild mint in steam. During the Middle Ages to kill the supposed tooth worms, the patient breathed in the fumes of heated seeds from the poisonous henbane plant, which might have helped relieve the pain, but it could also cause toxic effects and even death.

Scottish physician William Buchan (1729–1805) believed that "bad humors" needed to be drawn off to relieve the toothache and before the tooth was extracted. He administered purgatives, bled the patient, and frequently bathed their feet in warm water. These were not effective. Buchan also recommended chewing tobacco, mustard seeds, or other hot pungent herbs. They may have helped slightly as a counterirritant to dull the pain but caused nausea and vomiting. In the nineteenth century, a poultice made of herbs, such as ginger and ground mustard, were spread on a warm, moist cloth and applied to the face. A warm poultice would bring heat to the area and help to temporarily relieve inflammation and pain. Immobilizing the jaw with a cloth binding might have also helped relieve pain. To completely eliminate pain, American physician Edward John Waring (1819–1891) recommended that a cotton ball soaked in a mixture of chloroform and camphor be put into the cavity, which worked but over time could cause liver damage.

Folk cures for toothaches have been used for centuries. In one remedy, the patient inserts a clove or ground clove mixture into an infected tooth or cavity. Since cloves contain a substance that numbs skin and nerves, it provided temporary relief from the toothache. Today, clove oil is used as a folk treatment to relieve a toothache.

The primary method for eliminating a constant toothache was extracting the tooth, which in the past was a dangerous proposal as it could lead to a fractured jaw, infection, and even death. As the procedure was done without anesthesia until the late nineteenth century, it was excruciatingly painful. However, the technique for extraction has been common for centuries. Both Celsus in the first century BCE and Byzantine Greek physician Paulus Aegineta (c. 625–c. 690) gave similar instructions for pulling rotten teeth. First, the clinician or barber surgeon would loosen the gum from the tooth, and then the tooth was rocked back and forth until it moved easily, and finally forceps were used to pull it out. If the tooth was too rotten, lead was first put into the cavity to provide structure so the forceps could hold on to it.

Once teeth were pulled, artificial teeth were made for the rich. The Romans carved theirs from animal bone, agates, or ivory tusks, but they were only cosmetic and needed to be removed before eating. By the fifteenth century, human teeth were stolen by grave robbers and sold to barber dentists for dentures. However, they also were not usable for eating. In the nineteenth century, after the Battle of Waterloo in Belgium (1815), false teeth became known as "Waterloo teeth" due to the countless teeth taken from dead soldiers for use in dentures.

To keep teeth white, since it signaled youth, beauty, and wealth, ancient Egyptians developed a whitening paste that mixed ground pumice stone with white-wine vinegar that they rubbed on their teeth. Vinegar does have a bleaching effect. From ancient Rome through today, as folk medicine, both human and animal urine were used to whiten teeth as the ammonia in urine acts as a bleaching agency. During the seventeenth century, barber surgeons would file the teeth and apply nitric acid that would whiten them, but the procedure could cause serious gum ulcers. In the eighteenth century, some physicians started to use oxalic acid to bleach teeth. While all these practices made teeth whiter, they eroded the tooth enamel, which led to earlier tooth decay.

In the mid-twentieth century, hydrogen peroxide was found to bleach teeth. Today some alternative medical systems use herbs, such as turmeric, to whiten teeth or "oil pulling," in which a person swishes and holds oil in their mouth for 20 minutes. Proponents claim that the oil "pulls out" toxins from the body and whitens teeth. This has not been fully substantial by research but is generally harmless.

Treatment for gum disease has also been carried out throughout history. For example, the *Papyrus Egypt* for gum ulcers suggests that patients rinse their mouth for nine days with a mixture composed of cow's milk, fresh dates, and "nah-grain" that had been kept overnight in the dew. In addition, for gum disease it recommends that bran and sweet beer or celery and sweet beer be chewed and then spit out. These remedies were not very effective but were harmless. Celsus for ulcerated gums applied a hot iron to them, and then he rubbed honey and washed them with wine or water. The honey and wine, which have antibiotic properties, may have prevented infections. British physician Alexander Macaulay (1783–1868) applied roasted figs or onions to the inflamed gum ulcers along with milk and barley water, which likely had little effect. Folk treatment for gum boil abscesses today includes gargling and swishing the mouth with hydrogen peroxide and applying garlic, clove oil, tea tree oil, mustard oil, and turmeric to the gums. Some of these substances have antibiotic properties and sometimes may be effective.

Bad Breath (Halitosis)

For centuries through today, various herbs including mint and parsley have been chewed to take away bad breath. Ancient Egyptians (c. 3,000 BCE) combined frankincense, cinnamon, and myrrh into bits that were chewed to sweeten breath. They also crafted "chewing sticks" from twigs, in which they frayed the ends that acted like a brush to clean the teeth and reduce odor. Greek physician Hippocrates (450–357 BCE) recommended a mouth wash, which was made of alum, vinegar, and salt and would have helped sweeten breath. In the fifteenth century, the Chinese developed hog-bristle toothbrushes to reduce halitosis, but the invention was not adopted in the West until the eighteenth century. They also recommended elephant bile (which has antibiotic properties), diluted with water, to eliminate mouth odor, which may have helped by reducing mouth bacteria. Toothpicks until the Victorian era—when they began to be viewed as uncouth among the upper classes—made of various substances were used to remove food particles between the teeth, thus reducing mouth odor. From the fifteenth century to the nineteenth century, the affluent leisure class also scraped the tongue to remove bad breath. In the nineteenth century, commercial breath fresheners became popular.

Stuttering

Many remedies have been tried for stuttering, but most have only been temporarily successful, if at all. For example, the *Papyrus Ebers* mentions several recipes "to drive out disease of the tongue," which may, or may not, have referred to this speech issue. It suggests the patient first gargle with milk and spit it out. If that did not work, goose grease was added. If this was unsuccessful, then honey, incense, caraway seeds, and clay were mixed with water and chewed nine times. These may have worked by placebo effect and were largely harmless to the patient.

Greek orator Demosthenes put pebbles in his mouth and talked in front of a mirror and while walking up a hill to cure his stuttering, often with loud noises in the background such as crashing surf. Celsus in the first century massaged the tongue and suggested that the patient eat pungent substances, which were not harmful and may have worked by placebo effect. On the other hand, Romano-Greek physician Galen (129–210) advocated cauterizing the tongue. This and blistering the tongue were also used in the Middle Ages for stuttering; it was painful and not successful.

In the nineteenth century, several treatments were developed. French physician Itard maintained that stuttering was caused by muscular debility, so he developed a device made of gold or ivory—called Itard's Fork—and placed it under the stutterer's tongue to support it. He claimed it worked in several cases. English orator and writer Charles Canon Kingsley (1819–1875) recommended dumbbell exercises to help breathing. He also suggested placing a bit of cork between the back teeth when speaking and endorsed boxing, which he considered a healing art, for the person who stammers. These may have worked by placebo effect.

Also, in the nineteenth-century era, German surgeon Johann Friedrich Dieffenbach (1792–1847) to cure stuttering cut out a triangular wedge near the root of the tongue. Other surgeons cut out half of the tongue or cut the frenulum, the piece of tissue between the tongue and the floor of the mouth. These surgeries were not successful in relieving stuttering, and patients sometimes bled to death. Later treatments have involved electric shock, hypnosis, and raising the tip of the tongue to the roof of the mouth while speaking. Folk treatments involved smacking the child across the mouth with a greasy dish rag or holding a nutmeg under the tongue, not speaking for a whole year, or carrying marbles in the mouth. These may have sometimes temporarily worked due to placebo effect but could have been traumatic for the patient.

Tuberculosis

Tuberculosis (TB) comes in two forms. About 80 percent of patients suffer from "pulmonary TB." The active or "open" form of pulmonary TB usually has symptoms of coughing, blood-stained sputum, chest pain, weight loss, fever, chills, and night sweating. If the TB bacteria are enclosed in a tubercle in the lungs, the disease is considered "closed" and is not contagious. However, about 20 percent of TB patients suffer from "extra-pulmonary TB" where other organs are attacked. This form also is not contagious. Common sites for this type are the lymph nodes of the neck, which cause painful lumps formerly known as "scrofula." The infectious nature of TB was demonstrated in 1865, and in 1882, German physician Robert Koch (1843–1910) isolated the *Mycobacterium tuberculosis* that causes the disease. TB is usually spread through coughing or breathing bacteria into the air. However, infected domestic mammals such as cows can also spread the disease through unpasteurized milk.

Over the centuries, TB has been associated with a high death rate and has killed more people than any other infectious disease. Globally it is now one of the top 10 causes of death in developing nations. The World Health Organization (WHO) by 2020 estimated that TB resulted in almost 2 million deaths per year. Moreover, about 10 million people developed the disease annually. All ages are susceptible to TB, but people who are immunocompromised, such as those with HIV, and malnourished young adults have a higher death rate. In the absence of effective treatment, roughly two in three die within five years of diagnosis.

TB has been known by numerous names. Hippocrates in ancient Greece called it "phthisis," and the Romans called it "tabes." The pallor of TB sufferers was the origin of the term "white plague" in the 1700s. The name "consumption" was based upon the fact that the disease consumed the patient. Phthisis and consumption were the two most common terms in the 1600s through late 1800s for the respiratory form of the illness. Scrofula, however, was thought to be a separate disease. The term "tuberculosis" was first used in an English-language medical textbook in the 1860s and became common around the turn of the twentieth century. Its name is derived from the Latin *tubercle*, meaning "a small swelling."

The origin of this ancient disease is thought to be in East Africa. Early people spread it along trade routes as they migrated out of Africa. It was transmitted to domestic animals that spread it to seals and sea lions along the African coast. These creatures in turn brought the disease to the Americas prior to European exploration. DNA evidence and skeletal deformities typical of TB have been found in eastern Mediterranean, Egyptian, and Peruvian remains from around 7000 and 2400 BCE and 1000 CE, respectively. The first written documents describing TB are from India circa 1500 BCE.

The disease was widely known in the European Middle Ages and became more prevalent as people swarmed into crowded unsanitary towns and cities during the Industrial Revolution. From the 1600s through early 1900s, it was a leading cause of death. In London, for example, in 1632—other than for a higher infant death rate—consumption was the number one cause of death. In 1900, it was the fourth-leading cause of death in that city. From ancient Greece through the late 1890s, physicians proposed many causes for the condition. These included a contagion, miasmas (bad air), acid blood, heredity, or an innate predisposition as it ran in families. By the mid-1800s, sedentary indoor occupations and bad air—especially in factories—along with communicability, were thought to be the primary causes of the illness.

Various treatments were attempted over the ages for TB, but the first successful remedy against the disease in recent times was the sanatorium with fresh mountain air and good nutrition first described in 1854. This method was adopted by physicians on both sides of the Atlantic from the late 1800s through the 1930s and became known as the "sanatorium movement." During the late 1880s, the tuberculin test to detect exposure to TB was developed, and in 1895, the x-ray machine was invented, which enabled a clear diagnosis of the disease. In 1921, the Calmette–Guérin (BCG) vaccination was first used to immunize children against TB. The modern era of TB treatment and control was heralded by the discovery of streptomycin in 1944 and isoniazid in 1952, which cured 80 percent of sufferers. By this time, the disease was in decline in developed nations due to better hygiene, public health, and a higher standard of living.

TB was so much a part of life in the nineteenth and early twentieth centuries that it became a frequent subject in the arts. A widespread belief also held that the disease assisted artistic talent, and it became known as the "romantic disease." Many works featured TB with the tragic death of a principal character. For example, in Italian musician Giacomo Puccini's (1858–1924) most famous opera *La bohème* (c. 1894), Mimi the seamstress dies of the disease. Norwegian artist Edvard Munch's (1863–1944) painting *The Sick Child* (1885) illustrates his young sister while she is dying of the TB. German novelist Erich Maria Remarque's (1898–1970) *Three Comrades* (1936) and American playwright Eugene O'Neil's (1888–1953) *Long Day's Journey into Night* (1940s) both portray main characters with TB.

UNUSUAL TREATMENTS

Various treatments for TB have been tried since antiquity. Egyptian healers around 1500 BCE recited incantations to the gods and then lanced scrofula tumors on the neck. After this surgery, they placed a poultice containing a mixture of acacia seeds, the blood of the Nehur bird and wasps, lead, copper, sea salt, and honey on the wound. This did not cure the tumors, but copper and lead may have prevented infection. In Greece during the first century, physicians administered opium to help the patient sleep, along with bleeding to get rid of bad humors and a diet of barley water, fish, and fruit for strength, which was not effective in curing the condition but was harmless to the patient.

Religious rituals were also used as treatment in many cultures. For example, "laying-on of hands" was practiced by English and French monarchs from 1300 through the mid-1700s. These rulers claimed they had a divine gift of healing and employed the "king's touch" as a cure for the "king's evil"—another term for scrofula. The monarch touched the swollen lymph nodes and in England hung a gold coin talisman, called an Angel, around the sufferer's neck to wear constantly to ensure the success of the treatment. In some cases, the swellings abated as this form of TB sometimes goes into remission, leading patients to believe they had been cured by this royal touch. Laying-on of hands is still used by some religious groups today for various ills.

By the mid-1800s, most physicians gave patients cod liver oil, wine, beer, and tonics containing iron and quinine, as they were believed to help restore the patient's strength and nourish the body. Physicians also favored raw beef and brandy and raw eggs and milk for this purpose. Chaulmoogra oil from an Asian Indian nut, also a cure for leprosy discussed in another entry, was tried with little success. These did not cure the patient but were harmless.

Koch recommended treatment with tuberculin, an extract of supposedly killed TB bacteria, which he created in the 1880s; however, it made patients worse or led to death. A diluted concoction was later developed and was, and is still used, to test for exposure to TB. By the turn of the twentieth century, a few physicians collapsed or surgically removed parts of the patient's lungs, which cured some but also led to many deaths.

From the 1860s to early 1950s, physicians often sent patients to high mountains such as the Rockies or the Alps or dry areas such as Arizona or Egypt as a cure. They recommended sea voyages or travel in the winter to warmer climates such as Florida or southern France. In some cases, patient became better with a change in environment and reduced stress. In the early twentieth century, patent medicines were sold to cure TB, which contained creosote, digitalis (heart medication), chloroform, opium, mercury, gold salts, and Fowler's solution (contains arsenic). They were inhaled in vaporizers or taken as liquid and likely poisoned the patient. A skin liniment, which claimed to draw out poisons through the skin to cure a host of diseases, including TB, was marketed in the 1920s. It contained turpentine, ammonia, formaldehyde, mustard, and wintergreen. It often caused blisters but did not cure TB.

Typhoid Fever

Typhoid, also called "enteric," "continued fever," or "abdominal typhus," has a gradual onset of symptoms over several days. These include a sustained high fever, abdominal pain, constipation, severe headaches, and after the first week, a hard abdomen. Some people develop a skin rash with rose-colored spots, nosebleeds, and diarrhea. In severe cases, delirium, stomach and intestinal ulcers, bowel perforation, intestinal bleeding, and pneumonia develop, which often foreshadows death. Typhoid fever can be identified by smell as patients often have a peculiar musty odor characteristic of a freshly killed mouse. The disease is caused by *Salmonella typhi* bacteria, first identified in 1880, which are spread by water, food, and sometimes flies contaminated with fecal material. Paratyphoid, a similar disease, produces milder symptoms and is caused by the *S. paratyphoid* bacterium. Typhoid can last for weeks or months. Sometimes after the patient appears well, since the bacteria can live in the gall bladder, they can become re-infected and relapse, as the disease does not provide long-lasting immunity.

The disease affects all ages, but children and young adults are more likely to die from it. The fever has a fatality rate from 10 to 30 percent without treatment. The World Health Organization (WHO) in 2015 estimated there were about 12.5 million cases of typhoid and 149,000 deaths. The disease is most common in developing nations, in particular parts of Africa and India. Typhoid in the past was sometimes confused with typhus (discussed in another entry) as some of its symptoms are similar. Typhoid fever was first recognized as a separate disease in 1643, and in 1829, French physician Pierre C. A. Louis (1787–1872) first used the word *typhoide*, meaning "resembling typhus." Until the late 1800s, physicians believed the disease was caused by a contagion, *miasma* (foul air), fatigue, or anxiety.

Typhoid is one of the few diseases that can produce asymptomatic carriers. These individuals sometimes unknowingly infect others. One of the most infamous healthy carriers was Mary Mallon (1869–1938), an Irish American, who worked as a cook for wealthy New York families in the late 1800s. Known as "Typhoid Mary," she is thought to have infected 51 people, of which 3 died, throughout her cooking career. Because she would not give up cooking or get surgery to remove her infected gall bladder, the New York Health Commission quarantined her on an island near Manhattan, where she remained until her death.

Throughout the twentieth century, in developed nations typhoid fever steadily declined due to adequate sewage disposal, chlorination of drinking water, better hygiene, and immunization. Beginning in 1948, antibiotics were found to cure the disease. Although like other bacteria, some antibiotic resistant typhoid strains have been developed.

The illness has been found since antiquity. Egyptian, Greek, and Roman physicians observed fever and alimentary tract diseases and may have influenced history. Recent DNA analysis suggests that the devastating Athenian plague (430–424 BCE) was typhoid fever although not all medical historians agree. The plague killed one-third of the Athenian population, including its leader. Following this disaster, the balance of power shifted from Athens to Sparta and ended the Greek Golden Age and Athenian dominance of the ancient Greek world.

Some historians suggest that the Jamestown, Virginia, colony may have died out because of typhoid fever. In the English colonies overall, the fever killed more than 6,000 settlers between 1607 and 1624. Typhoid raged throughout Europe and North America from the seventeenth through early twentieth century and killed tens of thousands including many prominent men and women.

Until the early twentieth century, typhoid had been a deadly companion of war. In the nineteenth century, for example, during the American Civil War (1861–1865), over 80,000 soldiers died due to typhoid fever or dysentery. In the Shiloh campaign, the northern general, W. T. Sherman (1820–1891), mustered only half of his 10,000 troops because the other half were sick, which resulted in his troops being overrun by the Confederates. Troops in Spanish-American War (April–August 1898) encountered typhoid both on the field and in training camps. Among the recruits over 20,700 contracted the disease (1,590 died), which represented 72 percent of all sick soldiers. By the WWI era (1914–1918), the American, British, and European armies immunized soldiers against typhoid, so the disease was only a minor factor during this and succeeding wars.

UNUSUAL TREATMENTS

Various treatments for the typhoid fever have been tried over the centuries. In ancient Egypt before any treatment was given, healers performed an incantation to the goddess Isis. To cure a hard and distended abdomen—often indicative of typhoid—the physician gave the patient a drink of crushed emmer wheat, barley, grapes, figs, and onions added to sweet beer, honey, and goose fat. In addition, a poultice "to drive out the hardness of the abdomen," which contained cat dung, red lead, watermelon, sweet beer, and wine, was placed on the patient's abdomen; these were not effective. Ancient Greek physicians (c. 500 BCE) bled patients to reduce the fever and gave clay to patients to eat as a treatment for fevers and to absorb poisons. *Terra sigillata*, stamped clay tablets, were still used into the 1600s for typhoid and other fevers. The Romans treated typhoid and other fevers with cold baths.

In early 1800s, newly identified electricity, or "galvanism," was used to treat typhoid and other ills. A few physicians thought that passing electricity through a person energized the body and invigorated the patient. Most physicians, however, in the 1800s relied on standard treatments that had been used for centuries such as emetics (causes vomiting) containing the poisonous metal antimony along with purging and bloodletting to "clean the body of poisons." These did not cure typhoid, and antimony could have poisoned the patient. Leeching or cupping (hot cups placed on skin to draw up blood) was used on the back of the neck and abdomen even though new research of the time found these were not helpful. Calomel, containing mercury, was also used with little effect other than poisoning the patient.

When the fever began to decline, physicians gave the patient whey wine. For weaker individuals, they administered brandy or whiskey punch along with milk every few hours as it was thought to build up strength, which could have helped hydrate the patient. Some gave coffee and claimed it helped build strength while others administered quinine, although this drug was only effective for malaria and intermittent fevers discussed in another entry and had no effect upon typhoid.

By the 1890s, a few physicians—after it was realized that microorganisms caused disease—recommended destroying the typhoid bacilli in the stomach and intestine. They gave patients antiseptic drinks that included chlorine water (diluted bleach), eucalyptus and turpentine oil, and carbolic acid, which may have further sickened the patient but was unlikely to kill the bacteria. Some physicians placed patients in cold water tanks as a new treatment, although cold baths had been used in antiquity, claiming the treatment reduced mortality from typhoid. Physician also ordered opium enemas "to splint the bowels," which may have worked to stop diarrhea.

At the turn of the twentieth century, American physician John Harvey Kellogg (1852–1943) tried light baths in specially made boxes to cure typhoid at his Battle Creek Sanitarium, which had little effect. He also operated on intestinal perforations caused by the disease. These operations are still done; however, they have a high mortality rate. Many of the older treatments were discontinued as scientific evidence found them most ineffective.

Typhus

Typhus is caused by different species of the tiny *rickettsia* bacteria. Symptoms of the disease include the sudden appearance of a high fever, chills, headache, malaise, and severe pains in the bones and joints. A rash of small red spots appears first on the chest and spreads over the body except for the face, palms, and soles of the feet. As the disease progresses, some patients experience delirium, coma, and death.

Typhus is a potentially fatal illness and is found in two major forms. The major form of the disease, "epidemic typhus," is a louse-borne disease caused by the *Rickettsia prowazekii* bacterium. The disease is transmitted through infected lice feces by scratching bites or breathing in the feces from lice infested clothing, bedding, or the environment. It primarily emerges in crowded filthy conditions with little access to bathing and laundry facilities. It is more common during cold weather when people huddle together for warmth. Typhus affects all ages, but children are less likely to die from it. Epidemic typhus has a fatality rate from 10 to 60 percent without treatment. The other forms of the disease have few deaths and are transmitted by flea feces (murine or endemic typhus) or bites from mites (scrub typhus). Murine typhus is found in areas with rodent infestations, and scrub typhus is common in rural areas of Asia and Northern Australia.

The term "typhus" was based upon the Greek word *typhos*, meaning "smoky" or "hazy," since the disease often produced delirium. The term was first used around 1760 after a massive outbreak of the disease in England. One of the first conclusive descriptions of epidemic typhus was during the Spanish army's siege of Granada in 1489, where 17,000 Spanish troops contracted the disease. However, typhus was often confused with other diseases such as smallpox, measles, and typhoid fever (discussed in other entries), so it is possible that the disease existed earlier. A typhus-like disease, which was different from typhoid fever, was reported in 1641 in China. Since it caused few deaths, it was likely endemic or scrub typhus. However, in the West it was not until 1829 that typhus and typhoid were identified as separate diseases. Since typhus appeared to be a new disease in the early 1500s, physicians thought it had been imported from the New World like syphilis (see separate entry). But a fatal disease like epidemic typhus was rare among the indigenous peoples of the Americas.

In the sixteenth century, sailors on long sea voyages with little hygiene—and many lice—came down with "ship-fever," which was spread through the communities of the ports they visited. Jail or "gaol fever" in England often spread from jails into the general population through judges, court officials, and jailers when they acquired infected lice from prisoners. One such typhus outbreak between 1577 and 1579 killed about 10 percent of the English population. In 1759, English authorities estimated that about 25 percent of all prisoners died from typhus each year. For those on death row, more died of typhus than by hanging. It was also contracted by those nursing the sick.

"Famine fever," still another name for typhus, wiped out whole communities in the Irish Potato Famine of 1846–1849, and thousands died of the disease in English workhouses. In Canada, the 1847 typhus epidemic killed more than 20,000 people, mostly newly arrived Irish immigrants who were escaping famine and disease. Many had contracted it aboard the immigrant ships. In the 1800s, outbreaks also raged through Eastern Europe. Typhus was a factor in public health laws in the United States. In 1892, when typhus was found among Eastern European Jewish immigrants—who had arrived by ship to New York—they were placed in quarantine to keep the epidemic from spreading. A year later, procedures for the medical inspection of all immigrants were enacted due to fear of typhus, bubonic plague, and cholera discussed in other entries.

Throughout the mid-1800s, typhus was thought to be transmitted by "contagions" in bedding and clothing or by *miasmas* (bad air). By 1909, French physician Charles Nicolle (1866–1936) noticed that typhus patients were no longer infectious after receiving hot baths and clean clothes. He correctly deducted that the disease was transmitted by infected lice. In 1916, the rickettsia bacterium was identified. By the 1950s, insecticides and antibiotics eliminated or cured the disease in developed nations. However, it was still found in some poorer nations and among refugees and others living in squalor in the twenty-first century

Typhus has influenced history and the outcome of European wars. For example, during the English Civil War (1642–1651), around 180,000 people died. Over 50 percent were killed by wound infections and "camp fever," which particularly affected the king's conscripted troops. The outcome led to the execution of Charles I and the end of the monarch as the ruling entity. A major example of typhus and war was Napoleon's retreat from Moscow in 1812. It is estimated that several hundred thousand soldiers were killed by the "war plague," augmented by the Russian winter, which resulted in Napoleon's defeat. Curiously, during the American Civil War (1861–1855), typhus was not a major killer in POW camps, which generally offered an ideal breeding ground for the disease. Unlike wars in Europe, it was rarely transmitted to the surrounding populations. Some researchers suggest that something in the American environment was inhospitable to the louse and extensive spread of the disease.

During WWI and its immediate aftermath, at least 3 million deaths out of 20–30 million cases of typhus occurred among both soldiers and civilians—particularly on the eastern front. On the western front, delousing stations were in place. Likewise, during WWII, typhus was also responsible for hundreds of thousands of deaths in concentration camps, among refugees, and among German and eastern European troops. On the other hand, the Americans and British used insecticides to kill the lice, and few contracted typhus.

UNUSUAL TREATMENTS

In the late 1500s, physicians recommended blistering agents to the temples to release "poisons in the blood" as a cure for typhus based upon humoral theory. By the mid-1700s, they no longer recommended this treatment as it had little effect. In the 1700s, some doctors gave a broth made from poisonous snakes, milk, and eggs to patients as they thought the mixture reduced the effect of the disease. Depending upon the dose of the snake venom, it was harmless but did not cure the disease.

However, by the mid-1800s, physicians knew there was no known cure for typhus and that the disease, like many other diseases with fever, needed to "run its course." Supportive care was given to keep patients alive, which included bleeding (to reduce "inflamed blood"), cooling lime juice in water, a simple diet, and alcohol. Physicians found alcohol had both a medicinal and food value based upon years of use for many diseases. These liquids would also have hydrated the patient. Alcohol was especially thought beneficial for those with intermittent or a rapid pulse as alcohol slowed down the heart rate. Alcohol was given for delirium, when extremities were cold and the body hot, and for congested lungs. Doctors believed that sherry, port, and beer were the best drinks for typhus sufferers. "Old people and drunkards" were also given brandy, or "old rye whiskey" mixed with beef tea, milk, or eggs for nourishment. For severe cases of typhus, physicians recommended "a half bottle of fine old sherry" per day, the only effect of which was to put patients to sleep. To relieve thirst, 20 drops of diluted hydrochloric acid (what is found in the stomach) in an ounce of water was given several times a day. Although some physicians used quinine to reduce the fever, others did not find it effective as quinine was specific for intermittent or malaria-type fevers.

A very effective prevention method against the louse emerged in WWII. The Allies halted typhus by killing lice on human bodies and clothing with insecticides. In 1943, applications of MYL powder—made from the chrysanthemum flower—reduced louse infestations to a low level. This treatment was so popular that the powder became a black market item as it eliminated itching from lice bites. By 1944, the newly developed and more effective insecticide powder, DDT, was blown with a power sprayer on and under clothing. The military established delousing stations, and the resulting mass delousing effort eliminated typhus as a major problem throughout Western Europe. On the other hand, due to its toxic effect on the environment and humans, DDT was discontinued for most uses worldwide in the 1970s.

Urinary Tract Issues

Common urinary tract issues include lower and upper urinary tract infections (bladder and kidneys) that can be acute or chronic, bladder control issues, and "stones" that form in the kidneys and travel to the bladder. Urine is secreted in the kidneys and passes into the bladder via the ureters and then out of the bladder through the urethra.

Bladder and Kidney Inflammation and Infection

Symptoms of cystitis (bladder infection or inflammation) include pain or burning on urination (dysuria), frequent urination, and feeling the need to urinate despite having an empty bladder. It is often caused by microorganisms and can reduce the quality of life, particularly for women. If bacteria travel up to the kidneys, they can cause an acute kidney infection (pyelonephritis). Symptoms include fever and flank pain and generally cystitis symptoms and can be lethal. For example, until the development of antibiotics, "honeymoon cystitis" from frequent sexual intercourse sometimes killed young women due to "uremic poisoning" because of kidney failure. Urinary tract infections (UTIs) also become increasingly common as people age. Since women have a shorter urethra, they are more likely to have infections compared to men. Up to 10 percent of women have a UTI every year, and about 50 percent of women have at least one infection at some point in their life.

Chronic UTIs can lead to chronic kidney disease (CKD), also called nephritis, glomerulonephritis, or until the mid-twentieth century Bright's disease. Due to infection or injury of the glomerulus that filters the urine, kidney function slowly declines over time. This results in edema or swelling (called dropsy in the past) of the extremities and other areas of the body. Albumen, and often blood, is found in the urine. High blood pressure, fatigue, and loss of appetite are also common. CKD is associated with underlying conditions such as diabetes, obesity, high cholesterol levels, and heredity. It affected 753 million people globally in 2016, including 417 million females and 336 million males. It is the ninth leading cause of death in North America, seventh in the United Kingdom, and twelfth in Australia and the world.

UTI have likely been common since the emergence of early mammals. They were mentioned in the ancient Egyptian *Papyrus Ebers* circa 1500 BCE. Dropsy, a sign of nephritis, was described in the Christian Gospels circa 70–100 CE. The first use of the term "nephritis" in English for kidney inflammation was in 1566 derived from Late Latin, from the Greek *nephros* for kidney (*-itis* means "inflammation"). "Cystitis" for a bladder infection derived from Latin was used in 1783. The term "chronic kidney disease" was not coined until the 1940s as being more inclusive of several kidney issues. In the seventeenth century, deaths from UTI were initially recorded. In London, John Graunt (1620–1674) recorded 56 deaths from urinary tract diseases out of 9,584 deaths.

Causes of urinary disorders until the early nineteenth century were based upon those suggested by Greek physician Hippocrates (c. 460–370 BCE) and Greco-Roman physician Galen (129–210 CE), who believed that urinary disorders were caused by disharmony of the humors. Scottish physician William Buchan (1729–1805) in the eighteenth century thought that bladder and kidney inflammation was from intemperance and the lack of exercise. Even though infectious diseases were found to be caused by bacteria during the nineteenth century, many physicians did not recognize that UTIs were also caused by bacteria for decades. American physician Elijah B. Hammack (1826–1888), for example, believed that UTIs were caused by intemperance, cold, strain in the back, and the internal use of turpentine in addition to highly seasoned food and sexual indulgence.

Although UTI began to be treated with antibiotics in the 1950s, which saved many lives, many bacteria became resistant to them. No cure yet exists for CKD, and treatment is focused on reducing symptoms with diet and dialysis, which began in the 1940s.

Incontinence and Overactive and Underactive Bladder

Urinary incontinence is the loss of bladder control that negatively impacts a person's life. About 70 percent of urinary incontinence is due to overactive bladder, and its cause is largely unknown. "Urge incontinence" is a loss of bladder control when a person suddenly and frequently feels the need to urinate, and in men, it can be caused by an enlarged prostate. In "stress incontinence," urine leaks when a person coughs, sneezes, laughs, exercises, or lifts a heavy object. "Overflow incontinence" is due to poor bladder contraction or a blockage of the urethra. Incontinence is most common in women. Of women over 65 years of age, more than 50 percent have experienced some form of incontinence.

Moreover, overactive bladders and incontinence in Western cultures has been considered a shameful and taboo subject. This changed in the 1980s when a few paper manufacturing companies began marketing adult diapers for active older people. More acceptance of incontinence and using diapers and other devices to collect urine also occurred by the late twentieth century when military personnel and astronauts adopted them when stopping to urinate was impossible.

In contrast to an overactive bladder is the underactive bladder syndrome. The person has extreme difficulty in emptying the bladder and urination may cause burning and pain—strangury is an older term for this condition. In some cases, the individual is unable to urinate. Common causes include aging, diabetes, injury, paralysis, surgery, or obstruction in the bladder neck or kidneys by tumors or stones. These conditions have also been noted since antiquity. The words "incontinent" and "dysuria" were first used in the fourteenth century in the English language. Incontinent is from the Latin *incontinent*, meaning "not holding together," and dysuria also via Late Latin from the Greek *dysouria*, meaning "difficult" plus "urine."

Physicians have suggested various causes for these maladies. In the eighteenth century, Buchan, for example, believed that a slowing or stoppage of urine in addition to stones might be caused by inflammation of the bladder, swelling of hemorrhoid veins, hard feces lodged in the rectum, or "hysteric affections." Since antiquity to treat urine retention, catheters have been used. They have ranged from reeds to clay to metal tubes. By the late twentieth century, only sterile deposable plastic catheters were used.

Kidney and Bladder Stones

Hard calcium oxalate or uric acid stones (renal calculus or nephrolith) cause extreme painful spasms in the kidneys and when they pass into the bladder through the ureters. They have been described throughout history, and the earliest bladder stone has been found in an Egyptian mummy from around 4800 BCE. The disorder has increased since the late twentieth century likely from dietary changes. It affects about 12 percent of the world population and occurs more frequently in men than in women. Stones have been associated with an increased risk of kidney disease. In past centuries, bladder stones were common in children but were rare by the late twentieth century in Western nations. However, they are still a problem in the Middle East, South Asia, and the Far East among children.

Throughout history, descriptions of lithotomy surgery are found, but this surgery was considered the last resort. The lithotomy position for female pelvic exams comes from the position of the patient lying on the table for stone removal surgery. Hippocrates, in ancient Greece, was against cutting into the bladder for stones as it was generally lethal. Even today in the Hippocratic Oath, the new physician agrees not to "cut for the stone" but to leave it to the professionals. By the late twentieth century, ultrasound and laser treatments that crushed stones became the most common methods for stone removal. Ureteroscopy (a small catheter with a small camera and light), which is passed into the ureter and percutaneous (small incision) for stone removal, is also done. Today only about 4 percent of stone removal is done by invasive surgery.

Urinary tract stones and infection may have affected the course of French history. For example, Napoleon Bonaparte I (1769–1821) during the Russian campaign in 1812 was in great pain from a bladder stone and may not have been clearly thinking. In addition, Napoleon III (1808–1873), nephew of Napoleon I, surrendered to Germany in 1870 at the battle of Sedan during the Franco-Prussian war. Due to painful bladder and kidney stone and gallstone attacks, he was under the influence of opium, which was reported to make him lethargic, sleepy, and apathetic.

UNUSUAL TREATMENTS

Various remedies have been tried for urinary tract issues since antiquity. Some are still practiced today as both regular and folk medicine. Many were used out of tradition and were not effective.

Bladder and Kidney Inflammation and Infections

North American indigenous peoples ate the American cranberry (*Vaccinium macrocarpon*) to treat UTI. They are still used as part of folk medicine although there is a difference of opinion regarding their effectiveness. During the Renaissance (c. 1300–1600), eating clay pieces, or sigillata, was thought to help bladder and kidney complaints. In the early nineteenth century, English physician Richard Bright (1789–1858)—for which Bright's diseased is named—treated nephritis patients based upon humor theory and tradition. He drew out body fluids through bleeding and purging and used digitalis (foxglove plant), which may have helped reduced swelling. However, when the condition got worse, he ordered bleeding from a vein, more calomel (poisonous mercury compound), and an enema of colocynth (purgative drug from a plant). When the patient further deteriorated, he recommended additional bleeding and a cantharides (ground caustic beetle) plaster be placed on the neck. All these treatments likely led to an earlier death.

Later in the century to increase urination in dropsy, British physician Edward John Waring (1819–1891) prescribed the diuretic juniper (the berry used to make gin) that decreased some swelling. American physician Henry Hartshorne (1823–1897) suggested lancing the edema to reduce swelling and cupping (hot cups to create a vacuum place over skin that had been cut) the loins, which may have helped but could have caused infections. For acute nephritis, another American Elijah B. Hammack (1826–1888) suggested hot water soaks or a bottle at the feet, a hot pack of bitter herbs placed over the painful area, and Dover's powder (ipecac and opium) to cause sweating and relieve pain. In order "to keep the bowels open," based upon humoral theory, he believed that radishes and watermelons were effective.

Incontinence and Overactive and Underactive Bladder

The *Papyrus Ebers* suggests external urine-collection devices be used for male incontinence, which are still used today. It also recommended vaginal inserts to compress the bladder for women. Various concoctions were mixed to drink, rub on the penis, or given as enemas to regulate urination. Many contained milk or beer and were likely ineffective. For example, one remedy for an enema called for a mixture olive oil, honey, sweet beer, sea salt, and seeds-of-the-wonder fruit (pomegranate, which has antimicrobial properties).

For male incontinence in the eighteenth century, penal clamps were used to shut off unwanted urination. These devices are still advertised today and may help with dribbling. In the nineteenth century, physicians prescribed tincture of cantharides (caustic beetle mixed with alcohol) to relieve "paralysis of the bladder" and incontinence, as they believed it restored the bladder to a healthy tone thus relieving these problems. Physicians also suggested "syrup of the iodide of iron" first made in 1840 by combining iron wire with iodine and then adding honey or sugar. It acted as a purgative but was thought to prevent incontinence. By the late twentieth century, abdominal muscle training, Pilates, Tai Chi, breathing exercises, and generalized fitness were recommended to prevent incontinence, but evidence is insufficient to support these as effective.

For bladder retention, healers in ancient Egypt recommended cooking a papyrus in oil and smearing it on the body, particularly for children. Hippocrates, besides catheterization, administered a diuretic such as juniper in addition to bloodletting and opium for pain. In the thirteenth century, physicians injected various chemical and herbal preparations into the bladder or rubbed ointments onto the penis or kidney area. When these were not effective, they catheterized it and as a last resort sent the patient for surgery, which was likely lethal. In the nineteenth century based upon tradition from antiquity, in addition to catheterization, Hartshorne, for example, placed leeches on the perineum (area between the anus and genitals), which could have led to infections. He also recommended laudanum (alcohol and opium) enemas and inhalation of anesthesia such as ether or chloroform to relieve pain and relax the patient, which likely relieved pain. Some physicians recommended hot enemas to relieve urine retention, which may or may not have helped.

Kidney and Bladder Stones

Treatments for stones were mentioned in the *Papyrus Ebers.* Throughout the late eighteenth century, special barber surgeons often removed stones by fingers through incisions into the bladder through the perineum. However, most patients died from infection or bleeding. Various instruments in the eighteenth and nineteenth centuries were also developed to enter the urethra to crush and extract stones. Some surgeons described passing a nail up the urethra and then cracking the stone by striking the nail with a hammer, which would have been extremely painful and could have perforated the bladder. By the late nineteenth century when anesthesia and aseptic conditions were instituted, lithotomy surgery became safer.

Buchan in the eighteenth century recommended a teaspoonful of the "sweet spirits of nitre" (a distillation of alcohol and nitric and sulfuric acid that acts as a diuretic) for stones lodged in the ureter and for the patient to "exercise on horseback, or in a coach" to help pass the stones. In the nineteenth century, Hammack prescribed chloroform to relax the patient to help pass painful kidney stones. Once passed into the bladder, slippery elm tea with aspirin-like qualities was given to drink. It was also injected into the bladder. If the stone had not been passed, the dangerous lithotomy surgery was performed. Over the twentieth century, various methods to remove stones were advanced. By the 1980s, urologists used high-energy shock waves to destroy stones in the kidney or ureter.

Wounds and Wound Infections

Wounds can range from simple scrapes to deep lacerations and punctures that can lead to immediate death. Wounds are generally caused by battles, violence, or accidents. Several different types of wounds are found. An incised wound is produced by a sharp edge and is usually longer than it is deep, such as that caused by a sharp edge during surgery, a knife while cutting vegetables, or other activities. It causes a clean, straight cut, and there is often much bleeding. A laceration is a complex wound caused by tearing or crushing forces, such as an automobile accident, and often causes much damage to the surrounding tissues. In an abrasion, the wound is caused by scraping, such as a fall off a bicycle. It is not deep but may contain many foreign bodies such as dirt. A puncture is a deep wound caused by a sharp object such as a nail or sword. It can appear small but might bring in pathogenic bacteria. Another type of wound is an amputation. This can be caused by trauma, surgery to save the person's life with an uncontrolled infection, or tissue damage that is beyond repair. Finally, chronic wounds such as decubitus ulcers (bedsores), which are difficult to treat, are found among those with spinal cord injuries or diabetes.

Wound infections may occur in any wound. Since the skin is broken, it allows microorganisms to enter, which can lead to infections. The most serious is gangrene, which is the death of body tissue resulting from bacterial infection or lack of blood circulation. To prevent death, amputations were, and still are, performed if tissue death is spreading and the wound is not healing.

It is estimated that almost 2 million people in Europe suffer from acute or chronic wounds, and in the United States, it is estimated that from 1 to 2 percent of the total population will experience a chronic wound. Among Medicare beneficiaries, around 15 percent have some type of wound or wound infection. The largest group of infections are from surgical wounds and are found in roughly 4 percent of surgical patients. In addition, about 3.4 percent of diabetics have a chronic wound infection. Wounds and wound infections cause pain, distress, social isolation, job performance, quality of life, and even death.

Infections from wounds have been common since the first vertebrates evolved. From ancient Greek civilizations into the early nineteenth century, pus formation was seen as a good sign inasmuch as it was believed that the body was eliminating "toxins" and balancing the four humors based upon humoral theory. Infections were also thought to be caused by miasma or bad smells through the air. However, with little understanding of the nature of infections, people did not realize that the lack of hygiene transmitted pathogenic bacteria, causing many wounds to become infected and to discharge pus. Aseptic techniques did not begin until the late nineteenth century. The development of sulfa drugs in the 1930 and other antibiotics in the 1940s dramatically reduced wound infections, gangrene, and death.

The term "wound" was first used before the twelfth century from the Old English *wund*, from Old High German *wunta*, meaning "wound." Around the world, similar techniques have been used to treat wounds. The wound was first washed, and debris cleaned out. Then herbs, or other substances or material, were placed upon the wound before it was bandaged. However, many became infected.

Archaeological evidence suggests that serious wounds were sutured in ancient Egypt using linen and animal sinew to close wounds. The ancient Greeks refined suturing. Greco-Roman physician Galen (129–c. 210), for example, sutured severed tendons of gladiators and was the first to describe gut sutures. He also advocated the use of sutures made of silk. These substances are still used in the early twenty-first century, although synthetic material or even gluing an incision is now common. The Romans developed fine needles for suturing.

The most common surgical procedure performed for thousands of years for serious wound infections of limbs was amputation. For leg wounds with gangrene, it was the best chance for survival although the surgery had a 60 percent mortality rate. These amputations were most often carried out in military activities. Moreover, in all wars up to the twentieth century, more troops died of wound infections than from deaths in battle. Mortality from wounds then decreased into the twenty-first century due to personal upper-body armor, antibiotics, safer anesthesia, and various surgical techniques—including plastic surgery. By 2020, about 90 percent of wounded troops survived the Iraq and Afghanistan conflicts—the highest in history despite increasingly lethal weapons—but many returned home with permanent mental, emotional, and physical disabilities.

Wounds and infections, and sometimes their treatments, have been found in art and literature. For instance, in ancient Greece, numerous wars occurred. Greek author Homer (c. late eighth or early seventh BCE) describes about 150 different battle wounds made by arrows and spears in the epic poem the *Iliad*. Treatments of these injuries included bandaging and amputations. The *Iliad* concerns the conflict between the Greeks and Trojans. In Western cultures, thousands of illustrations of Christ with wounds from his crucifixion have been created. Terrible wounds have also been described in stories and film. The 1952 film *The Snows of Kilimanjaro*, based upon American novelist Ernest Hemingway's (1899–1961) short story of the same name (1936), concerns the reminiscence of a writer who is on an African safari and who is dying of an infected leg wound. He lies outside of his tent as the vultures and hyenas watch him. His female companion nurses him and dresses his gangrenous wound, which is infested with maggots. He is finally rescued and taken back to civilization.

UNUSUAL TREATMENTS

Since antiquity, healers have attempted to treat wounds. Folk treatments also arose that sometimes were successful, and many remedies came out of military battles. These therapies generally were based upon trial and error. In ancient Egypt, the *Papyrus Ebers* (c. 1500 BCE) reported many recipes for different types of wounds. For simple wounds, such as a splinter, worm's blood was cooked and crushed in oil. Another recipe called for killing a mole, cooking it in oil, and a third was mixing ass's dung with fresh milk. These concoctions were applied to the wound opening to draw out the splinter. It may or may not have worked, but the dung could have caused an infection.

For an open wound, fresh cow's meat was placed on it until it festered. Then oil, fats, and fruit, and finally a plaster of fresh ivory powder mixed with honey were applied. Another recipe called for human excrement crushed with beer yeast, oil, and honey to be applied as a poultice. For a huge gaping laceration wound, fresh meat was placed on it the first day, and it was held together by stitching. If the stitches came loose, it was treated with grease and honey and bound with strips of linen. Some of these remedies may have helped, such as honey that has antiseptic properties, but serious infections from raw meat or tetanus from manure also likely occurred, leading to death.

The ancient Greeks placed honey, vinegar, rainwater, and medicinal spices on open wounds. Some of these items in addition to oregano, cinnamon, and garlic have been found to have anti-inflammatory or antiseptic properties, were harmless to the patient, and may have helped to prevent infection. When pus oozed out of an infected wound, healers put willow leaves and thyme on it and covered it with a bandage. These herbs may have helped reduce inflammation and pain and were harmless to the patient. Greek physician Hippocrates (c. 460–370 BCE) placed clay on open lacerations to heal them, which could have introduced bacteria leading to infections. He also treated wounds with wine, which acts as an antiseptic, and then kept them dry.

The Romans, besides wine, also used hot oil or boiling water, which may have prevented infection. These treatments were adopted by Greco-Roman physician Galen (129–c. 210) and influenced care of injuries and wounds into the Middle Ages. Greek physician, pharmacologist, botanist, and author of *De materia medica*, Dioscorides (c. 40–90 CE) noted he used frankincense resin to treat ulcers and wounds. Frankincense has been found to have antiseptic properties and was largely harmless to the patient. Byzantine Greek physician Paulus Aegineta (c. 625–c. 690) cut away the flesh to remove the gangrene and sometimes used cautery. These techniques were used into the nineteenth century but could cause blood poisoning (septicemia) before the widespread use of aseptic techniques.

During the Middle Ages, cauterization with a hot metal rod was a common procedure for serious wounds. The physician burned lacerations to prevent infection and help clot the blood to improve healing. However, it was extremely painful to the patient and often resulted in disfigurement. From the Middle Ages until the late eighteenth century, physicians put "man's grease" on wounds. This "cure-all" substance was the boiled-down fat cut from an executed criminal. It was not effective but usually did not harm the patient as most pathogens were killed by boiling. Barber surgeons into the eighteenth century also practiced bloodletting to restore the balance of humors based upon the theories of Hippocrates and Galen. They would place leeches near the wound, draw blood from a vein, or sometimes use "wet cupping" (a heated cup was placed over incised skin, which produced a vacuum and drew out blood). This likely weakened the patient and did little to stop the spread of infection.

Bloodletting by cutting a vein was common for many illnesses and conditions into the nineteenth century. It was also used for infected wounds to draw out poisonous blood. (Wellcome Collection. Attribution 4.0 International (CC BY 4.0))

From antiquity, analgesics such as opium, mandrake, and henbane have been put on wounds to relieve pain. Sometimes they were put on a sponge, and the patient inhaled the vapors and even ingested the mixtures. These were often used for anesthesia prior to surgery and limb amputations. Dominican bishop and physician Albertus Magnus (c. 1193–1280), for example, used a mixture of mandrake and wine as a narcotic for amputations. However, mandrake and henbane are extremely toxic, and mixed with opiates and alcohol, they likely led to some deaths. During some later wars, such as the American Civil War (1861–1865), little anesthesia other than alcohol and opiates was used prior to amputations. Opiates such as morphine were also used for pain relief, which led to postwar opiate addiction being called the "soldiers' disease" among numerous wounded soldiers who survived.

Some successful therapies, however, have emerged out of wartime. French barber surgeon Ambroise Paré (1509–1590), for example, like most physicians cauterized gunshot wounds as it was believed that gunpowder was poisonous. At one point, he ran out of elderberry oil for cauterization, so he instead used the post-cauterization treatment of egg yolk, rose oil, and turpentine on the wounds. These soldiers developed little pain or problem, but the ones who had the usual treatment of cauterization had excruciating painful wounds. Paré and others also found that tying off vessels in amputations worked better than cautery. However, this knowledge became unknown, or was ignored, for wound management, and cautery was still used for gunshot wounds during the American Civil War.

A wound treatment used by the military that was successful was maggot debridement therapy. During WWI (1914–1918), it was found that troops with severe wound infections that became infested with maggots had fewer problems and were more likely to survive compared to other troops. This was because the maggots ate the dead tissue. This therapy was first recorded by American physician William Stevenson Baer (1872–1931). After the war, he experimented with bot fly maggots. The therapy became popular in the 1930s and 1940s but was discontinued with the introduction of sulfa drugs and other antibiotics. In addition, if sterile maggots were not used, patients could develop infections such as tetanus. However, when antibiotic resistant bacteria evolved in the late twentieth century, sterile maggots were approved as a medical device in 2005 by the U.S. Food and Drug Administration and are used in hard-to-treat chronic wound infections.

Yellow Fever

Most people who are infected with yellow fever are asymptomatic (do not have symptoms). Those with symptoms may exhibit three stages. In the first stage, fever, headache, muscle aches, loss of appetite, and vomiting are found. Jaundice, from where the disease gets its name due to yellowed skin and eyes, is also seen. This period usually lasts several days. Then the disease goes into remission and most recover. However, about 15 percent of people go into a dangerous third stage, when the mortality rate can be up to 50 percent. The patient has a rapidly rising temperature, becomes agitated, bleeds from the gums and nose, and has black vomit and diarrhea, and organ damage occurs. Those who survive are usually immune for life. All ages are susceptible to the disease, and it is estimated that yellow fever now causes about 200,000 cases and 30,000 deaths worldwide primarily in tropical and subtropical areas of Africa and South America.

Some researchers estimate that the disease evolved in central and western sub-Saharan Africa around 3000 BCE. The disease is caused by the *flavivirus* first identified in 1927. It is spread from humans to humans though the bite of an infected female *Aedes aegypti* mosquito. The mosquito can also transfer the virus to her eggs. When they hatch and mature into mosquitoes, their bites can spread the infection.

The disease was carried to the New World on slave ships beginning in the sixteenth century. An outbreak of the fever in the Yucatan with many deaths was documented by the Spanish in 1648. Between 1668 and 1699, the disease was carried to port cities along the Atlantic, where thousands died. Over the next century, widespread epidemics were recorded. In Philadelphia (1793), for example, the fever killed 10 percent of the population, and hundreds of thousands were killed in the Caribbean and in Central and South America. Into the late 1880s, serious outbreaks occurred in the southern United States and along the Ohio and Mississippi rivers. However, this "yellow jack," another name for the fever, was not found in the East Indies, the eastern shores of Africa, the Pacific coast of America, Australia, and Asia. The reasons for this are still unknown.

During the eighteenth century, this "American plague" spread to southern Europe. Between 1800 and 1828, several epidemics devastated Spanish and Portuguese ports. Later the fever was found in French and British seaports. Due to the disease, ships were quarantined or prevented from entering many European ports, which caused considerable economic hardships for international traders.

Many theories arose as to the cause of the disease. In the 1700s, some physicians claimed it was caused by foul odors from rotten cargo and garbage, which were carried by the air. Other physicians thought it was spread by contact with infected individuals or contaminated objects. Until the late 1800s, another theory suggested the disease was caused by microscopic vegetation from newly upturned earth, decaying vegetation around wharfs, or even high heat and humidity. A few physicians, as late as the 1880s, believed it was triggered by the alignment of the planets which produced more carbon in the atmosphere, which in turn created microbes. There is no cure for the disease—only supportive care and prevention with immunization and mosquito control, which has eliminated it from much of the world.

The "yellow plague" influenced the history of the Americas. Native peoples, European settlers, and their indentured servants became sick or died from the fever. However, enslaved Africans from areas where the disease was endemic generally only had a very mild illness and thus were considered valuable workers, which fueled the slave trade. Yellow fever like malaria, typhus, and typhoid fevers (discussed in other entries) also influenced conquests and wars in the New World. For example, a British expedition of 27,000 troops attempted to annex Mexico in 1741. They were forced to withdraw as the number of troops was reduced to about 7,000 by this fever and other diseases. Likewise, about 60 years later, Napoleon (1769–1821) attempted to regain control of the lucrative sugar trade in what is now Haiti. However, due to the yellow plague, only one-third of his troops sent to Haiti survived. Napoleon consequently gave up his plans for re-conquest and sold the entire Louisiana territory to the United States in 1803.

During the Spanish-American War (1898), more American troops died from diseases, including yellow fever, than from combat. After the war, the fever continued to ravage both the American occupation force and the Cubans. Meanwhile, the French had attempted to build the Panama Canal and found it impossible due to the fever, and they abandoned the project. Because the United States was now interested in building the canal, in 1900 the Yellow Fever Commission headed by army surgeon Major Walter Reed (1851–1902) was established to investigate the cause and prevention of the fever. Cuban physician Carlos Juan Finlay (1833–1915) had suggested that the disease was transmitted by mosquito bites, and Reed's research supported this theory. Therefore, to eliminate mosquitoes, swamps were drained, or oil was poured on them, and people slept under mosquito nets, which dramatically reduced the disease, resulting in the canal being completed in 1914 and thus changing international shipping.

Yellow fever has appeared in the arts. For example, French writer Victor Hugo (1802–1885) mentioned the fever in some of his literary works. A young adult historical novel, *Fever 1793* by American writer Laurie Halse Anderson (1961–), looks at the 1793 Philadelphia epidemic through the eyes of a teenage girl who attempts to survive among this public health crisis and death and destruction of this unknown illness.

UNUSUAL TREATMENTS

By the early 1800s, although many physicians realized there was no cure for yellow fever, they felt they should do something for their patients. Therefore, into the late 1800s, based upon humoral theory, they used standard treatments for fevers, which included bleeding by vein or leeches, cupping (hot cups placed on skin to draw up blood), or blistering (chemicals that caused blisters) on the abdomen to "draw out the poisons." These remedies were often followed by quinine administered by mouth or rectum, which was not effective as the drug was specific for malaria only. These treatments were not effective.

When the patient had a high fever and vomiting, a few physicians recommended mineral water, frequent drinks of ice champagne, charcoal water, and hot coffee, which may have kept the patient hydrated. In addition, a mustard plaster was placed over the epigastrium (stomach area) to stop the vomiting. Physicians also recommended cold water enemas to cool the patient along with cold water cloths placed on the head and limbs. They kept patients' bowels loose through enemas made of "good beef tea and brandy" and gave them lime water to drink. Several physicians claimed these treatments reduced mortality. Other physicians recommended tepid baths followed by soaking the feet in hot water. A purgative of calomel (contains mercury), jalap (root of the Mexican climbing plant), and ginger was given. If this remedy did not clear the bowels, an enema of castor oil, molasses, and warm water was used to eliminate poisons, which could have caused dehydration.

After the fever had abated, physicians gave opium with calomel to drink. If the patient was not vomiting, many recommended a "stout glass of rum," wine, brandy, or whiskey as a stimulant. Some physicians experimented with antiseptic substances, such as chlorine, sulfites, and carbolic acid, but they had no effect upon the disease. Around the turn of the twentieth century, American physician Harvey Kellogg (1852–1943) placed yellow fever patients in specially made boxes for light baths at his Battle Creek Sanitarium, which may have helped decrease jaundice but did not cure the disease.

At a loss of how to prevent the yellow plague, various communities into the late 1800s tried different public health methods to get rid of the "contagion" that sometimes were successful, although they were not aware of the reason for the success. For example, Mobile, Alabama, paved its streets with oyster shells, which eliminated muddy roads and the fever almost disappeared. If inhabitants were removed from an infected area to another place, the epidemic would stop. Some communities sprayed sulfur and lime mixtures into houses of the infected, which may have killed some mosquitoes and slowed the spread of the disease. On the other hand, other communities burned barrels of tar in the street to disinfect the air and get rid of the contagion, which was not successful as this did not eliminate disease-carrying mosquitoes. By 1910, most of the old treatments and some public health measures were discontinued as scientific evidence found them ineffective.

Z

Zoonotic Diseases

Zoonosis refers to diseases that can be passed from animals to humans. The animal acts as a reservoir (where pathogen lives) or vector (spreads infections) and transfers the illness to a susceptible human or animal host. Hundreds of zoonotic illnesses exist and can be caused by viruses, bacteria, parasites, worms, and fungi. Some result in uncomfortable maladies and others serious illnesses or death. They are transmitted by bites, contact with animal products, respiratory droplets, insect vectors, and infected food products or water. Discussion of zoonotic diseases such as equine encephalitis or West Nile virus transmitted by mosquitoes or Lyme disease from ticks are discussed in the "Bites and Stings: Insects and Arachnids" entry. Plague transmitted by fleas, malaria and yellow fever from mosquitoes, and influenza—which sometimes can originate in swine or birds—are detailed in their own entries. A greatly feared zoonotic scourge for centuries has been rabies or hydrophobia. Dogs and wild animals transmit the rabies virus in their saliva, and the affliction is discussed in detail under the "Bites: Mammals and Reptiles" entry. It causes inflammation in the brain and death.

An ancient potentially deadly zoonotic disease is anthrax, which is spread through close contact with infected animals or their products. It is caused by the *Bacillus anthracis* bacteria that forms spores. These spores usually enter the animal or human body through skin abrasions, ingesting tainted food or water, or inhaling infected dust from hides. However, in the twenty-first century, some IV drug users in Europe were sickened by street drugs contaminated with spores. Once the spores are in the body, they are activated by fluids and are transformed into live bacteria. About 95 percent of all anthrax infections are cutaneous that, in the past, was identified as a carbuncle or woolsorters' disease. A small blister forms on the skin with swelling around the sore that often turns into a painless ulcer with a black center. Around 10–40 percent of untreated cutaneous cases may result in death. The intestinal form causes nausea and vomiting, bloody diarrhea, and abdominal pains; left untreated, it has a mortality rate of 25–75 percent. The respiratory form causes fever, chest pain, and shortness of breath and has a fatality rate that is 80 percent or higher. However, if diagnosed early, anthrax can be medically managed with antibiotics and prevented with immunizations.

Anthrax is more common in the agricultural regions of developing countries and is rare in developed nations. Hunters, agricultural workers, and tanners are most likely to be infected. German physician Robert Koch (1843–1910) discovered the anthrax bacterium in 1875. In Britain in 1878, woolsorters' disease and anthrax were shown to be one and the same. The first use of the term "anthrax" in English was in 1398. It is derived from the Greek word for "coal" due to the black center of the lesion. Anthrax is thought to have originated in Egypt or Mesopotamia. During biblical times, Christian, Islamic, and Judaic scriptures relate that when the Egyptian pharaoh would not allow the enslaved Israelites to leave Egypt, 10 plagues arose. Some scholars suggest that the fifth plague of Egypt might have been anthrax as numerous domestic animals died.

Various causes of anthrax have been given over the centuries. For example, Greek physician Galen (129–c. 210) who practiced in the Roman Empire and whose theories were used into the nineteenth century believed that the black carbuncle was caused by black blood rising from the blister. In the seventeenth century, the French thought anthrax was from contaminated fields that infected animals and then people. Several breakouts of anthrax among domestic animals and humans have occurred over the centuries. A major epidemic occurred in southern Europe in 1613 when 60,000 died. In 1979, an outbreak of inhalation anthrax occurred at a Soviet military microbiology facility, where at least 68 people died. Anthrax spores can survive for centuries in contaminated soil. During 2016, for example, a number of people and reindeer were sickened or died in Siberia by spores released from melting permafrost, where reindeer in the past had died from the disease.

Another ancient zoonotic scourge is trichinosis. Trichinosis is transmitted from ingesting the *Trichinella spiralis* worm in undercooked infected pork or wild game. It produces gastrointestinal complaints and muscle aches and is generally not fatal. In the past, it may have been associated with eating infected meat. Globally, approximately 10,000 cases of trichinosis are diagnosed per year. Trichinosis is rare in developed nations since there are strict laws for meat processing and animal feed. The first use of the term "trichinosis," now called "trichinellosis," was in 1835 when the worm was discovered in Britain in a cadaver. The word is derived from the Greek *thrix*, meaning "hair."

The prohibition of pork in Islam and Judaism may have come from awareness in biblical times that eating meat from non-cud-chewing ruminant animals such as pigs was associated with disease, so pork was deemed unclean. Shellfish, which are also considered unclean by some of Abrahamic religions, such as Judaism, can carry certain hepatitis viruses. Hepatitis is discussed in the "Liver and Gallbladder Issues" entry. Although chicken and other fowl meat and their eggs can be a source of human zoonotic pathogens—such as salmonella and campylobacter bacteria discussed in "Gastroenteritis, Dysentery, and Diarrhea"—many cultures and religions accept chickens and other birds as food.

Another bird disease, Psittacosis, which causes flu-like symptoms, results in 50–200 infections per year in the United States. It is transmitted by inhaling the parasite *Chlamydia psittaci* from the discharges of infected birds. Bird and bat guano can also transmit histoplasmosis, a respiratory illness with flu-like symptoms, caused by inhaling the fungus *Histoplasma capsulatum*. The disease is highest in the Midwest of the United States, with an estimated 6.1 cases per 100,000 people.

Similarly, inhaling dried infected cat feces or soil or eating infected undercooked meat can transmit the *Toxoplasma gondii* parasite. The parasite can also be passed from a pregnant woman to her developing fetus and cause mental disability. In the United States, of the 750 deaths attributed to toxoplasmosis each year, about 50 percent are believed to be caused by eating contaminated meat. In addition, from 400 to 4,000 cases of congenital toxoplasmosis occur each year. The parasite was discovered 1908, but its importance in causing birth defects was not known until mid-century. The term is derived from the Latin *toxicum*, "poison," and *plasma*, "something molded."

New emerging zoonotic diseases fatal to humans such as Middle East Respiratory Syndrome (MERS) from camels and severe acute respiratory syndrome (SARS) perhaps originating in bats and Ebola (probably originating in monkeys) are transmitted by close contact with the infected animal or person. Bats are thought to transmit these diseases by bites along with airborne particles from their infected saliva. The COVID-19 pandemic of 2020–2022 that sickened and killed hundreds of thousands likely originated from bats, or their airborne fecal material. The virus was spread to humans, or other animals. Once humans were infected, it easily spreads to others mostly through respiratory droplets. Most zoonotic diseases have only been identified in recent years, and medical personnel attempt to prevent or manage them with immunizations if available, supportive therapy, or newly developed antimicrobial, viral, or parasitic drugs.

Some of these diseases have been associated with war and terrorist activities in an attempt to influence the course of history. For example, during WWI, anthrax was used as a biological weapon in animal feed. The Germans attempted to infect allied horses, but it did not alter the course of the war. The Japanese, who occupied Manchuria, China, from 1932 to 1945, deliberately infected the prisoners with the disease, which killed at least 10,000. It was not used during WWII, but in the postwar years, international concerns of more virulent anthrax spores developed as a military weapon led to the *Biological Weapons Convention* in 1972 that prohibited biological and toxic weapons. During the 1991 Gulf War, American military personnel were immunized against the disease due to fear that the substance might be used. Furthermore, in 2001, an American terrorist mailed anthrax powder in packages to people in Washington, D.C., New York City, and West Palm Beach, Florida, which sickened 22 people and killed 5.

UNUSUAL TREATMENTS

Because only a few diseases in the past were associated with animals, the symptoms of most afflictions were often treated based upon trial or error or the prevailing medical philosophy. Anthrax, however, was treated throughout the ages. Roman philosopher Celsus (c. 25–c. 50) burned and cauterized the black lesion of the common cutaneous form of the disease to eliminate the poisoned black blood of the eruption. Similarly, Galen, in the second century, applied substances whose "properties resembled fire." These include arsenic, iron sulfate, quicklime, and sandarach (gum resin from cypress). Other healers applied mustard and figs; none were effective in curing the disease.

Byzantine Greek physician Paulus Aegineta (c. 625–c. 690), in his work based upon the manuscripts of others, recommended bloodletting, cutting into the lesion, and using a poultice made of the "tender part of bread baked in an earthen pan."

Physicians placed other items on the pustule, such as a mixture of black seaweed, passionflower, wine must, vinegar, and other herbs and substances.

In the nineteenth century, some physicians employed ether and carbolic acid (phenol) poultices on oiled silk, which had a soothing effect on the anthrax lesion. Many favored laxatives to "keep the bowels open" to balance the humors and used wine on the lesion to prevent gangrene. In terms of most other zoonotic diseases, specific treatments were not formulated until recent times as the causes of the maladies were unknown. Generalized treatment usually included bloodletting, purging, and vomiting to eliminate toxic poisons and rebalance the humors of these unknown illnesses.

Glossary

Acupuncture
The insertion of needles at points along channels in the body to aid energy called *qi* to flow when it is "blocked."

Age of Enlightenment
Eighteenth and early nineteenth centuries when the scientific method emerged.

Blistering
Placing ground cantharides beetles on the skin to cause a blister and fluid drainage to draw off "bad humors" and balance the body's humoral system.

Bloodletting
Drawing off blood from a vein by cupping or leeching to balance the body humors. No scientific data supports this procedure other than a few rare genetic diseases.

Case Fatality Rate
The number of people who die from the number of people with the disease.

Clyster
An enema where fluid is injected into the rectum to cause a fecal discharge.

Congenital Anomalies
Birth defects.

Consumption
Tuberculosis.

Contagion Theory
Belief that disease is spread by direct contact, indirect contact, or from a distance. Substances or "seeds of the disease" are found on the person or in the clothing of sick or dead people and are spread to the healthy. The seeds enter the body through the respiratory system or injuries lead to putrefaction in the blood or skin.

COPD
Chronic obstructive pulmonary disease.

Cupping
Heated glass cups are placed on the skin to draw blood to the skin's surface. In wet cupping, the heated cups are placed over the skin that has been scratched to draw

out blood. In dry cupping, they are not. By the twenty-first century, rubber cups to create a vacume were used. Cupping has been used in Western medicine throughout the mid-nineteenth century and in Eastern and alternative medicine into present time for centuries.

CVA
Cerebral vascular disease or strokes.

Debridement
Cleaning out damaged tissue or containments in a wound such as battle wounds.

Early Middle Ages: 500–1000
Decline of classical science, learning, and arts.

Early Modern Period
Late fifteenth century to the late eighteenth century that included the Protestant Reformation and the Industrial Revolution in the West.

Emetics
A substance that causes vomiting; it is generally used to get rid of bad humors.

Endemic
Disease is always found in a population, such as malaria in tropical areas.

Epidemic
Outbreak of a disease that occurs suddenly with a rapid increase in the number of cases above what is expected, such as measles, due to lack of vaccination in a population or a new diseases such as COVID-19.

FDA
Federal Drug Administration in the United States that approves treatments and medical devices.

High Middle Ages: 1000 CE–1250
European cities and wealth began to rise again out of the darker times of the early Middle Ages.

HIV
Human immunodeficiency virus.

Homer
Ancient Greek author may be a group of authors and not one individual who wrote the epic poems the *Iliad* and the *Odyssey* in the late eighth or early seventh centuries BCE.

Humoral Theory
Suggested the body was made up of four fluids (blood, phlegm, black bile, and yellow bile) or humors. When they were out of balance, disease arose. Methods such as bloodletting, purging, emetics, and blistering were carried out to rebalance the humors. There is no scientific basis for this theory.

Ischemic Heart Disease
Heart disease that often leads to heart attacks.

Late Middle Ages: 1300–1500
Out of the bleak fourteenth century and Black Death, classical learning arose again to form the Renaissance.

Lower Respiratory Infections
Pneumonia and tuberculosis.

Miasma Theory
Disease was thought to be caused by bad air or odors that emerge from rotting garbage or swamps. There is no scientific basis for this theory.

Microcephaly
Abnormal smallness of the head.

Morbidity
Illness.

Mortality
Death.

Natron
A mixture of sodium carbonate (soda ash), sodium bicarbonate (baking soda), and small quantities of sodium chloride and sodium sulfate used for embalming in ancient Egypt.

Pandemic
An outbreak of a disease that spreads worldwide such as COVID-19 first reported in late 2019.

Paracelsus
Noted Swiss physician and alchemist of the German Renaissance whose real name was Theophrastus Phillippus Aureolus Bombastus von Hohenheim (1493–1541).

Placebo Effect
A positive effect produced by a remedy that is not attributed to the properties of the substance or treatment but is due to the patient's belief in the remedy. Sometimes called a fake treatment.

Plaster
Mustard or other herbs ground and added to warm water to make a paste and spread on the chest or other areas of the body. Sometimes a piece of thick cloth was put between the skin and the mixture to prevent irritation.

Poultice
Herbs such as ginger or turmeric spread on a warm, moist cloth and applied to the skin. It supposedly draws out inflammation.

Psychotropic Drugs
Mind-altering drugs such as cannabis, alcohol, and opiates.

Purgative
A laxative that causes diarrhea.

Renaissance
Revival of classical learning and wisdom that began in Italy in the fourteenth century and spread throughout Europe into the seventeenth century.

Saltpeter (Nitre, Potassium Nitrate)
It was used to prevent asthma, arthritis, and promote urination. It can cause kidney damage. There is no evidence that it quells sexual urges.

Schizophrenia
Mental illness that has faulty perception, hallucinations, and delusion.

Seton
A thread placed under the skin to cause festering and keep the wound open and to allow drainage. Its ends protrude, and the thread is often moved back and forth. It also acts as a counterirritant.

Spanish Fly
A blistering beetle that contains *terpenoid cantharidin*, a toxic blistering agent. It was used as an aphrodisiac and as a diuretic. But it can cause organ failure and death.

Spirit of Nitre
A distillation of alcohol with nitric and sulfuric acids with a sweet odor.

Sudorifics
A substance that induces sweating.

Tincture
A substance mixed with ethyl alcohol such as tincture of cannabis.

Trepanation
Cutting or scraping a hole in the head to allow evil spirits to depart from the person, also used to cure headaches, epilepsy, and mental illness.

Venesection
Cutting a vein to cause bleeding as a way of drawing out bad humors.

Victorian Era
The years when Queen Victoria reigned in Great Britain from 1837 to 1901.

WWI
World War I, Great War (1914–1918)

WWII
World War II (1939–1945)

Bibliography

Abou El-Soud, Neveen. "Herbal Medicine in Ancient Egypt." *Journal of Medicinal Plants Research* 4, no. 2 (February 2010): 82–86. https://www.researchgate.net/publication/228634623_Herbal_medicine_in_ancient_Egypt

Abou-Ghazi, Dia' M. "Ahmed Kamal 1849–1923." In *Mélanges Ahmed Kamal,* edited by Dia' Abou-Ghazi, 1–14. Annales du Service des Antiquités de L'Egypte 64. Cairo: Al-Shaab Printing House, 1981. https://archive.org/stream/ASAE-64-1981/ASAE_64_1981_djvu.txt

Ahmed, Awad M. "History of Diabetes Mellitus." *Saudi Medical Journal* 23, no. 4. (April 2002): 373–78. https://smj.org.sa/index.php/smj/article/download/4029/1803

Aitken, John. *Principles of Midwifery or Puerperal Medicine*. 3rd ed. Enlarged and Illustrated with Engravings. For the Use of Students. Edinburgh: Printed for J. Murray, Fleet-Street, 1786.

Al Awar, Omar, and Gytis Sustickas. "Landmarks in the History of Traumatic Head Injury." March 2017. https://www.researchgate.net/publication/314004316_Landmarks_in_the_History_of_Traumatic_Head_Injury

Al Binali, H. A. "Night Blindness and Ancient Remedy." *Heart Views* 15, no. 4 (October–December 2014): 136–39. https://www.ncbi.nlm.nih.gov/pmc/articles/PMC4348990/

Alelign, Tilahun, and Beyene Petros. "Kidney Stone Disease: An Update on Current Concept." *Advances in Urology* (February 4, 2018): 3068365. https://www.hindawi.com/journals/au/2018/3068365/

Allan, David B., and Gordon Waddell. "An Historical Perspective on Low Back Pain and Disability." *Acta Orthopaedica Scandinavica* 60, no. 234 (1989): 1–23.

American Academy of Sleep Medicine. *The International Classification of Sleep Disorders*. 3rd ed. Darien, IL: American Academy of Sleep Medicine, 2014.

American Psychiatric Association. *Diagnostic and Statistical Manual of Mental Disorders*. 5th ed. Arlington, VA: American Psychiatric Association, 2013.

Ancient Egyptian Medicine: The Papyrus Ebers. Translated from the German version by Cyril P. Bryan with an introduction by G. Elliot Smith. Chicago: Ares Publishers Inc., 1974.

Andersen, S. Ry. "History of Ophthalmology: The Eye and Its Diseases in Ancient Egypt." *Acta Ophthalmologica Scandinavica* 75 (1997): 338–44.

Anderson, D. Mark, Daniel I. Rees, and Tianyi Wang. *The Phenomenon of Summer Diarrhea and Its Waning, 1910–1930.* IZA Institute of Labor Economics. IZA DP No. 12232. Bonn, Germany: IZA Institute of Labor Economics, 2019. http://ftp.iza.org/dp12232.pdf

Angelakis, Emmanouil, Yassina Bechah, and Didier Raoult. "The History of Epidemic Typhus." *Microbiology Spectrum* 4, no. 4 (August 2016). https://www.ncbi.nlm.nih.gov/pubmed/27726780

Arderne, John. *De Arte Phisicali et de Cirurgia of Master John Arderne, Surgeon of Newark, Dated 1412.* Translated by Sir D'Arcy Power, K.B.E., M.B.Oxon, F.R.C.S., from a Transcript Made by Eric Millar, M.A. Oxon, from the Replica of the Stockholm Manuscript in the Wellcome Historical Medical Museum. New York: William Wood & Co., 1922.

Arnold, Melina, Mónica S. Sierra, Mathieu Laversanne, Isabelle Soerjomataram, Ahmedin Jemal, and Freddie Bray. "Global Patterns and Trends in Colorectal Cancer Incidence and Mortality." *Gut*, 66, no. 4 (2017): 683–91. https://gut.bmj.com/content/66/4/683

Ashby, Steven P. "Grooming the Face in the Early Middle Ages." *Internet Archaeology* 42 (2016). https://intarch.ac.uk/journal/issue42/6/9.cfm

Asrani, Sumeet K., Harshad Devarbhavi, John Eaton, and Patrick S. Kamath. "Burden of Liver Diseases in the World." *Journal of Hepatology* 70, no. 1 (January 2019): 151–71.

Astbury, Leah. "Being Well, Looking Ill: Childbirth and the Return to Health in Seventeenth-Century England." *Social History of Medicine* 30, no. 3 (August 2017): 500–19.

Atiyeh, B. S., S. W. A. Gunn, and S. N. Hayek. "Military and Civilian Burn Injuries during Armed Conflicts." *Annals of Burns and Fire Disasters* 20, no. 4 (December 2007): 203–15.

Attur, M. G., M. Dave, M. Akamatsu, M. Katoh, and A. R. Amin. "Osteoarthritis or Osteoarthrosis: The Definition of Inflammation Becomes a Semantic Issue in the Genomic Era of Molecular Medicine." *Osteoarthritis and Cartilage* 10, no. 1 (2002): 1–4. https://www.oarsijournal.com/article/S1063-4584(01)90488-1/pdf

Australia Government. Department of Health. *Guidelines for the Public Health Management of Gastroenteritis Outbreaks due to Norovirus or Suspected Viral Agents in Australia.* Last modified 2010. https://www1.health.gov.au/internet/publications/publishing.nsf/Content/cda-cdna-norovirus.htm-l

Azari, Amir A., and Neal P. Barney. "Conjunctivitis: A Systematic Review of Diagnosis and Treatment." *JAMA* 310, no. 16 (October 2013): 1721–29.

Babkin, Igor V., and Irina N. Babkina. "The Origin of the Variola Virus." *Viruses* 7, no. 3 (March 2015): 1100–12. https://www.ncbi.nlm.nih.gov/pmc/articles/PMC4379562/

Baer, William S. "The Treatment of Chronic Osteomyelitis with the Maggot (Larva of the Blow Fly)." *The Journal of Bone and Joint Surgery* 13, no. 3 (July 1931): 438–75.

Baker, A. B. "Artificial Respiration, the History of an Idea." *Medical History* 15, no. 4 (1971): 336–51. https://www.ncbi.nlm.nih.gov/pmc/articles/PMC1034194/

Ballenger, Jesse F. "Dementia, Society, and History." *AMA Journal of Ethics* 19, no. 7 (2017): 713–19. doi: 10.1001/journalofethics.2017.19.7.mhst1-1707.

Bandaranayake, Ilian, and Paradi Mirmirani. "Hair Loss Remedies—Separating Fact from Fiction." *Cutis* 73, no. 2 (2004): 107–14.

Barry, John. *The Great Influenza: The Story of the Deadliest Pandemic in History*. New York: Viking Penguin, 2004.

Beebe, M. J., J. M., Bauer, and H. R. Mir. "Treatment of Hip Dislocations and Associated Injuries: Current State of Care." *The Orthopedic Clinics of North America* 47, no. 3 (July 2016): 527–49.

Bell, John. *A Treatise on Baths: Including Cold, Sea, Warm, Hot, Vapour, Gas, and Mud Baths; Also, on Hydropathy, and Pulmonary Inhalation; with a Description of Bathing in Ancient and Modern Times*. 2nd ed. Philadelphia: Lindsay & Blakiston, 1859.

Berchtold, Nicole C., and Carl W. Cotman. "Evolution in the Conceptualization of Dementia and Alzheimer's Disease: Greco-Roman Period to the 1960s." *Neurobiology of Aging* 19, no. 3 (May 1998): 173–89.

Berg, Friedrich Paul. "Typhus and the Jews." *The Journal of Historical Review* 8, no. 4 (Winter 1988): 433–81. http://www.ihr.org/jhr/v08/v08p433_Berg.html

Bhattacharya, Tanya, Mark A. Strom, and Peter A. Lio. "Historical Perspectives on Atopic Dermatitis: Eczema through the Ages." *Pediatric Dermatology* 33, no. 4 (July/August 2016): 375–79.

Bianucci, Raffaella, Antonio Perciaccante, Philippe Charlier, Otto Appenzeller, and Donatella Lippi. "Earliest Evidence of Malignant Breast Cancer in Renaissance Paintings." *The Lancet Oncology* 19, no. 2 (February 1, 2018): 166–67.

Bikbov, Boris, Norberto Perico, and Giuseppe Remuzzi. "Disparities in Chronic Kidney Disease Prevalence among Males and Females in 195 Countries: Analysis of the Global Burden of Disease 2016 Study." *Nephron* 139, no. 4 (May 23, 2018): 313–18.

Blanchette, Christopher M., Melissa H. Roberts, Hans Petersen, Anand A. Dalal, and Douglas W. Mapel. "Economic Burden of Chronic Bronchitis in the United States: A Retrospective Case-Control Study." *International Journal of Chronic Obstructive Pulmonary Disease* 6 (January 2013): 73–81.

Bland, Edward F. "Rheumatic Fever: The Way It Was." *Circulation* 76, no. 6 (December 1987): 1190–95. https://www.ahajournals.org/doi/pdf/10.1161/01.CIR.76.6.1190

Blomstedt, Patric. "Orthopedic Surgery in Ancient Egypt." *Acta Orthopaedica* 85, no. 6 (December 2014): 670–76. https://www.ncbi.nlm.nih.gov/pmc/articles/PMC4259025/

Bos, Kirsten I., Kelly M. Harkins, Alexander Herbig, Mireia Coscolla, Nico Weber, Iñaki Comas, Stephen A. Forrest, et al. "Pre-Columbian Mycobacterial Genomes Reveal Seals as a Source of New World Human Tuberculosis." *Nature* 514, no. 7523 (October 23, 2014): 494–97.

Bou Assi, Tarek, and Elizabeth Baz. "Current Applications of Therapeutic Phlebotomy." *Blood Transfusion* 12, no. 1 (January 2014): s75–s83. https://www.ncbi.nlm.nih.gov/pmc/articles/PMC3934278/

Bourne, Rupert R. A., Gretchen A. Stevens, Richard A. White, Jennifer L. Smith, Seth R. Flaxman, Holly Price, Jost B. Jonas et al. "Causes of Vision Loss Worldwide, 1990–2010: A Systematic Analysis." *The Lancet Global Health* 1, no. 6 (December 1, 2013): E339–49. https://www.thelancet.com/journals/langlo/article/PIIS2214-109X(13)70113-X/fulltext

Bowers, Elizabeth Shimer. "Weird Food Remedies for Diabetes." Medically reviewed by Farrokh Sohrabi, MD. Everyday Health. Last modified December 11, 2012. https://www.everydayhealth.com/diabetes/weird-food-remedies-for-diabetes.aspx

Boylston, Arthur. "The Origins of Inoculation." *Journal of the Royal Society of Medicine* 105, no. 7 (July 2012): 309–13. https://www.ncbi.nlm.nih.gov/pmc/articles/PMC3407399/

Breasted, James Henry. *The Edwin Smith Surgical Papyrus: Published in Facsimile and Hieroglyphic Transliteration with Translation and Commentary in Two Volumes.* Chicago: University of Chicago Press, 1930.

Brorson, Stig. "Management of Fractures of the Humerus in Ancient Egypt, Greece, and Rome: An Historical Review." *Clinical Orthopedics and Related Research* 467, no. 7 (July 2009): 1907–14. https://www.ncbi.nlm.nih.gov/pmc/articles/PMC2690737/

Brun, Rita, and Branden Kuo. "Functional Dyspepsia." *Therapeutic Advances in Gastroenterology* 3, no. 3 (May 2010): 145–64.

Brunton, Deborah. "19th Century." In *Health and Medicine through History: From Ancient Practices to 21st-Century Innovations*, edited by Ruth Clifford Engs, vol. 2, pt. 3. Santa Barbara, CA: Greenwood, 2019.

Bryan, Charles S. "New Observations Support William Osler's Rationale for Systemic Bloodletting." *Baylor University Medical Center Proceedings* 32, no. 3 (July 2019): 372–76. https://www.ncbi.nlm.nih.gov/pmc/articles/PMC6650279/#CIT0058

Bryan, Leon S., Jr. "Blood-letting in American Medicine, 1830–1892." *Bulletin of the History of Medicine* 38 (October 21, 1964): 516–29.

Buchan, William. *Domestic Medicine; Or, The Family Physician: Being an Attempt to Render the Medical Art More Generally Useful, by Shewing People What Is In Their Own Power Both With Respect to the Prevention And Cure of Diseases. Chiefly Calculated to Recommend a Proper Attention to Regimen And Simple Medicines.* 2nd American ed. Philadelphia: Printed by Joseph Crukshank, for R. Aitken, 1774.

Bush, Jeffrey S., and Simon Watson. "Trench Foot." In *StatPearls* [Internet]. Treasure Island, FL: StatPearls Publishing, 2019. https://www.ncbi.nlm.nih.gov/books/NBK482364/

Butler, J. V., E. C. Mulkerrin, and S. T. O'Keeffe. "Nocturnal Leg Cramps in Older People." *Postgraduate Medical Journal* 78 (October 2002): 596–98. https://pmj.bmj.com/content/78/924/596

Butter, William. *A Treatise on the Kinkcough: With an Appendix Containing an Account of Hemlock and Its Preparations*. London: Printed for T. Cadell, in the Strand, 1773.

Bynum, Bill. "Ptomaine Poisoning." *The Lancet* 357, no. 9261 (March 31, 2001): 1050.

Byrne, Joseph P. *Daily Life during the Black Death*. Westport, CT: Greenwood Press, 2006.

Byrne, Joseph P. *Encyclopedia of the Black Death*. Santa Barbara, CA: ABC-CLIO, 2012.

Byrne, Joseph P. "15th through 18th Centuries." In *Health and Medicine through History: From Ancient Practices to 21st-Century Innovations*, edited by Ruth Clifford Engs, vol. 1 pt. 2. Santa Barbara, CA: Greenwood, 2019.

Canning, Brendan J. "The Cough Reflex in Animals: Relevance to Human Cough Research." *Lung* 186, no. 1 (2008): S23–28. https://www.ncbi.nlm.nih.gov/pmc/articles/PMC2882536/

Cardan, Jerome. *The Book of My Life* (*de vita propria liber*). Translated by Jean Stoner. New York: E. P. Dutton & Co., Inc., 1930. http://djm.cc/library/cardan-book-of-my-life-1930.pdf

Carter, A. J. "Dwale: An Anaesthetic from Old England." *BMJ* 319, no. 7225 (December 18, 1999): 1623–26. https://www.ncbi.nlm.nih.gov/pmc/articles/PMC1127089/

Carter, K. Codell. "Leechcraft in Nineteenth Century British Medicine." *Journal of the Royal Society of Medicine* 94 (January 2001): 38–42.

Caulfield, Ernest. "The 'Throat Distemper' of 1735–1740." Part II. *Yale Journal of Biology and Medicine* 11, no. 4 (March 1939): 277–335. https://www.ncbi.nlm.nih.gov/pmc/articles/PMC2602120/pdf/yjbm00530-0001.pdf

Cayton, Evangeline T. "The Phenomenon of Polio. Part I: From Antiquity to the Twenty-first Century." *Baylor University Medical Center Proceedings* 2, no. 1 (January 1989): 5–14. https://www.tandfonline.com/doi/pdf/10.1080 08998280.1989.11929689

Celsus, Aulus Cornelius. *Of Medicine: In Eight Books*. Translated with Notes Critical and Explanatory by James Greive. Edinburgh: Printed at the University Press for Dickinson and Company, 1814.

Channing, William F. *Notes on the Medical Application of Electricity*. Boston: Daniel Davis, Jr., and Joseph M. Wightman, 1849.

Chase, A. W. *Dr. Chase's Recipes; or, Information for Everybody: An Invaluable Collection of about Eight Hundred Practical Recipes*. 26th ed. Ann Arbor, MI: The author, 1865.

Chin, T., and P. D. Welsby. "Malaria in the UK: Past, Present, and Future." *Post Graduate Medical Journal* 80, no. 949 (2004): 663–66. https://pmj.bmj.com/content/80/949/663

Choi, Charles Q. "Ancient, Unknown Strain of Plague Found in 5,000-Year-Old Tomb in Sweden." Live Science. December 6, 2018. https://www.livescience.com/64246-ancient-plague-swedish-tomb.html

Cirillo, Vincent J. "Fever and Reform: The Typhoid Epidemic in the Spanish-American War." *Journal of the History of Medicine and Allied Sciences* 55, no. 4 (October 1, 2000): 363–97.

Cohen, Henry. "The Evolution of the Concept of Disease." *Proceedings of the Royal Society of Medicine* 48 (October 26, 1953): 155–60.

Cohen, Sheldon, David A. J. Tyrrel, and Andrew P. Smith. "Psychological Stress and Susceptibility to the Common Cold." *New England Journal of Medicine* 325 (August 29, 1991): 606–12. https://www.nejm.org/doi/full/10.1056/NEJM199108293250903

A Collection of the Yearly Bills of Mortality, From 1657 to 1758 Inclusive: Together with Several Other Bills of an Earlier Date. London: A. Millar, 1759.

Coventry, Charles Brodhead. *Epidemic Cholera: Its History, Causes, Pathology, and Treatment*. Buffalo, NY: Geo. H. Derby and Co., 1849.

Cox, F. E. G. "History of Human Parasitology." *Clinical Microbiology Reviews* 15, no. 4 (2002): 595–612. doi:10.1128/cmr.15.4.595-612.2002

Creighton, Charles. *A History of Epidemics in Britain. Volume II. From the Extinction of Plague to the Present Time*. Cambridge: The University Press, 1894.

Crocq, Marc-Antoine. "A History of Anxiety: From Hippocrates to DSM." *Dialogues in Clinical Neuroscience* 17, no. 3 (September 2015): 319–25. https://www.ncbi.nlm.nih.gov/pmc/articles/PMC4610616/

Crocq, Marc-Antoine. "The History of Generalized Anxiety Disorder as a Diagnostic Category." *Dialogues in Clinical Neuroscience* 19, no. 2 (June 2017): 107–16. https://www.ncbi.nlm.nih.gov/pmc/articles/PMC5573555/

Crocq, Marc-Antoine, and Louis Crocq. "From Shell Shock and War Neurosis to Posttraumatic Stress Disorder: A History of Psychotraumatology." *Dialogues in Clinical Neuroscience* 2, no. 1 (March 2000): 47–55. https://www.ncbi.nlm.nih.gov/pmc/articles/PMC3181586/

Crutcher, James M., and Stephen L. Hoffman. "Malaria." In *Medical Microbiology*. 4th ed., edited by Samuel Baron, 995–1008. Galveston: University of Texas Medical Branch at Galveston, 1996.

Culpepper, Nicholas. *The English Physician Enlarged: With Three Hundred and Sixty-Nine Medicines, Made of English Herbs, That Were Not in Any Impression until This*. London: Printed for the Booksellers, 1785.

Cunha, Cheston B. "Prolonged and Perplexing Fevers in Antiquity: Malaria and Typhoid Fever." *Infectious Disease Clinics of North America* 21, no. 4 (December 2007): 857–66.

Cunha, Cheston B., and Burke A. Cunha. "Brief History of the Clinical Diagnosis of Malaria: From Hippocrates to Osler." *Journal of Vector Borne Diseases* 45 (September 2008): 194–99.

Daley, Mary Doreen. "Pseudocyesis." *Postgraduate Medical Journal* 22, no. 254 (December 1, 1946): 395–99. https://pmj.bmj.com/content/postgradmedj/22/254/395.full.pdf

Dapling, Amy C. "Juvenile Mortality Ratios in Anglo-Saxon and Medieval England: A Contextual Discussion of Osteoarchaeological Evidence for Infanticide and Child Neglect." PhD diss., University of Bradford, 2010. https://bradscholars.brad.ac.uk/bitstream/handle/10454/5381/Dapling.%20Juvenile%20Mortality%20Ratios%20in%20Anglo-Saxon%20and%20Medie.pdf?sequence=12

David, Henry P. *From Abortion to Contraception: A Resource to Public Policies and Reproductive Behavior in Central and Eastern Europe from 1917 to the Present.* Westport, CT: Greenwood Press, 1999.

Davis, Geetha. "The Evolution of Cataract Surgery." *Missouri Medicine* 113, no. 1 (January–February 2016): 58–62. https://www.ncbi.nlm.nih.gov/pmc/articles/PMC6139750/

Davis, Larry E. "Unregulated Potions Still Cause Mercury Poisoning." *Western Journal of Medicine* 173, no. 1 (July 2000): 19. https://www.ncbi.nlm.nih.gov/pmc/articles/PMC1070962/

De Jong, Paulus T. V. M. "A Historical Analysis of the Quest for the Origins of Aging Macula Disorder, the Tissues Involved, and Its Terminology." *Ophthalmology and Eye Disease* 8, no. 1 (2016): 5–14. https://www.ncbi.nlm.nih.gov/pmc/articles/PMC5091095/

Demaitre, Luke. *Leprosy in Premodern Medicine: A Malady of the Whole Body.* Baltimore, MD: The Johns Hopkins University Press, 2007.

DeSesso, John M. "The Arrogance of Teratology: A Brief Chronology of Attitudes throughout History." *Birth Defects Research* 111, no. 3 (December 4, 2018): 123–41.

Diaz, Denisse, Vivian Fonseca, Yamil W. Aude, and Gervasio A. Lamas. "Chelation Therapy to Prevent Diabetes-Associated Cardiovascular Events." *Current Opinion in Endocrinology & Diabetes and Obesity* 25, no. 4 (August 2018): 258–66. https://www.ncbi.nlm.nih.gov/pmc/articles/PMC6058685/#:~:text=Diabetic%20patients%20assigned%20

Drake, Daniel. *A Systematic Treatise, Historical, Etiological and Practical, on the Principal Diseases of the Interior Valley of North America as They Appear in the Caucasian, African, Indian and Esquimaux Varieties of Its Population.* Cincinnati, OH: Winthrop B. Smith & Co., 1850.

Dubos, René Jules. *The White Plague: Tuberculosis, Man, and Society.* New York: Little, Brown, and Company, 1952.

Dumas dos Santos, Fernando Sergio, Letícia Pumar Alves de Souza, and Antonio Carlos Siani. "Chaulmoogra Oil as Scientific Knowledge: The Construction of a Treatment for Leprosy." English translation by Diane Grosklaus Whitty. *História, Ciências, Saúde-Manguinhos* 15, no. 1 (2008): 29–46. https://www.scielo.br/pdf/hcsm/v15n1/en_03.pdf

Dundes, Lauren. "The Evolution of Maternal Birthing Position." *American Journal of Public Health* 77, no. 5 (May 1987): 636–41.

Dunn, P. M. "Aristotle (384–322 BCE): Philosopher and Scientist of Ancient Greece." *Archives of Disease in Childhood. Fetal and Neonatal Edition* 91, no.1 (January 2006): F75–77.

Edwards, Martha L. "Deaf and Dumb in Ancient Greece." In *The Disability Studies Reader,* edited by Lennard J. Davis, 29–51. New York: Routledge, 1997.

Ehrlich, George E. "Low Back Pain." *Bulletin of the World Health Organization* 81, no. 9 (2003): 671–76. https://www.who.int/bulletin/volumes/81/9/Ehrlich.pdf

Eknoyan, Garabed. "A History of Obesity, or How What Was Good Became Ugly and Then Bad." *Advances in Chronic Kidney Disease* 13, no. 4 (October 1,

2006): 421–27. https://www.ackdjournal.org/article/S1548-5595%2806%2900106-6/fulltext

El Bcheraoui, Charbel, Ali H. Mokdad, Laura Dwyer-Lindgren, Amelia Bertozzi-Villa, Rebecca W. Stubbs, Chloe Morozoff, Shreya Shirude, Mohsen Naghavi, and Christopher J. L. Murray. "Trends and Patterns of Differences in Infectious Disease Mortality among US Counties, 1980–2014." *JAMA* 319, no. 12 (2018): 1248–60.

Ellis, Harold. "A History of Bladder Stone." *Journal of the Royal Society of Medicine* 72, no. 4 (April 1979): 248–51. https://www.ncbi.nlm.nih.gov/pmc/articles/PMC1437036/?page=1

The Embryo Project. "The Embryo Project Encyclopedia." The Embryo Project at Arizona State University, 2019. https://embryo.asu.edu/search

Engs, Ruth C. *Alcohol and Other Drugs: Self-responsibility.* Bloomington, IN: Tichenor Publishing, 1987.

Engs, Ruth C. *Clean Living Movements: American Cycles of Health Reform.* Westport, CT: Praeger Press, 2000.

Engs, Ruth C. *Controversies in the Addiction's Field.* Dubuque, IA: Kendall-Hunt, 1990.

Engs, Ruth C. "Do Traditional Western European Drinking Practices Have Origins in Antiquity?" *Addiction Research* 2, no. 3 (1995): 227–39. http://hdl.handle.net/2022/17485

Engs, Ruth C. *The Eugenics Movement: An Encyclopedia.* Westport, CT: Greenwood Publishing Co., 2005.

Engs, Ruth C. "20th Century and Beyond." In *Health and Medicine through History: From Ancient Practices to 21st-Century Innovations*, edited by Ruth Clifford Engs, vol. 2 pt. 4. Santa Barbara, CA: Greenwood, 2019.

Engs, Ruth C. "Women, Alcohol, and Health: A Drink a Day Keeps the Heart Attack Away?" *Current Opinion in Psychiatry* 9, no. 3 (May 1996): 217–20. http://hdl.handle.net/2022/17473

Epifano, L. D., and R. D. Brandstetter. "Historical Aspects of Pneumonia." In *The Pneumonias*, edited by Monroe Karetzky, Burke A. Cunha, and Robert D. Brandstetter, 1–14. New York: Springer, 1993.

Epstein, Randi Hutter. *Get Me Out: A History of Childbirth from the Garden of Eden to the Sperm Bank.* New York: Norton, 2010.

Estes, Stephen A., and Jane Estes. "Therapy of Scabies: Nursing Homes, Hospitals, and the Homeless." *Seminars in Dermatology* 12, no. 1 (March 1993): 26–33.

Eveleth, Rose. "To Get Rid of Body Hair, Renaissance Women Made Lotions of Arsenic, Cat Dung and Vinegar." *Smithsonian Magazine*, March 5, 2014. https://www.smithsonianmag.com/smart-news/get-rid-body-hair-renaissance-women-made-lotions-arsenic-cat-dung-and-vinegar-180949977/

Fallon, Cara Kiernan, and Jason Karlawish. "It's Time to Change the Definition of 'Health.'" Stat. First Opinion, July 17, 2019. https://www.statnews.com/2019/07/17/change-definition-health/

Faria, Miguel A., Jr. "Violence, Mental Illness, and the Brain—A Brief History of Psychosurgery: Part 1—From Trephination to Lobotomy." *Surgical Neurology International* 4 (April 5, 2013): 49. https://www.ncbi.nlm.nih.gov/pmc/articles/PMC3640229/

Feig, Milton. "Diarrhea, Dysentery, Food Poisoning, and Gastroenteritis: A Study of 926 Outbreaks and 49,879 Cases Reported to the United States Public Health Service (1945–1947)." *American Journal of Public Health* 40, no. 11 (November 1950): 1372–94. https://www.ncbi.nlm.nih.gov/pmc/articles/PMC1528983/?page=1

Feldmann, H. "Nasenbluten in der Geschichte der Rhinologie; Bilder aus der Geschichte der Hals-Nasen-Ohren-Heilkunde, dargestellt an Instrumenten aus der Sammlung im Deutschen Medizinhistorischen Museum in Ingolstadt" ["Nosebleed in the History of Rhinology. Images of the History of Otorhinolaryngology Presented by Instruments from the Collection of the Ingolstadt Medical History Museum"]. *Laryngo-Rhino-Otologie* 75, no. 2 (February 1996): 111–20.

Fenner, Frank, Donald Ainslie Henderson, Isao Arita, Zdenek Jezek, and Ivan D. Ladnyi. "The History of Smallpox and Its Spread around the World." In *Smallpox and Its Eradication*, 209–244. Geneva: WHO, 1988.

Ferrari, Roberto, Cristina Balla, and Alessandro Fucili. "Heart Failure: An Historical Perspective." *European Heart Journal Supplements* 18, no. G (December 28, 2016): G3–10.

Ferretti, Joseph, and Werner Köhler. "History of Streptococcal Research." In *Streptococcus Pyogenes: Basic Biology to Clinical Manifestations* [Internet], edited by J. J. Ferretti, D. L. Stevens, and V. A. Fischetti. Oklahoma City: University of Oklahoma Health Sciences Center, 2016. https://www.ncbi.nlm.nih.gov/books/NBK333430/?report=reader

Finsen, Niels R. "The Red Light Treatment of Smallpox." *British Medical Journal* 2, no. 1823 (December 7, 1895): 1412–14. https://www.ncbi.nlm.nih.gov/pmc/articles/PMC2509016/pdf/brmedj08798-0013b.pdf

Freeman, Jennifer. "RA Facts: What Are the Latest Statistics on Rheumatoid Arthritis?" Rheumatoid Arthritis Support Network. Last modified October 27, 2018. https://www.rheumatoidarthritis.org/ra/facts-and-statistics/

Frith, John. "The History of Plague—Part 1. The Three Great Pandemics." *Journal of Military and Veterans' Health* 20, no. 2 (April 2012): 8–12. http://jmvh.org/article/the-history-of-plague-part-1-the-three-great-pandemics/

Furman, Yury, Sheldon M. Wolf, and David S. Rosenfeld. "Shakespeare and Sleep Disorders." *Neurology* 49, no. 4 (October 1, 1997): 1171–72.

Furuse, Yuki, Akira Suzuki, and Hitoshi Oshitani. "Origin of Measles Virus: Divergence from Rinderpest Virus between the 11th and 12th Centuries." *Virology Journal* 7, no. 52 (March 4, 2010). https://www.ncbi.nlm.nih.gov/pmc/articles/PMC2838858/

Galanaud, J.-P., J.-P. Laroche, and M. Righini. "The History and Historical Treatments of Deep Vein Thrombosis." *Journal of Thrombosis and Haemostasis* 11 (2013): 402–11.

Galkowski, Victoria, Brad Petrisor, Brian Drew, and David Dick. "Bone Stimulation for Fracture Healing: What's All the Fuss?" *Indian Journal of Orthopaedics* 43, no 2 (April–June 2009): 117–20. https://www.ncbi.nlm.nih.gov/pmc/articles/PMC2762251/

GBD 2015 Mortality and Causes of Death Collaborators. "Global, Regional, and National Life Expectancy, All-Cause Mortality, and Cause-Specific Mortality for 249 Causes of Death, 1980–2015: A Systematic Analysis for the Global Burden of Disease Study 2015." *The Lancet* 388, no. 10053 (October 2016): 1459–1544.

Gensini, Gian Franco, and Andrea A. Conti. "The Evolution of the Concept of 'Fever' in the History of Medicine: From Pathological Picture per se to Clinical Epiphenomenon (and Vice Versa)." *Journal of Infection* 49, no. 2 (August 2004): 85–87.

Gjonej, Arben, Risida Gjonej, and Edvin Selmani. "Clubfoot since Ancient Time up to Now." *Journal of Osteoarthritis* 1, no. 1 (2016): 1000108. https://www.omicsonline.org/open-access/clubfoot-since-ancient-time-up-to-now-joas-1000108.php?aid=68612

Goddard, Allison L., and Peter A. Lio. "Alternative, Complementary, and Forgotten Remedies for Atopic Dermatitis." *Evidence-Based Complementary and Alternative Medicine* 2015 (2015): 676897. https://www.ncbi.nlm.nih.gov/pmc/articles/PMC4518179/

Goer, Henci. "Dueling Statistics: Is Out-of-Hospital Birth Safe?" *Journal of Perinatal Education* 25, no. 2 (2016): 75–79. https://www.ncbi.nlm.nih.gov/pmc/articles/PMC4944459/

Goetz, Christopher G. "The History of Parkinson's Disease: Early Clinical Descriptions and Neurological Therapies." *Cold Spring Harbor Perspectives in Medicine* 1, no. 1 (2011): a008862.

Gould, Tony. *A Summer Plague: Polio and Its Survivors*. New Haven, CT: Yale University Press, 1995.

Graham, David Y. "History of *Helicobacter pylori*, Duodenal Ulcer, Gastric Ulcer and Gastric Cancer." *World Journal of Gastroenterology* 20, no. 18 (May 14, 2014): 5191–204. https://www.ncbi.nlm.nih.gov/pmc/articles/PMC4017034/

Graunt, John. *Natural and Political Observations Made upon the Bills of Mortality*. Edited with an introduction by Walter Francis Willcox. Baltimore, MD: The Johns Hopkins Press, 1939.

Gregg, Rollin R. *Diphtheria: Its Cause, Nature, and Treatment*. Buffalo, NY: Printing House of Matthews Bros. & Bryant, 1880.

Griffith, F. L., ed. *The Petrie Papyri: Hieratic Papyri from Kahun and Gurob (Principally of the Middle Kingdom)*. London: Bernard Quaritch, 1898.

Gross, Robert A. "A Brief History of Epilepsy and Its Therapy in the Western Hemisphere." *Epilepsy Research* 12, no. 2 (1992): 65–74.

Gruber, Franjo, Jasna Lipozenčić, and Tatjana Kehler. "History of Venereal Diseases from Antiquity to the Renaissance." *Acta Dermatovenerologica Croatica* 23, no. 1 (2015): 1–11.

Gupta, Subash C., Sridevi Patchva, and Bharat B. Aggarwal. "Therapeutic Roles of Curcumin: Lessons Learned from Clinical Trials." *AAPS Journal* 15, no. 1 (January 2013): 195–218. https://www.ncbi.nlm.nih.gov/pmc/articles/PMC3535097/pdf/12248_2012_Article_9432.pdf

Guterman, Mark A., Payal Mehta, and Margaret S. Gibbs. "Menstrual Taboos among Major Religions." *The Internet Journal of World Health and Societal Politics* 5, no. 2 (2007). http://ispub.com/IJWH/5/2/8213

Guttmacher Institute. "Induced Abortion Worldwide." Fact Sheet, March 2018. http://dl.icdst.org/pdfs/files3/3ff4ac7f7b67079d0f352d8fb8c5924b.pdf

Haddon, John. "The Natural History and Treatment of Pneumonia." *British Medical Journal* 2, no. 2190 (December 20, 1902): 1932. https://www.ncbi.nlm.nih.gov/pmc/articles/PMC2401909/?page=1

Hajar, Rachel. "Congestive Heart Failure: A History." *Heart Views* 23, no. 3 (July–September 2019): 129–32. https://www.ncbi.nlm.nih.gov/pmc/articles/PMC6791096/

Hajar, Rachel. "Rheumatic Fever and Rheumatic Heart Disease a Historical Perspective." *Heart Views* 17, no. 3 (July–September 2016): 120–26. https://www.ncbi.nlm.nih.gov/pmc/articles/PMC5105226/

Hamilton, Brady E., Joyce A. Martin, Michelle J. K. Osterman, and Lauren M. Rossen. "Births: Provisional Data for 2018." Vital Statistics Rapid Release Report No. 007. May 2019. https://www.cdc.gov/nchs/data/vsrr/vsrr-007-508.pdf

Hammack, Elijah B. *The Family Physician and Guide to Health; A System of Domestic Medicine. Including a Treatise on Midwifery and the Diseases Peculiar to Women*. St. Louis, MO: Southwestern Book and Publishing Co., 1869.

Hanson, Ann Ellis. "Hippocrates: The 'Greek Miracle' in Medicine." Medicina Antiqua. Accessed February 3, 2020. https://www.ucl.ac.uk/~ucgajpd/medicina%20antiqua/sa_hippint.html

Harold, John Gordon. "Historical Perspectives on Hypertension." *Cardiology Magazine*, November 20, 2017. https://www.acc.org/latest-in-cardiology/articles/2017/11/14/14/42/harold-on-history-historical-perspectives-on-hypertension

Harris, Bridget, Peter J. D. Andrews, Gordon D. Murray, John Forbes, and Owen Moseley. "Systematic Review of Head Cooling in Adults after Traumatic Brain Injury and Stroke." *Health Technology Assessment (Winchester, England)* 16, no. 45 (2012): 1–175. https://www.ncbi.nlm.nih.gov/pmc/articles/PMC4781040/

Hartshorne, Henry. *Essentials of the Principles and Practice of Medicine. A Handy-Book for Students and Practitioners*. Philadelphia: H. C. Lea, 1867.

Hauck, Fern R., and Kawai O. Tanabe. "International Trends in Sudden Infant Death Syndrome: Stabilization of Rates Requires Further Action." *Pediatrics* 122, no. 3 (September 2008): 660–66.

Hawass, Zahi, Yehia Z. Gad, Somaia Ismail, Rabab Khairat, Dina Fathalla, Naglaa Hasan, Amal Ahmed, et al. "Ancestry and Pathology in King Tutankhamun's Family." *JAMA* 303, no 7. (2010): 638–47.

Hawkins, Summer Sherburne, Marco Ghiani, Sam Harper, Christopher F. Baum, and Jay S. Kaufman. "Impact of State-Level Changes on Maternal Mortality: A Population-Based, Quasi-Experimental Study." *American Journal of Preventive Medicine* 58, no. 2 (February 1, 2020): 165–74.

Hecker, J. F. C. *The Epidemics of the Middle Ages.* Translated by B. G. Babington. London: George Woodfall and Son, 1844.

Heinerman, John. *Heinerman's Encyclopedia of Fruits, Vegetables, and Herbs.* West Nyack, NY: Parker Publishing Co., 1988.

Hendrie, Colin A., and Gayle Brewer. "Evidence to Suggest that Teeth Act as Human Ornament Displays Signaling Mate Quality." *PLoS One* 7, no.7 (July 31, 2012): e42178. https://www.ncbi.nlm.nih.gov/pmc/articles/PMC3409146/

Hernigou, Philippe, Maxime Huys, Jacques Pariat, and Sibylle Jammal. "History of Clubfoot Treatment, Part I: From Manipulation in Antiquity to Splint and Plaster in Renaissance before Tenotomy." *International Orthopaedics* 41, no. 8 (2017): 1693–704.

Hippocrates. *The Aphorisms of Hippocrates, from the Latin Version of Verhoofd, with a Literal Translation on the Opposite Page, and Explanatory Notes by Elias Marks.* New York: Collins & Co., 1817.

Hippocrates. *The Genuine Works of Hippocrates Translated from the Greek, with a Preliminary Discourse and Annotations, by Francis Adams.* 2 vols. London: Printed for the Sydenham Society, 1849.

Hippocrates. *Of the Epidemics.* Translated by Francis Adams. The Internet Classics Archive. http://classics.mit.edu//Hippocrates/epidemics.html

Hoblyn, Richard D., and John A. P. Price. *Hoblyn's Dictionary of Terms Used in Medicine and the Collateral Sciences.* 15th ed. Revised Throughout with Numerous Additions by John A. P. Price. London: G. Bell & Sons, 1912.

Holmes, Bayard, and P. Gad Kitterman. *Medicine in Ancient Egypt: The Hieratic Material.* Cincinnati, OH: The Lancet-Clinic Press, 1914.

Home, Francis. *Clinical Experiments, Histories, and Dissection.* 2nd ed. corrected. London: Printed for J. Murray, 1782.

Homei, Aya, and Michael Worboys. "Athlete's Foot: A Disease of Fitness and Hygiene." In *Fungal Disease in Britain and the United States 1850–2000: Mycoses and Modernity*, Chapter 2. Basingstoke, UK: Palgrave Macmillan, 2013.

Hume, Jennifer C. C., Emily J. Lyons, and Karen P. Day. "Malaria in Antiquity: A Genetics Perspective." *World Archaeology* 35, no. 2 (2003): 180–92.

Humphreys, Margaret. "A Stranger to Our Camps: Typhus in American History." *Bulletin of the History of Medicine* 80 (2006): 269–90.

International Headache Society. "The International Classification of Headache Disorders." 3rd ed. *Cephalalgia* 38, no. 1 (2018): 1–211. https://ichd-3.org/

International Osteoporosis Foundation. "Facts and Statistics." Accessed January 8, 2020. https://www.iofbonehealth.org/facts-statistics#category-23

The Internet Classics Archive. *Works by Hippocrates.* Accessed December 10, 2019. http://classics.mit.edu/Browse/index-Hippocrates.html

Iqbal, Javed, Banzeer Ahsan Abbasi, Tariq Mahmood, Sobia Kanwal, Barkat Ali, Sayed Afzal Shah, and Ali Talha Khalil. "Plant-derived Anticancer

Agents: A Green Anticancer Approach." *Asian Pacific Journal of Tropical Biomedicine* 7, no. 12 (December 2017): 1129–50.

Johnson, James Rawlins. *A Treatise on the Medicinal Leech; Including Its Medical and Natural History, with A Description of its Anatomical Structure: Also, Remarks Upon the Diseases, Preservation and Management of Leeches.* London: Longman, Hurst, Rees, Orme, and Brown, 1816.

Johnston, Hardee. "Tetanus Antitoxin in Poliomyelitis." *Medical Record* 90, no. 7 (August 12, 1916): 292–93.

Jones, Peter Murray, and Lea T. Olsan. "Performative Rituals for Conception and Childbirth in England, 900–1500." *Bulletin of the History of Medicine* 89, no. 3 (Fall 2015): 406–33.

Jones, W. H. S. "Ancient Roman Folk Medicine." *Journal of the History of Medicine and Allied Sciences* 12, no. 4 (October 1957): 459–72.

Jones, W. H. S. *Malaria, a Neglected Factor in the History of Greece and Rome, with an Introduction by Major R. Ross, F.R.C.S., C.B. and A Concluding Chapter by G. G. Ellett, M.B.* Cambridge: Macmillan & Bowes, 1907.

Joudrey, Alan D., and Jean E. Wallace. "Leisure as a Coping Resource: A Test of the Job Demand-Control-Support Model." *Human Relations* 62, no. 2 (2009): 195–217.

Joy, Gracie. "Obstetrics and Gynecology in the Ancient World." *The Histories* 3, no. 1 (2016): Article 7. http://digitalcommons.lasalle.edu/the_histories/vol3/iss1/7

Kang, Lydia, and Nate Pedersen. *Quackery: A Brief History of the Worst Ways to Cure Everything.* New York: Workman Publishing, 2017.

Karamanou, Marianna, Emmanovil Agapitos, Antonis Kousoulis, and George Androutsos. "From the Humble Wart to HPV: A Fascinating Story throughout Centuries." *Oncology Reviews* 4 (2010): 133–35.

Karenberg, Axel. "Stroke in Literary Works around the World." *World Neurology: The Official Newsletter of the World Federation of Neurology.* Posted on June 3, 2014. https://worldneurologyonline.com/article/stroke-literary-works-around-world/

Katz, Arnold M., and Phyllis B. Katz. "Disease of the Heart in the Works of Hippocrates." *British Heart Journal* 24, no. 3 (May 1962): 257–64. https://www.ncbi.nlm.nih.gov/pmc/articles/PMC1017881/

Kellaway, Peter. "The Part Played by Electric Fish in the Early History of Bioelectricity and Electrotherapy." *Bulletin of the History of Medicine* 20, no. 2 (1946): 112–37.

Kellogg, E. W. "The Use of Turpentine in the Treatment of Diphtheria." *Journal of the American Medical Association* 28, no 4 (July 1894): 151–53.

Kellogg, John Harvey. *Light Therapeutics: A Practical Manual of Phototherapy for the Student and the Practitioner.* Battle Creek, MI: The Good Health Publishing Co., 1910.

Kempińska-Mirosławska, Bogumiła, and Agnieszka Wozniak-Kosek. "The Influenza Epidemic of 1889–90 in Selected European Cities—A Picture Based on the Reports of Two Poznań Daily Newspapers from the Second Half of the Nineteenth Century." *Medical Science Monitor* 19 (December 10, 2013): 1131–41. https://www.ncbi.nlm.nih.gov/pmc/articles/PMC3867475/

Kenealy, Tim. "Sore Throat." *BMJ Clinical Evidence* 2007 (2007): 1509. https://www.ncbi.nlm.nih.gov/pmc/articles/PMC2943825/

Kent, Susan K. *The Influenza Pandemic of 1918–1919: A Brief History with Documents.* Boston: Bedford/St. Martin's, 2013.

Khan, Ahsan A., and Gregory Y. H. Lip. "The Prothrombotic State in Atrial Fibrillation: Pathophysiological and Management Implications." *Cardiovascular Research* 115, no. 1 (2019): 31–45.

Khosla, S. N. *Typhoid Fever: Its Cause, Transmission and Prevention.* New Delhi, India: Atlantic Publishers & Distributers, 2008.

Klaver, Elizabeth. "Erectile Dysfunction and the Post War Novel: *The Sun Also Rises* and *In Country*." *Literature and Medicine* 30, no. 1 (Spring 2012): 86–102.

Kleisiaris, Christos F., Chrisanthos Sfakianakis, and Ioanna V. Papathanasiou. "Health Care Practices in Ancient Greece: The Hippocratic Ideal." *Journal of Medical Ethics and the History of Medicine* 7, no. 6 (March 2014). https://www.ncbi.nlm.nih.gov/pmc/articles/PMC4263393/

Klingbeil, G. M. "The Historical Background of the Modern Speech Clinic." *Journal of Speech Disorders* 4 (1939): 115–32.

Knapp, Vincent J. "Major Medical Explanations for High Infant Mortality in Nineteenth-Century Europe." *Canadian Bulletin of Medical History* 15, no. 2 (Fall 1998): 317–36. https://www.utpjournals.press/doi/pdf/10.3138/cbmh.15.2.317

Komorowski, Andrzej L. "History of the Inguinal Hernia Repair." In *Inguinal Hernia*, edited by Silvestro Canonico, 3–16. Rijeka, Croatia: InTech, 2014.

Kothari, Shyam S. "Of History, Half-Truths, and Rheumatic Fever." *Annals of Pediatric Cardiology* 6, no. 2 (July–December 2013): 117–20. https://www.ncbi.nlm.nih.gov/pmc/articles/PMC3957438/

Kugeler, Kiersten J., J. Erin Staples, Alison F. Hinckley, Kenneth L. Gage, and Paul S. Mead. "Epidemiology of Human Plague in the United States, 1900–2012." *Emerging Infectious Diseases* 21, no. 1 (January 2015): 16–22. https://www.ncbi.nlm.nih.gov/pmc/articles/PMC4285253/

Kunjumoideen, Kottapurath. "History of Cancer." MedicineWorld.Org: History of Cancer. August 4, 2005. https://medicineworld.org/cancer/history.html

Lakhtakia, Ritu. "The History of Diabetes Mellitus." *Sultan Qaboos University Medical Journal* 13, no. 3 (August 2013): 368–70. https://www.ncbi.nlm.nih.gov/pmc/articles/PMC3749019/

Lambert, Agnes. "Leprosy: Present and Past. II. Past." *The Nineteenth Century.* 16 (September 1884): 467–89.

Lang, Ursula, and Sabine Anagnostou. "Combating Rotting Flesh and Putrid Smells: The History of Antisepsis from Antiquity to the Nineteenth Century." *Pharmaceutical Historian* 48, no. 1 (2018): 1–11. https://www.ingentaconnect.com/content/bshp/ph/2018/00000048/00000001/art00001?crawler=true

Lee, Chia-Lin, Lien-Chai Chiang, Li-Hung Cheng, Chih-Chuang Liaw, Mohamed H. Abd El-Razek, Fang-Rong Chang, and Yang-Chang Wu. "Influenza A

(H1 N1) Antiviral and Cytotoxic Agents from *Ferula assa-foetida*." *Journal of Natural Products* 72, no. 9 (2009): 1568–72.

Lee, Kwang Chear, Kavita Joory, and Naiem S. Moiemen. "History of Burns: The Past, Present and the Future." *Burn & Trauma* 2, no. 4 (2014): 169–80. https://www.ncbi.nlm.nih.gov/pmc/articles/PMC4978094/

Leonardo, Richard A. *History of Gynecology*. New York: Froben Press, 1944.

Lescarboura, Austin C. "Our Abrams Verdict. The Electronic Reactions of Abrams and Electronic Medicine in General Found Utterly Worthless." *Scientific American*, September 1924.

Lewin, Philip. *The Foot and Ankle: Their Injuries, Diseases, Deformities and Disabilities, with Special Application to Military Practice*. 2nd ed. Philadelphia: Lea & Febiger, 1941.

Liddell, Keith. "Skin Diseases in Antiquity." *Clinical Medicine (London, England)* 6, no. 1 (January–February 2006): 81–86. https://www.ncbi.nlm.nih.gov/pubmed/16521362

Lipton, May. "The History and Superstitions of Birth Defects: Part I." *Journal of School Health* 39, no. 8 (October 1969): 579–82.

Liu, Jung-Tai, Horng-Ming Yeh, Shyun-Yeu Liu, and Kow-Tong Chen. "Psoriatic Arthritis: Epidemiology, Diagnosis, and Treatment." *World Journal of Orthopedics* 5, no.4 (September 18, 2014): 537–43.

Liu, Wei-T'ung, S. H. Zia, H. L. Chung, and C. W. Wang. "Typhus Fever in Peiping: Epidemiological Considerations." *American Journal of Epidemiology* 35, no. 2 (March 1, 1942): 231–50.

London, Kathleen. "The History of Birth Control." In *The Changing American Family: Historical and Comparative Perspectives*. Curricular Resources. 1982. Volume VI. Unit 3. Section 1. New Haven, CT: Yale University Yale-New Haven Teachers Institute, 1982.

Loriaux, D. Lynn. "Diabetes and the Ebers Papyrus 1552 B.C." *The Endocrinologist* 16, no. 2 (March/April 2006): 55–56.

Loudon, Irvine. "General Practitioners and Obstetrics: A Brief History." *Journal of the Royal Society of Medicine* 101, no. 11 (November 2008): 531–35. https://www.ncbi.nlm.nih.gov/pmc/articles/PMC2586862/

Loudon, Irvine. "Maternal Mortality in the Past and its Relevance to Developing Countries Today." *The American Journal of Clinical Nutrition* 72, no. 1 (July 1, 2000): 241S–46S.

Lurie, Samuel. "The History of the Diagnosis and Treatment of Ectopic Pregnancy: A Medical Adventure." *European Journal of Obstetrics & Gynecology and Reproductive Biology* 43 (January 1992): 1–7. https://www.ejog.org/article/0028-2243(92)90235-Q/pdf

Macaulay, Alexander. *A Dictionary of Medicine, Designed for Popular Use. Containing an Account of Diseases and Their Treatment, Including Those Most Frequent in Warm Climates*. 2nd ed. Edinburgh: Adam Black, 1831.

MacDorman, Marian F., T. J. Mathews, and Eugene Declercq. "Trends in Out-of-Hospital Births in the United States, 1990–2012." NCHS Data Brief no. 144, March 2014. https://www.cdc.gov/nchs/products/databriefs/db144.htm

Magiorkinis, Emmanouil, Aristidis Diamantis, Kalliopi Sidiropoulou, and Christos Panteliadis. "Highights [sic] in the History of Epilepsy: The Last 200 Years." *Epilepsy Research and Treatment* 2014 (2014): 582039. https://www.hindawi.com/journals/ert/2014/582039/

Magner, Lois N. *A History of Medicine*. New York: Marcel Dekker, 1992.

Majori, Giancarlo. "Short History of Malaria and Its Eradication in Italy with Short Notes on the Fight against the Infection in the Mediterranean Basin." *Mediterranean Journal of Hematology and Infectious Diseases* 4, no. 1 (March 10, 2012): e2012016. https://www.ncbi.nlm.nih.gov/pmc/articles/PMC3340992/

Maltby, Maryanne Tate. "Ancient Voices on Tinnitus: The Pathology and Treatment of Tinnitus in Celsus and the Hippocratic Corpus Compared and Contrasted." *The International Tinnitus Journal* 17, no. 2 (2012): 140–45.

Mammas, Ioannis N., and Demetrios A. Spandidos. "Paediatric Virology in the Hippocratic Corpus." *Experimental and Therapeutic Medicine* 12, no. 2 (August 2016): 541–49. https://www.ncbi.nlm.nih.gov/pmc/articles/PMC4950906/

Marineli, Filio, Gregory Tsoucalas, Marianna Karamanou, and George Androutsos. "Mary Mallon (1869–1938) and the History of Typhoid Fever." *Annals of Gastroenterology* 26, no. 2 (2013): 132–34.

Markel, Howard. *Quarantine! East European Jewish Immigrants and the New York City Epidemics of 1892.* Baltimore, MD: Johns Hopkins University Press, 1997.

Mascarenhas, Maya N., Seth R. Flaxman, Ties Boerma, Sheryl Vanderpoel, and Gretchen A. Stevens. "National, Regional, and Global Trends in Infertility Prevalence since 1990: A Systematic Analysis of 277 Health Surveys." *PloS Medicine* 9, no. 12 (2012): e1001356. https://journals.plos.org/plosmedicine/article?id=10.1371/journal.pmed.1001356

Mathews, Joan J., and Kathleen Zadak. "The Alternative Birth Movement in the United States: History and Current Status." *Women & Health* 17, no. 1 (1991): 39–56.

Mattern, Susan. "Galen and His Patients." *The Lancet* 378, no. 9790 (August 6, 2011): 478–79.

Mattick, A., and J. P. Wyatt. "From Hippocrates to the Eskimo—History of Techniques Used to Reduce Anterior Dislocation of the Shoulder." *Journal of the Royal College of Surgeons of Edinburgh* 45, no. 5 (October 2000): 312–16.

Maughan, Karen L., and Blake Reid Boggess. "Achilles Tendinopathy and Tendon Rupture." UpToDate. Last modified June 6, 2019. https://www.uptodate.com/contents/achilles-tendinopathy-and-tendon-rupture

McCarty, Arthur C. "History of Appendicitis Vermiformis: Its Diseases and Treatment." Presented to the Innominate Society, May 1927. https://www.innominatesociety.com/19261929

McClintock, Charles T., and Willard H. Hutchings. "The Treatment of Tetanus." *The Journal of Infectious Diseases* 13, no. 2 (September 1913): 309–20.

Mckay, W. J. Stewart. *The History of Ancient Gynaecology.* New York: William Wood & Company, 1901.

Merk Manual. Professional Version. Kenilworth, NJ: Merck & Co., Inc., 2020. https://www.merckmanuals.com/professional

Merriam-Webster Online. https://www.merriam-webster.com

Meyer, Christian G., Florian Marks, and Jürgen May. "Editorial: Gin Tonic Revisited." *Tropical Medicine and International Health* 9, no. 12 (December 2004): 1239–40.

Meynell, G. G., ed. *Thomas Sydenham's (1624–1689) Observationes Medicae (London, 1676) and his Medical Observations Manuscript (Manuscript 572 of the Royal College of Physicians of London), with New Transcripts of Related Locke MSS. in the Bodleian Library.* Folkestone: Winterdown, 1991.

Michaels, Paula A. "Childbirth and Trauma, 1940s–1980s." *Journal of the History of Medicine and Allied Sciences* 73, no. 1 (January 2018): 52–72.

Minguillon, Jesus. "Blue Lighting Accelerates Post-Stress Relaxation: Results of a Preliminary Study." *PLoS One* 12, no. 10 (2017): e0186399. https://www.ncbi.nlm.nih.gov/pmc/articles/PMC5648169/#:~:text=The%20results%20reported%20in%20the,comparison%20with%20conventional%20white%20lighting

Morillon, M., B. Mafart, and T. Matton. "Yellow Fever in Europe in 19th Century." In *Ecological Aspects of Past Settlement in Europe*, edited by P. Bennike, E. B. Bodzsar, and C. Suzanne, 211–22. European Anthropological Association, *2002 Biennal Yearbook.* Eötvös Budapest: University Press, Budapest, 2002. http://bertrand.mafart.free.fr/paleoanthropology_paleopathology_full_text_mafart/Yellowfever_history_europe_mafart.pdf

Morrow, Prince Albert. *Leprosy.* New York: Wood, 1899.

Moscucci, Ornella. "Gender and Cancer in Britain, 1860–1910: The Emergence of Cancer as a Public Health Concern." *American Journal of Public Health* 95, no. 8 (August 2005): 1312–21. https://ajph.aphapublications.org/doi/full/10.2105/AJPH.2004.046458

Mukhtar, Kashif, Hasham Nawaz, and Shahab Abid. "Functional Gastrointestinal Disorders and Gut-Brain Axis: What Does the Future Hold?" *World Journal of Gastroenterology* 25, no. 5 (2019): 552–66. https://www.wjgnet.com/1007-9327/full/v25/i5/552.htm

National Safety Council. *Injury Facts.* 2015 ed. Itasca, IL: National Safety Council. Accessed November 8, 2020. [Internet Archive] https://web.archive.org/web/20170926231142/http://www.nsc.org/Membership%20Site%20Document%20Library/2015%20Injury%20Facts/NSC_InjuryFacts2015Ed.pdf

Nerlich, Andreas G., Bettina Schraut, Sabine Dittrich, Thomas Jelinek, and Albert R. Zink. "*Plasmodium falciparum* in Ancient Egypt." *Emerging Infectious Diseases* 14, no. 8 (August 2008): 1317–19. https://www.ncbi.nlm.nih.gov/pmc/articles/PMC2600410/

Neumann, Donald A. "Polio: Its Impact on the People of the United States and the Emerging Profession of Physical Therapy." *Journal of Orthopedic Sports and Physical Therapy* 34, no. 8 (August 2004): 479–92. https://www.jospt.org/doi/pdfplus/10.2519/jospt.2004.0301

Neville, J. "Rabies in the Ancient World." In *Historical Perspective of Rabies in Europe and the Mediterranean Basin: A Testament to Rabies*, edited by Arthur A. King, Anthony R. Fooks, Michel Aubert and Alex Wandeler, 1–13. Weybridge, UK: World Organisation for Animal Health (OIE) in conjunction with the World Health Organisation (WHO) Collaborating Centre for the Characterisation of Rabies and Rabies-related Viruses at the Veterinary Laboratories Agency, Weybridge, UK, World Organisation for Animal Health, 2004. https://www.oie.int/doc/ged/d11246.pdf

North, Carol S. "The Classification of Hysteria and Related Disorders: Historical and Phenomenological Considerations." *Behavioral Sciences* 5, no. 4 (November 2015): 496–517. https://www.mdpi.com/2076-328X/5/4/496

Nuki, George, and Peter A. Simkin. "A Concise History of Gout and Hyperuricemia and Their Treatment." *Arthritis Research & Therapy* 8, no. 1 (April 12, 2006). https://arthritis-research.biomedcentral.com/articles/10.1186/ar1906

Ogoina, Dimie. "Fever, Fever Patterns and Diseases Called 'Fever'—A Review." *Journal of Infection and Public Health* 4, no. 3 (August 2011): 108–24.

Oldstone, Michael B. A. *Viruses, Plagues, and History: Past, Present and Future*. Revised and updated ed. Oxford; New York: Oxford University Press, 2010.

Online Etymology Dictionary. https://www.etymonline.com

Osler, William. "The War and Typhoid Fever: An Address Delivered before the Society of Tropical Medicine and Hygiene, November 20th, 1914." *British Medical Journal* 2, no. 2813 (November 28, 1914): 909–13.

Osler, William, Henry A. Christian, and Thomas McCrae. *The Principles and Practice of Medicine: Originally Written by Sir William Osler, Bart.* 14th Semicentennial (1892–1942) ed. New York: D. Appleton-Century Company, Incorporated, 1942.

Osler, William, and Thomas McCrae. *The Principles and Practice of Medicine: Designed for the Use of Practitioners and Students of Medicine*. 9th ed., thoroughly revised. New York and London: D. Appleton and Company, 1921.

Oswald, Leopold. "The Remedies of Nature: Catarrh.—Pleurisy.—Croup." *Popular Science Monthly*, March 1, 1884.

The Oxyrhynchus Papyri, vol. 80. Extant Medical Texts (5219–5229); New Medical Texts (5230–5253); Doctors' Reports (5254–5257); Indexes. Edited with translations and notes by M. Hirt, D. Leith and W. B. Henry, with contributions by D. Colomo, N. Gonis, and L. Tagliapietra. London: The Egypt Exploration Society, 2014.

Papagrigorakis, Manolis J., Christos Yapijakis, Philippos N. Synodinos, and Effie Baziotopoulou-Valavani. "DNA Examination of Ancient Dental Pulp Incriminates Typhoid Fever as a Probable Cause of the Plague of Athens." *International Journal of Infectious Diseases* 10, no. 3 (2006): 206–14. https://www.ncbi.nlm.nih.gov/pubmed/16412683

Papavramidou, Niki S., and Helen Christopoulou-Aletras. "Treatment of 'Hernia' in the Writings of Celsus (First Century AD)." *World Journal of Surgery* 29, no. 10 (October 2005): 1343–47. https://www.ncbi.nlm.nih.gov/pubmed/16151666

Pappas, Georgios, Ismene J. Kiriaze, and Matthew E. Falagas. "Insights into Infectious Disease in the Era of Hippocrates." *International Journal of Infectious Diseases* 12, no. 4 (July 1, 2008): 347–50.

The Papyrus Ebers. Translated from the German version by Cyril P. Bryan with an introduction by Professor G. Elliot Smith. London: Geoffrey Bles, 1930.

The Papyrus Ebers: The Greatest Egyptian Medical Document. Translated by Bendix Ebbell. Copenhagen: Levin & Munksgaard, 1937.

Parish, Lawrence C. "An Historical Approach to the Nomenclature of Rheumatoid Arthritis." *Arthritis and Rheumatism* 6, no. 2 (April 1963): 138–58.

Parkinson, James. *An Essay on the Shaking Palsy*. London: Printed by Whittingham and Rowland, for Sherwood, Neely, and Jones, 1817.

Pasero, Giampiero, and Piero Marson. "Hippocrates and Rheumatology." *Clinical and Experimental Rheumatology* 22 (2004): 687–89. https://pdfs.semanticscholar.org/2fcd/8f657d7eab4a213e2aa61c0a98af14e81e5a.pdf

Patel, Puja, and Solomon L. Moshé. "The Evolution of the Concepts of Seizures and Epilepsy: What's in a Name?" *Epilepsia Open* 5, no. 1 (March 2020): 22–35. https://www.ncbi.nlm.nih.gov/pmc/articles/PMC7049807/#:~:text=The%20word%20seizure%20is%20derived,around%202500%20BC%20from%20Mesopotamia

Paul, John R., and Horace T. Gardner. "Viral Hepatitis. Evolution of Concepts of Hepatitis." In *Preventive Medicine in World War II*, vol. 5. *Communicable Diseases Transmitted Through Contact or by Unknown Means*, chapter 17, 411–462. Washington, DC: Office of the Surgeon General, Department of the Army, 1960 [U.S. Army Medical Department, Office of Medical History]. https://history.amedd.army.mil/booksdocs/wwii/communicablediseasesV5/DEFAULT.htm

Paulus, Aegineta. *The Seven Books of Paulus Ægineta. Translated from the Greek. With a Commentary Embracing a Complete View of the Knowledge Possessed by the Greeks, Romans, and Arabians on All Subjects Connected with Medicine and Surgery.* By Francis Adams. 3 vols. London: Printed for the Sydenham Society, 1844–1847.

Pearce, J. M. S. "A Brief History of the Clinical Thermometer." *QJM: An International Journal of Medicine* 95, no. 4 (April 2002): 251–52.

Pećanac, Marija, Zlata Janjić, Aleksandar Komarčević, Miloš Pajić, Dušanka Dobanovački, and Sanja Skeledžija Mišković. "Burns Treatment in Ancient Times." *Medicinski pregled* 66, no. 5–6 (May–June 2013): 263–67. https://scindeks-clanci.ceon.rs/data/pdf/0025-8105/2013/0025-81051306263P.pdf

Peitzman, Steven J. "From Dropsy to Bright's Disease to End-stage Renal Disease." *The Milbank Quarterly* 67, no. 1 (1989): 16–32.

Petersen, Emily E., Nicole L. Davis, David Goodman, Shanna Cox, Nikki Mayes, Emily Johnston, Carla Syverson, et al. "Vital Signs: Pregnancy-Related

Deaths, United States, 2011–2015, and Strategies for Prevention, 13 States, 2013–2017." *Morbidity and Mortality Weekly Report* 68, no. 18 (2019): 423–29.

Phillips, Howard, and David Killingray, eds. *The Spanish Influenza Pandemic of 1918–1919: New Perspectives*. Routledge Studies in the Social History of Medicine, 12. Abingdon-on-Thames, UK: Routledge, 2003.

Pierce, John, and Jim Writer. *Yellow Jack: How Yellow Fever Ravaged America and Walter Reed Discovered Its Deadly Secrets*. Hoboken, NJ: John Wiley & Sons, 2005.

Pliny the Elder. *Natural History*. Chapter 39. (29.39). Accessed October 22, 2019. http://perseus.uchicago.edu/perseus-cgi/citequery3.pl?dbname=PerseusLatinTexts&getid=1&query=Plin.%20Nat.%2029.39

Pliny the Elder. *The Natural History of Pliny*. Translated, with Copious Notes and Illustrations by the Late John Bostock, and H. T. Riley. London: George Bell & Sons, 1893. https://archive.org/details/naturalhistoryp00bostgoog/page/n5

Podolsky, Scott H. *Pneumonia before Antibiotics: Therapeutic Evolution and Evaluation in Twentieth-Century America*. Baltimore, MD: The Johns Hopkins University Press. 2006.

Pott, Percivall. *The Chirurgical Works of Percivall Pott, F.R.S. and Surgeon to St. Bartholomew's Hospital*. A new edition In Three Volumes. London: Printed for T. Lowndes, J. Johnson, G. Robinson, T. Cadell, T. Evans, W. Fox, J. Bew and S. Hayes, 1779.

Pott, Percivall. *Remarks on that Kind of Palsy of the Lower Limbs, Which Is Frequently Found to Accompany A Curvature of the Spine, and Is Supposed to Be Caused by It. To Which are Added, Observations on the Necessity and Propriety of Amputation, in Certain Cases,* London: Printed for J. Johnson, 1779.

Pott, Percivall. *Treatise on Ruptures*. London: C. Hitch and L. Hawes, 1756.

Purnell, Jonathan Q. "Definitions, Classification, and Epidemiology of Obesity." MDText.com, Inc., South Dartmouth, MA. Last modified April 12, 2018. https://www.ncbi.nlm.nih.gov/books/NBK279167/

Rafferty, J., A. Tsikoudas, and B. C. Davis. "Ear Candling: Should General Practitioners Recommend It?" *Canadian Family Physician* 53, no. 12 (December 2007): 2121–22. https://www.ncbi.nlm.nih.gov/pmc/articles/PMC2231549/

Ranganathan, S., and T. Mukhopadhyay. "Dandruff: The Most Commercially Exploited Skin Disease." *Indian Journal of Dermatology* 55, no. 2 (April–June 2010): 130–34. https://www.ncbi.nlm.nih.gov/pmc/articles/PMC2887514/

Raoult, Didier, Theodore Woodward, and J. Stephen Dumler. "The History of Epidemic Typhus." *Infectious Disease Clinics of North America* 18, no. 1 (March 2004): 127–40.

Rasnake, Mark S., Nicholas G. Conger, C. Kenneth McAllister, King K. Holmes, and Edmund C. Tramont. "History of U.S. Military Contributions to the Study of Sexually Transmitted Diseases." *Military Medicine* 170, no. 4 (2005): 61–65.

Rathee, Manu, and Amit Sapra. "Dental Caries." *StatPearls* [Internet]. Treasure Island, FL: StatPearls Publishing. Last modified January 2020. https://www.ncbi.nlm.nih.gov/books/NBK551699/?report=reader

Rāzī, Abū Bakr Muḥammad ibn Zakarīyā (commonly called Rhazes). *A Treatise on the Small-Pox and Measles.* Translated from the Original Arabic by William Alexander Greenhill. London: Sydenham Society, 1848.

Rehemtulla, Alnawaz. "Dinosaurs and Ancient Civilizations: Reflections on the Treatment of Cancer." *Neoplasia* 12, no. 12 (December 2010): 957–68. https://www.ncbi.nlm.nih.gov/pmc/articles/PMC3003131/

Rennie, Claire. "The Treatment of Whooping Cough in Eighteenth-Century England." *Ex Historia* 8 (2016): 1–33. https://humanities.exeter.ac.uk/media/universityofexeter/collegeofhumanities/history/exhistoria/volume8/Rennie_1-33.pdf

Rezaei, Kourous A., and Gary W. Abrams. "The History of Retinal Detachment Surgery." In *Primary Retinal Detachment: Options for Repair*, edited by Ingrid Kreissig, 1–24. Berlin; Heidelberg: Springer, 2005.

Riddle, John M. *Contraception and Abortion from the Ancient World to the Renaissance.* Cambridge, MA: Harvard University Press, 1992.

Rider, Catherine. "Men and Infertility in Late Medieval English Medicine." *Social History of Medicine* 29, no. 2 (May 2016): 245–66.

Riedel, Stefan. "Edward Jenner and the History of Smallpox and Vaccination." *Baylor University Medical Center Proceedings* 18, no. 1 (2005): 21–25. https://www.ncbi.nlm.nih.gov/pmc/articles/PMC1200696/

Risse, Guenter B. "The Limits of Medical Science: Hospitals in Fin-de-Siècle Europe and America; Typhoid Fever and Johns Hopkins Hospital, Baltimore, 1891." In *Mending Bodies, Saving Souls: A History of Hospitals*, 399–422. Oxford: Oxford University Press, 1999.

Robb, R. L., J. V. Bean, and S. Lucretia Robb. *Robb & Co's Family Physician: A Work on Domestic Medicines Designed to Show the Causes, Symptoms and Treatment of Disease: For the Use of the People.* 4th ed. carefully revised and enlarged. Burlington, IA: Robb & Co., Book Publishers, 1880.

Robins, Gay. *Women in Ancient Egypt.* Cambridge: Harvard University Press, 1993.

Robinson, Sean M., and Bryon Adinoff. "The Classification of Substance Use Disorders: Historical, Contextual, and Conceptual Considerations." *Behavioral Sciences (Basel, Switzerland)* 6, no. 3 (August 18, 2016): 18. doi: 10.3390/bs6030018

Rogers, Naomi. *Dirt and Disease: Polio before FDR.* New Brunswick, NJ: Rutgers University Press, 1992.

Rogers, Leonard. "Chaulmoogra Oil in Leprosy and Tuberculosis: The Successful Treatment of Leprosy by Injections of Soluble Preparations of the Fatty Acids of Chaulmoogra and Other Oils and Its Bearing on the Tuberculosis Problem." *The Lancet* 197, no. 5101 (June 4, 1921): 1178–80.

Roguin, Ariel. "Rene Theophile Hyacinthe Laënnec (1781–1826): The Man behind the Stethoscope." *Clinical Medicine and Research* 4, no. 3 (September 2006): 230–35. https://www.ncbi.nlm.nih.gov/pmc/articles/PMC1570491/

Rolleston, J. D. "The History of Scarlet Fever." *British Medical Journal* 2, no.3542 (1928): 926–29. https://www.ncbi.nlm.nih.gov/pmc/articles/PMC2456687/

Rosenberg, Charles E. *The Cholera Years: The United States in 1832, 1849, and 1866*. Chicago: The University of Chicago Press, 1962.

Roser, Max, Hannah Ritchie, and Bernadeta Dadonaite. "Child and Infant Mortality." Our World in Data, 2013. Last modified November 2019. https://ourworldindata.org/child-mortality

Rueff, Jakob. *The Expert Midwife, Or an Excellent and most Necessary Treatise of the Generation and Birth of Man Wherein is Contained Many very Notable and Necessary Particulars Requisite to be Knovvne and Practised*. London: Printed by E. G. for S. B. and are to be sold by Thomas Aleborn Sign of the Greene Dragon in Saint Paul's Church-Yard, London, 1637.

Russell-Jones, D. L. "Sudden Infant Death in History and Literature." *Archives of Disease in Childhood* 60 (March 1985): 278–81. https://adc.bmj.com/content/archdischild/60/3/278.full.pdf

Saitoh, H. ["The Descriptions on Disorders of Urination in the Hippocratic Collection."] *Nihon Hinyokika Gakkai Zasshi* 96, no. 3 (March 2005): 432–41 [Article in Japanese]. https://www.ncbi.nlm.nih.gov/pubmed/15828260

Sanders, D. L., and A. N. Kingsnorth. "From Ancient to Contemporary Times: A Concise History of Incisional Hernia Repair." *Hernia* 16, no. 1 (February 2012): 1–7.

Sangaré, Abdoul Karim, Ogobara K. Doumbo, and Didier Raoult. "Management and Treatment of Human Lice." *BioMed Research International* 2016 (2016): 8962685. https://www.ncbi.nlm.nih.gov/pmc/articles/PMC4978820/

Santoro, Rosa. "Skin over the Centuries. A Short History of Dermatology: Physiology, Pathology and Cosmetics." *Medicina Historica* 1, no. 2 (2017): 94–102.

Scurlock, Jo Ann. "Baby-Snatching Demons, Restless Souls and the Dangers of Childbirth: Medico-Magical Means of Dealing with Some of the Perils of Motherhood in Ancient Mesopotamia." *Incognita* 2 (1991): 137–85.

Sen, Chandan K., Gayle M. Gordillo, Sashwati Roy, Robert Kirsner, Lynn Lambert, Thomas K. Hunt, Finn Gottrup, Geoffrey C. Gurtner, and Michael T. Longaker. "Human Skin Wounds: A Major and Snowballing Threat to Public Health and the Economy." *Wound Repair and Regeneration*, 17, no. 6 (2009): 763–71. https://www.ncbi.nlm.nih.gov/pmc/articles/PMC2810192/

Senkova, Michaela. "Male Infertility in Classical Greece: Some Observations." *Graeco-Latina Brunensia* 20, no. 1 (2015): 121–31. https://digilib.phil.muni.cz/bitstream/handle/11222.digilib/133970/1_GraecoLatinaBrunensia_20-2015-1_10.pdf

Shah, J. "Erectile Dysfunction through the Ages." *BJU International* 90, no. 4 (September 2002): 433–41. https://bjui-journals.onlinelibrary.wiley.com/doi/full/10.1046/j.1464-410X.2002.02911.x

Sinclair, Upton. *Unseen Upton Sinclair: Nine Unpublished Stories, Essays and Other Works*. Edited by Ruth Clifford Engs. Jefferson, NC: McFarland, 2009.

Singh, Vibha. “Medicinal Plants and Bone Healing.” *National Journal of Maxillofacial Surgery* 8, no. 1 (January–June 2017): 4–11.

Smith, Dale C. “Extremity Injury and War: A Historical Reflection.” *Clinical Orthopaedics and Related Research* 473 (2015): 2771–76.

Snow, John. *On the Mode of Communication of Cholera*. 2nd ed., Much Enlarged. London: John Churchill, 1855.

Stafford, M. A., P. Peng, and D. A. Hill. “Sciatica: A Review of History, Epidemiology, Pathogenesis, and the Role of Epidural Steroid Injection in Management.” *British Journal of Anaesthesia* 99, no. 4 (October 2007): 461–73.

Staples, J. Erin, and Thomas P. Monath. “Yellow Fever: 100 Years of Discovery.” *JAMA* 300, no. 8 (August 27, 2008): 960–62.

Starbuck, John M. “On the Antiquity of Trisomy 21: Moving Towards a Quantitative Diagnosis of Down Syndrome in Historic Material Culture.” *Journal of Contemporary Anthropology* 2, no. 1 (2011): 18–44. https://docs.lib.purdue.edu/cgi/viewcontent.cgi?article=1019&context=jca

Sternbach, George. “The History of Anthrax.” *Journal of Emergency Medicine* 24, no. 4 (May 2003): 463–67.

Stevens, Emily E., Thelma E. Patrick, and Rita Pickler. “A History of Infant Feeding.” *Journal of Perinatal Education* 18, no. 2 (Spring 2009): 18(2): 32–39. https://www.ncbi.nlm.nih.gov/pmc/articles/PMC2684040/

Strach, E. H. “Club-foot through the Centuries.” In *Historical Aspects of Pediatric Surgery*, edited by P. P. Rickham, 215–37. Progress in Pediatric Surgery, Vol. 20. Berlin; New York: Springer, 1986.

Stripling, Joshua, and Martin Rodriguez. “Current Evidence in Delivery and Therapeutic Uses of Fecal Microbiota Transplantation in Human Diseases—*Clostridium difficile* Disease and Beyond.” *American Journal of the Medical Sciences* 356, no. 5 (November 2018): 424–32.

Sun, Zhifei, and John Migaly. “Review of Hemorrhoid Disease: Presentation and Management.” *Clinics in Colon and Rectal Surgery* 29, no. 1 (March 2016): 22–29. https://www.ncbi.nlm.nih.gov/pmc/articles/PMC4755769/

Swedlund, Alan C., and Alison K. Donta. “Scarlet Fever Epidemics of the Nineteenth Century: A Case of Evolved Pathogenic Virulence?” In *Human Biologists in the Archives: Demography, Health, Nutrition and Genetics in Historical Populations*, edited by D. Ann Herring and Alan C. Swedlund, 159–177. Cambridge: Cambridge University Press, 2003.

Sydenham, Thomas. *The Entire Works of Dr. Thomas Sydenham: Newly Made English From the Originals: Wherein the History of Acute and Chronic Diseases, and the Safest And Most Effectual Methods of Treating Them, Are Faithfully, Clearly, and Accurately Delivered. To Which Are Added, Explanatory and Practical Notes, From the Best Medicinal Writers*. Translated by John Swan. London: Printed for Edward Cave at St. John’s Gate, 1742.

Sydenham, Thomas. *Selected Works of Thomas Sydenham, M.D. With a Short Biography and Explanatory Notes*. Edited by John D. Comrie. London: John Bale, Sons & Danielsson, Ltd., 1922.

Sydenham, Thomas. *Thomas Sydenham’s Observationes medicae (London, 1676) and his Medical observations (Manuscript 572 of the Royal College of*

Physicians of London), with new transcripts of related Locke MSS. in the Bodleian Library. Edited by G. G. Meynell. Folkestone: Winterdown, 1991.

Sydenham, Thomas. *The Whole Works of That Excellent Practical Physician, Dr. Thomas Sydenham: Wherein not only the History and Cures of Acute Diseases are Treated of, after a New and Accurate Method; but also the Shortest and Fastest Way of Curing Most Chronical Diseases*. 9th ed. Corrected from the Orginal Latin, by John Pechey, M.D. London: Printed for J. Darby in Bartholomew-Close, A. Bettesworth in Pater-noster-Row, and F. Clay without Temple-Bar in trust for Richard, James, and Bethel Wellington, 1729.

Sydenham, Thomas. *The Works of Thomas Sydenham, M.D., Translated from the Latin Edition of Dr. Greenhill, with a Life of the Author, by R.G. Latham, M.D.* 2 vols. London: Printed for the Sydenham Society, 1848–1850.

Talty, Stephan. *The Illustrious Dead: The Terrifying Story of How Typhus Killed Napoleon's Greatest Army*. New York: Crown, 2009.

Tanner, Thomas Hawkes. *On the Signs & Diseases of Pregnancy*. 2nd and enl. London ed. London: H. Renshaw, 1867.

Tefekli, Ahmet, and Fatin Cezayirli. "The History of Urinary Stones: In Parallel with Civilization." *ScientificWorldJournal* 2013 (November 20, 2013): 423964. https://www.ncbi.nlm.nih.gov/pmc/articles/PMC3856162/

Theeler, Brett J., Renee Mercer, and Jay C. Erickson. "Prevalence and Impact of Migraine among US Army Soldiers Deployed in Support of Operation Iraqi Freedom." *Headache* 48, no. 6 (June 2008): 876–82.

Thomas, D. H. "The Demise of Bloodletting." *Journal of the Royal College of Physicians of Edinburgh* 44, no. 1 (2014): 72–77. http://www.rcpe.ac.uk/sites/default/files/thomas_0.pdf

Thomas, George. "Coronary Artery Disease—Need for Better Terminology." *British Journal of Cardiology* 16 (July 2009): 192–93. https://bjcardio.co.uk/2009/07/coronary-artery-disease-need-for-better-terminology/

Tountas, Yannis. "The Historical Origins of the Basic Concepts of Health Promotion and Education: The Role of Ancient Greek Philosophy and Medicine." *Health Promotion International* 24, no. 2 (June 2009): 185–92.

Trevelyan, Barry, Matthew Smallman-Raynor, and Andrew D. Cliff. "The Spatial Dynamics of Poliomyelitis in the United States: From Epidemic Emergence to Vaccine-Induced Retreat, 1910–1971." *Annals of the Association of American Geography* 95, no. 2 (June 2005): 269–93. https://www.ncbi.nlm.nih.gov/pmc/articles/PMC1473032/

Trubuhovich, Ronald V. "History of Mouth-to-Mouth Rescue Breathing. Part 1." *Critical Care and Resuscitation* 7 (May 2005): 250–57. https://www.cicm.org.au/CICM_Media/CICMSite/CICM-Website/Resources/Publications/CCR%20Journal/Previous%20Editions/September%202005/20_2005_Sep_History-of-Medicine.pdf

Trubuhovich, Ronald V. "History of Mouth-to-Mouth Rescue Breathing. Part 2. the 18th Century." *Critical Care and Resuscitation* 8, no 2 (June 2006): 157–71. https://www.cicm.org.au/CICM_Media/CICMSite/CICM-Website

/Resources/Publications/CCR%20Journal/Previous%20Editions/June%20 2006/17_2006_Jun_History-of-mouth-to-mouth.pdf

Tsoucalas, Gregory, and Markos Sgantzos. "Hippocrates, on the Infection of the Lower Respiratory Tract among the General Population in Ancient Greece." *General Medicine (Los Angeles)* 4, no. 5 (2016): 1000272. https://www.longdom.org/open-access/hippocrates-on-the-infection-of-the-lower-respiratory-tract-among-the-generalpopulation-in-ancient-greece-2327-5146-1000272.pdf

Tubby, Alfred Herbert. *Deformities, Including Diseases of the Bones and Joints: A Text-book of Orthopaedic Surgery.* 2 vols. 2nd ed. London: Macmillan, 1912.

Turliuc, Mihaela Dana, Serban Turliuc, Andrei Ionut Cucu, Camelia Tamas, Alexandru Carauleanu, Catalin Buzduga, Anca Sava, Gabriela Florenta Dumitrescu, and Claudia Florida Costea. "Through Clinical Observation: The History of Priapism after Spinal Cord Injuries." *World Neurosurgery* 109 (January 2018): 365–71.

Underwood, Michael. *A Treatise On the Diseases of Children, with Directions for the Management of Infants From the Birth; Especially Such as are Brought Up by Hand.* London: Printed for J. Mathews, 1784.

United Nations Population Fund. *Trends in Maternal Mortality: 2000 to 2017. Estimates by WHO, UNICEF, UNFPA, World Bank Group and the United Nations Population Division. Executive Summary.* Geneva: World Health Organization, 2019. https://www.unfpa.org/resources/trends-maternal-mortality-2000-2017-executive-summary

United Nations Population Fund. "Unsafe, Unreliable: Dangerous Pregnancy-Prevention Methods." July 9, 2018. https://www.unfpa.org/dangerous-pregnancy-prevention

U.S. Bureau of the Census. *Mortality Statistics of the Seventh Census of the United States, 1850: Embracing—I.—the Cause of Death, II.—the Age and Sex, III.—the Color and Condition, IV.—the Nativity, V.—the Season of Decease, VI.—the Duration of Illness, VII.—the Occupation, of the Persons Reported to Have Died In the Twelve Months Preceding the First of June of That Year, with Sundry Comparative And Illustrative Tables.* Washington, DC: Bureau of the Census, 1855.

U.S. Department of Defense. *Emergency War Surgery.* 3rd United States Revision. Washington, DC: Department of Defense, 2004. https://apps.dtic.mil/dtic/tr/fulltext/u2/a428731.pdf

U.S. Department of Health and Human Services. Centers for Disease Prevention and Prevention. "Achievements in Public Health, 1900-1999: Control of Infectious Diseases." *Morbidity and Mortality Weekly Report* 48, no. 29 (July 30, 1999): 621–29.

U.S. Department of Health and Human Services. Centers for Disease Control and Prevention. "Acute Rheumatic Fever." Last modified July 12, 2018. https://www.cdc.gov/groupastrep/diseases-hcp/acute-rheumatic-fever.html

U.S. Department of Health and Human Services. Centers for Disease Control and Prevention. "Bats Lead in U.S. Rabies Risk. Awareness of Rabies Threats

Crucial to Preventing Deadly Disease." CDC Newsroom, Press Release. June 12, 2019. Last modified June 11, 2019. https://www.cdc.gov/media/releases/2019/p0611-bats-rabies.htm

U.S. Department of Health and Human Services. Centers for Disease control and Prevention. "CDCs Abortion Surveillance System FAQ's." Reproductive Health, Data and Statistics. Last modified November 25, 2019. https://www.cdc.gov/reproductivehealth/data_stats/abortion.htm

U.S. Department of Health and Human Services. Centers for Disease Control and Prevention. "Chain of Infection." In *Principles of Epidemiology in Public Health Practice*: *An Introduction to Applied Epidemiology and Biostatistics.* Lesson 1: "Introduction to Epidemiology." Section 10. 3rd ed. Last modified May 18, 2012. https://www.cdc.gov/csels/dsepd/ss1978/lesson1/section10.html

U.S. Department of Health and Human Services. Centers for Disease Control and Prevention. "Cholera—*Vibro cholerae* Infection." Last modified May 3, 2018. https://www.cdc.gov/cholera/index.html

U.S. Department of Health and Human Services. Centers for Disease Control and Prevention. "Chronic Sinusitis." FastStats. Last modified February 21, 2020. https://www.cdc.gov/nchs/fastats/sinuses.htm

U.S. Department of Health and Human Services. Centers for Disease Control and Prevention. "Epidemic Typhus." Typhus Fevers Home. Last modified March 7, 2017. https://www.cdc.gov/typhus/epidemic/index.html

U.S. Department of Health and Human Services. Centers for Disease Control and Prevention. "Fungal Diseases. Histoplasmosis Statistics." Last modified August 13, 2018. https://www.cdc.gov/fungal/diseases/histoplasmosis/statistics.html

U.S. Department of Health and Human Services. Centers for Disease Control and Prevention. "A History of Anthrax." Last modified January 31, 2017. https://www.cdc.gov/anthrax/resources/history/index.html

U.S. Department of Health and Human Services. Centers for Disease Control and Prevention. "The History of Malaria, an Ancient Disease." Last modified November 14, 2018. https://www.cdc.gov/malaria/about/history/index.html

U.S. Department of Health and Human Services. Centers for Disease Control and Prevention. "History of Smallpox." Last modified August 20, 2016. https://www.cdc.gov/smallpox/history/history.html

U.S. Department of Health and Human Services. Centers for Disease Control. "History Timeline Transcript." Suppl. to *Yellow Fever: History, Epidemiology, and Vaccination Information Lesson* [Internet Archive] December 2, 2017. https://web.archive.org/web/20170505164156/https://www.cdc.gov/travel-training/local/HistoryEpidemiologyandVaccination/HistoryTimelineTranscript.pdf

U.S. Department of Health and Human Services. Centers for Disease Control and Prevention. "Insects and Scorpions." Workplace Safety and Health Topics. Last modified May 31, 2018. https://www.cdc.gov/niosh/topics/insects/

U.S. Department of Health and Human Services. Centers for Disease Control and Prevention. "Mass Casualties: Burns." Emergency Preparedness and Response. Last modified July 18, 2006 [Internet Archive] https://web.archive.org/web/20111223141728if_/http://www.bt.cdc.gov/masscasualties/burns.asp

U.S. Department of Health and Human Services. Centers for Disease Control and Prevention. "1918 Pandemic Influenza: Three Waves." Last modified May 11, 2018. https://www.cdc.gov/flu/pandemic-resources/1918-commemoration/three-waves.htm

U.S. Department of Health and Human Services. Centers for Disease Control and Prevention. "Parasites." Last modified November 20, 2020. https://www.cdc.gov/parasites/index.html

U.S. Department of Health and Human Services. Centers for Disease Control and Prevention. "Plague Vaccine." *Morbidity and Mortality Weekly Report* 31, no. 22 (June 11, 1982): 301–4.

U.S. Department of Health and Human Services. Centers for Disease Control and Prevention. "Poliomyelitis." *Epidemiology and Prevention of Vaccine-Preventable Diseases: The Pink Book.* Course Textbook. 13th ed., 2015, chapter 18. Last modified July 17, 2020. https://www.cdc.gov/vaccines/pubs/pinkbook/polio.html

U.S. Department of Health and Human Services. Centers for Disease Control and Prevention. "Pregnancy Mortality Surveillance System." Reproductive Health. Last modified February 4, 2020. https://www.cdc.gov/reproductive-health/maternal-mortality/pregnancy-mortality-surveillance-system.htm

U.S. Department of Health and Human Services. Centers for Disease Control and Prevention. "Preventing Congenital Toxoplasmosis." *Morbidity and Mortality Weekly Report* 49, RR02 (March 31, 2000): 57–75.

U.S. Department of Health and Human Services. Centers for Disease Control and Prevention. "Sexually Transmitted Diseases (STDs)." CDC Fact Sheets. Last modified November 4, 2016. https://www.cdc.gov/std/healthcomm/fact_sheets.htm

U.S. Department of Health and Human Services. Centers for Disease Control and Prevention. National Center for Health Statistics. "Maternal Mortality." Last modified November 20, 2019. https://www.cdc.gov/nchs/maternal-mortality/index.htm

U.S. Department of Health and Human Services. Centers for Disease Control and Prevention. National Institute for Occupational Safety and Health. "Venomous Snakes." Workplace Safety and Health Topics. Last modified May 31, 2018. https://www.cdc.gov/niosh/topics/snakes/

U.S. Department of Health and Human Services. National Institutes of Health. Eunice Kennedy Shriver National Institute of Child Health and Human Development. "What Are The Stages Of Labor?" Accessed October 24, 2019. https://www.nichd.nih.gov/health/topics/labor-delivery/topicinfo/stages

U.S. Department of Health and Human Services. National Institutes of Health. National Institute of General Medical Science. "Burns." Content revised

January 2018. https://www.nigms.nih.gov/education/pages/factsheet_burns.aspx

U.S. Department of Health and Human Services. National Institutes of Health. National Institute on Deafness and Other Communication Disorders. "About Hearing: Quick Statistics." Last modified December 15, 2016. https://www.nidcd.nih.gov/health/statistics/quick-statistics

U.S. Public Health Service. *Public Health Reports* 32, no. 27 (July 6, 1917). https://www.ncbi.nlm.nih.gov/pmc/articles/PMC1999754/

Vasiliadis, Elias S., Theodoros B. Grivas, and Angelos Kaspiris. "Historical Overview of Spinal Deformities in Ancient Greece." *Scoliosis* 4, no. 6 (February 2009). https://scoliosisjournal.biomedcentral.com/articles/10.1186/1748-7161-4-6

Vaughan, Warren T. *Influenza: An Epidemiologic Study.* Monographic Series No. 1. Baltimore, MD: American Journal of Hygiene, 1921.

Veeresha, K. L., M. Bansal, and V. Bansal. "Halitosis: A Frequently Ignored Social Condition." *Journal of International Society of Preventive and Community Dentistry* 1, no. 1 (January–June 2011): 9–13.

Velasco-Villa, Andres, Matthew R. Mauldin, Mang Shi, Luis E. Escobar, Nadia F. Gallardo-Romero, Inger Damon, Victoria A. Olson, Daniel G. Streicker, and Ginny Emerson. "The History of Rabies in the Western Hemisphere." *Antiviral Research* 146 (October 2017): 221–32. https://www.ncbi.nlm.nih.gov/pmc/articles/PMC5620125/

Ventura, Hector O., and Mandeep R. Mehra. "Bloodletting as a Cure for Dropsy: Heart Failure Down the Ages." *Journal of Cardiac Failure* 11, no. 4 (May 2005): 247–52.

Vergun, David. "Survival Rates Improving for Soldiers Wounded in Combat, Says Army Surgeon General." U.S. Army. August 24, 2016. https://www.army.mil/article/173808/survival_rates_improving_for_soldiers_wounded_in_combat_says_army_surgeon_general

Verkuyl, Douwe Arie Anne. "Think Globally Act Locally: The Case for Symphysiotomy." *PLoS Medicine* 4, no. 3 (March 2007): e71. https://www.ncbi.nlm.nih.gov/pmc/articles/PMC1831724/

Volk, Anthony A., and Jeremy A. Atkinson. "Infant and Child Death in the Human Environment of Evolutionary Adaptation." *Evolution and Human Behavior* 34, no. 3 (May 2013): 182–92.

Wagner, David M., Jennifer Klunk, Michaela Harbeck, Alison Devault, Nicholas Waglechner, Jason W. Sahl, Jacob Enk, et al. "*Yersinia pestis* and the Plague of Justinian 541–543 AD: A Genomic Analysis." *The Lancet* 14, no. 4 (April 2014): 319–26.

Walter, Michael J., and Michael J. Holtzman. "A Centennial History of Research on Asthma Pathogenesis." *American Journal of Respiratory Cell and Molecular Biology* 32, no. 6 (June 2005): 483–89. https://www.ncbi.nlm.nih.gov/pmc/articles/PMC2715318/

Wardleworth, Thos. H. "Effects of Ergot in Producing Abortion." *Provincial Medical and Surgical Journal* 8, no. 6 (May 8, 1844): 78–79. https://www.ncbi.nlm.nih.gov/pmc/articles/PMC2558386/?page=2

Waring, Edward John. *A Manual of Practical Therapeutics, Considered With Reference to Articles of the Materia Medica.* Edited by Dudley W. Buxton. 4th revised ed. Philadelphia: P. Blakiston, Son & Co., 1886.

Whorton, James. "Civilization and the Colon: Constipation as the 'Disease of Diseases'." *The British Medical Journal* 321, no. 7276 (December 23-30, 2000): 1586–89. https://www.ncbi.nlm.nih.gov/pmc/articles/PMC1119264/

Williams, David R., and Irene Litvan. "Parkinsonian Syndromes." *Continuum (Minneapolis, Minn.)* 19, no. 5 Movement Disorders (October 2013): 1189–212. https://www.ncbi.nlm.nih.gov/pmc/articles/PMC4234134/

Williams, G. Rainey. "Presidential Address: A History of Appendicitis with Anecdotes Illustrating Its Importance." *Annals of Surgery* 197, no. 5 (May 1983): 495–506. https://www.ncbi.nlm.nih.gov/pmc/articles/PMC1353017/pdf/annsurg00135-0007a.pdf

Williamson, Laila. "Infanticide: An Anthropological Analysis." In *Infanticide and the Value of Life,* edited by Marvin Kohl, 61–75. Amherst, NY: Prometheus Books, 1978.

Wilson, Michelle, and Jason Stacy. "Shock Wave Therapy for Achilles Tendinopathy." *Current Reviews in Musculoskeletal Medicine* 4, no. 1 (March 2011): 6–10. https://www.ncbi.nlm.nih.gov/pmc/articles/PMC3070010/

Withers, Mellissa, Nina Kharazmi, and Esther Lim. "Traditional Beliefs and Practices in Pregnancy, Childbirth and Postpartum: A Review of the Evidence from Asian Countries." *Midwifery* 56 (January 2018): 158–70.

Wong, Michael. "Shining Light on Epilepsy: Optical Approaches for Treating Seizures." *Epilepsy Currents* 13, no. 2 (March/April 2013): 95–96.

Wood, Horatio C., Jr., and Joseph P. Remington. *The Dispensatory of the United States of America.* 20th ed. Philadelphia: Lippincott, 1918.

Wood, John M., Theodore Athanasiadis, and Jacqui Allen. "Laryngitis." *BMJ* 349 (October 9, 2014): 5827.

World Health Organization. "Cancer." Fact Sheets, September 12, 2018. https://www.who.int/news-room/fact-sheets/detail/cancer

World Health Organization. "The Case for Midwifery." Maternal, Newborn, Child and Adolescent Health. Accessed October 28, 2019. https://www.who.int/maternal_child_adolescent/topics/quality-of-care/midwifery/case-for-midwifery/en/

World Health Organization. "Depression." Fact Sheets, January 30, 2020. https://www.who.int/news-room/fact-sheets/detail/depression

World Health Organization. "Diarrhoeal Disease." Fact Sheets, May 2, 2017. https://www.who.int/en/news-room/fact-sheets/detail/diarrhoeal-disease

World Health Organization. "Global Leprosy Update, 2016: Accelerating Reduction of Disease Burden." *Weekly Epidemiological Record* 92, no. 35 (September 1, 2017): 501–20. https://apps.who.int/iris/bitstream/handle/10665/258841/WER9235.pdf?sequence=1

World Health Organization. "Influenza (Seasonal)." Fact Sheets, November 6, 2018. https://www.who.int/news-room/fact-sheets/detail/influenza-(seasonal)

World Health Organization. "Insect-Borne Diseases." World Health Report, 1996. Executive Summary. Accessed January 23, 2020. https://www.who.int/whr/1996/media_centre/executive_summary1/en/index9.html

World Health Organization. "Leprosy." Fact Sheets, February 9, 2018. https://www.who.int/en/news-room/fact-sheets/detail/leprosy

World Health Organization. "Leprosy Elimination." Classification of Leprosy. Accessed October 26, 2019. https://www.who.int/lep/classification/en/

World Health Organization. *Managing Prolonged and Obstructed Labor*. Geneva: World Health Organization, 2008.

World Health Organization. "The Ottawa Charter for Health Promotion." First International Conference on Health Promotion, Ottawa, 21 November 1986. Accessed November 22, 2019. https://www.who.int/teams/health-promotion/enhanced-wellbeing/first-global-conference

World Health Organization. "Pertussis." Immunization, Vaccines and Biologicals. August 14, 2020. http://www.who.int/immunization/monitoring_surveillance/burden/vpd/surveillance_type/passive/pertussis/en/

World Health Organization. "Plague." Fact Sheets, October 31, 2017. https://www.who.int/news-room/fact-sheets/detail/plague

World Health Organization. "Pneumonia." Fact Sheets, November 7, 2016. https://www.who.int/news-room/fact-sheets/detail/pneumonia

World Health Organization. "Poliomyelitis." Fact Sheets, March 1, 2019. https://www.who.int/news-room/fact-sheets/detail/poliomyelitis

World health Organization. *Yellow Fever*. Fact Sheets, May 1 2018. https://www.who.int/en/news-room/fact-sheets/detail/yellow-fever

World Health Organization. Regional Office for Europe. "History of Malaria Elimination in the European Region." Fact Sheet, Copenhagen. April 20, 2016. http://www.euro.who.int/__data/assets/pdf_file/0003/307272/Facsheet-malaria-elimination.pdf

Worrall, Graham. "Common Cold." *Canadian Family Physician* 57, no. 11 (November 1, 2011): 1289–90. https://www.cfp.ca/content/57/11/1289.short

Wyatt, Harold V. "Before the Vaccines: Medical Treatments of Acute Paralysis in the 1916 New York Epidemic of Poliomyelitis." *The Open Microbiology Journal* 8, no. 1 (2014): 144–47. https://www.ncbi.nlm.nih.gov/pmc/articles/PMC4293735/

Yakoot, Ashraf Ahmed. "The Journey of the Evolution of Rhinology." *The Egyptian Journal of Otolaryngology* [serial online] 29, no. 2 (2013): 136–42. http://www.ejo.eg.net/article.asp?issn=1012-5574;year=2013;volume=29;issue=2;spage=136;epage=142;aulast=Yakoot

Yang, Hyun Duk, Do Han Kim, Sang Bong Lee, and Linn Derg Young. "History of Alzheimer's Disease." *Dementia and Neurocognitive Disorders* 15, no. 4 (December 2016): 115–21. https://dnd.or.kr/DOIx.php?id=10.12779/dnd.2016.15.4.115

Yasuyuki, Miyano, Koyama Kunihiro, Sreekumari Kurissery, Nandakumar Kanavillil, Yoshiro Sato, and Yasushi Kikuchi. "Antibacterial Properties of Nine Pure Metals: A Laboratory Study Using Staphylococcus Aureus and Escherichia Coli." *Biofouling* 26, no. 7 (October 2010): 851–58. https://pubmed.ncbi.nlm.nih.gov/20938849/

York, William H. "Antiquity through the Middle Ages." In *Health and Medicine through History: From Ancient Practices to 21st-Century Innovations,* edited by Ruth Clifford Engs, vol. 1 pt. 1. Santa Barbara, CA: Greenwood, 2019.

Young, James Harvey. *The Medical Messiahs: A Social History of Health Quackery in Twentieth-Century America.* Princeton, NJ: Princeton University Press, 1967.

Zeb, Irfan, Naser Ahmadi, Khurram Nasir, Jigar Kadakia, Vahid Nabavi Larijani, Ferdinand Flores, Dong Li, and Matthew J. Budoff. "Aged Garlic Extract and Coenzyme Q10 Have Favorable Effect on Inflammatory Markers and Coronary Atherosclerosis Progression: A Randomized Clinical Trial." *Journal of Cardiovascular Disease Research* 3, no. 3 (July–September 2012): 185–90. https://www.ncbi.nlm.nih.gov/pmc/articles/PMC3425023/#!po=70.6522

Zinsser, Hans. *Rats, Lice, and History.* Boston: Little, Brown and Company, 1963.

Index

Note: Numbers in **bold** indicate main entries.

ABOUT THE AUTHOR

Ruth Clifford Engs, RN, EdD, is professor emerita of applied health science in the School of Public Health, Indiana University, Bloomington. She is the author of numerous articles, chapters, and books. Her publications with ABC-CLIO and its imprints include *Health and Medicine through History: From Ancient Practices to 21st-Century Innovations, The Eugenics Movement: An Encyclopedia, The Progressive Era's Health Reform Movement: A Historical Dictionary*, and *Clean Living Movements: American Cycles of Health Reform.*